‘Powerful . . . If we’re to stop
itself when the next pandemic pathogen emerges, books such as *Preventable* are very much welcome’
Oliver Barnes, *Financial Times*

‘One of the most sought-after expert voices . . . [*Preventable*] sets out a blueprint for what could be done differently when the next pandemic inevitably rears its ugly head . . . the book shines with irrepressible and defiant cheer’
Elsa Maishman, *Scotsman*

‘Sridhar’s prescience transformed her into one of Britain’s most prominent commentators . . . Sridhar distils the lessons of the time’
George Eaton, *New Statesman*

‘I always read and listen to what she has to say and I hope you will too’
Chelsea Clinton

## ABOUT THE AUTHOR

**Devi Sridhar** is Professor and Chair of Global Public Health at the University of Edinburgh. She has served as a policy advisor for the WHO, UNICEF, UNESCO and the Scottish, UK and German governments. Devi writes for the *Guardian* and regularly appears on broadcast media. This is her first book for a general readership.

[illegible] inquiry from repeating [illegible]

# Preventable

## *How a Pandemic Changed the World and How to Stop the Next One*

DEVI SRIDHAR

PENGUIN BOOKS

PENGUIN BOOKS

UK | USA | Canada | Ireland | Australia
India | New Zealand | South Africa

Penguin Books is part of the Penguin Random House group of companies
whose addresses can be found at global.penguinrandomhouse.com.

First published by Viking 2022
Published in Penguin Books 2023
001



Typeset by Jouve (UK), Milton Keynes
Printed and bound in Great Britain by Clays Ltd, Elcograf S.p.A.

The authorized representative in the EEA is Penguin Random House Ireland,
Morrison Chambers, 32 Nassau Street, Dublin D02 YH68

A CIP catalogue record for this book is available from the British Library

ISBN: 978-0-241-51055-1

*For the two bright stars in my life. You know who you are.*

# Contents

# Prologue: Grim News from Wuhan

On the evening of the 5th of January 2020, just as she was leaving work in Seoul, an email from the World Health Organization (WHO) landed in Dr Cho Chung's inbox. She wavered on whether to look at it, exhausted after a long shift in Yangji Hospital, but decided to give it a quick skim. The news was grim. She felt sick to her stomach. Rushing to inform her superiors in the hospital, she found that they had already heard the news, and a heavy silence filled the air. The hospital's Director began to race through what the next few days would entail. He would have to act quickly and decisively – and it was not going to be easy.

The email read: 'On 31 December 2019, the WHO China Country Office was informed of cases of pneumonia of unknown aetiology [unknown cause] detected in Wuhan City, Hubei Province of China.' Only forty-four cases were reported, and they were all linked to a so-called 'wet' market in Wuhan. These markets slaughter and sell live animals on site in front of buyers. They're found throughout China and are visited on an almost daily basis by those looking for fresh meat and fish, including specialities like snake, baby crocodile, pangolin, beaver, porcupine and civet cat.

No detail was given on what had caused this cluster, nor on whether the virus had the means to transmit not just from animal to human, but from human to human – a development that would make this outbreak much more difficult to control. Flashbacks to 2015 caused ripples through the South Korean medical community. Dr Chung remembered well the price of not having responded quickly to a previous coronavirus outbreak: that of MERS (Middle East respiratory syndrome).

The South Korean government had been accused of mishandling MERS in 2015. At that time it didn't have plans in place to stop infected people entering the country. It hadn't conducted rapid testing to identify cases when they did arrive. And there was a lack of clear messaging to the public on the pathogen, which added fuel to the fire.

The South Korean government was determined not to repeat past mistakes. Upon receiving the news of a novel pathogen in Wuhan, it rapidly implemented its updated outbreak prevention procedures to the letter. These included ignoring the advice of WHO and moving ahead to introduce travel restrictions, such as screening all passengers from Wuhan, contacting diagnostic manufacturers to see how quickly testing could be scaled up, and alerting the public to the new pathogen and advising caution in crowds and in mixing.

Seoul is one of the busiest cities on the planet, with over 10 million people living in the main city and surrounding areas. Markets such as Namdaemun, Dongdaemun and Gyeongdong swarm with thousands of people shopping, eating and mingling. From the 12th to the 15th of May 2015 a 68-year-old man visited four different hospitals in the city, trying to get appropriate treatment for his cough, breathing difficulties and wheezing. He was referred from doctor to doctor, with an initial diagnosis of simple pneumonia. During these stressful few days he sat in crowded emergency rooms, unknowingly infecting dozens of other people.

One of them, a 35-year-old man, left the hospital where he was infected and went to another medical centre, where he in turn infected even more people. The disease began to spread rapidly through crowded hospitals and waiting rooms before the underlying cause was identified. Five days later, on the 20th of May 2015, after he had been referred to the larger hospital in Seoul, the Samsung Medical Center, doctors learnt that the 68-year-old had recently visited the Middle East, and they quickly isolated him. The next challenge was finding a suitable test-kit for what the doctors rightly suspected was MERS. Knowing the danger of MERS, Dr Kim Lee alerted the South Korean government, which closed nearly 2,000 kindergartens and schools and cancelled mass gatherings but withheld details from the public on the extent of the outbreak in hospitals.

MERS, caused by a coronavirus (MERS-CoV), has one of the highest fatality rates of all diseases, with 35 per cent of patients dying from it. Coronaviruses are a family of viruses that cause disease in animals and are usually hosted in bats. Seven have made the jump from animals to humans, with four simply causing the common cold, and three others causing more severe illness: MERS, SARS and SARS-CoV-2. The last is the virus that causes the illness known as COVID-19. Coronaviruses get their name from their distinctive appearance, with crown-like spikes on their surface.

At that time in 2015 there were no vaccines or specific treatments for any coronaviruses, and clinical diagnosis proved difficult, given it can present like flu, with initial symptoms being fever, cough and breathing difficulties. Although there were delays in identifying the symptoms of this small cluster of patients as MERS, the South Korean government succeeded in stopping transmission in hospitals and ended the outbreak with 185 confirmed cases and 38 deaths. The public, however, saw these deaths as preventable. 'Never again' was the clear message that emerged, resulting in an overhaul of their entire approach to similar outbreaks. Leadership changed, legislation was passed, and preparations were made for the next inevitable arrival of a deadly pathogen.

In early January 2020 Foreign Minister Kang Kyung-wha – the first woman to serve as the Foreign Minister for South Korea – sat in a high-level meeting, fully aware of what the next few months would hold for her country and for the world, and the tragic significance of the WHO memo. She had worked hard to make South Korea an engaged global partner and leader, including improving relationships with powerful countries like the US; she wanted the world to see South Korea as responsible and strong. Hearing this news about a potential pandemic, she felt strangely calm. She knew the steps that had to be taken; and, unlike last time with MERS, they had had time to prepare and prevent a crisis. She turned to her fellow cabinet members: 'Let us begin . . .'

# Introduction: Warnings of a Global Pandemic

On a day in July 2021, like most others during the COVID-19 pandemic, I picked up my work mail and opened an envelope amidst the usual plethora of letters and packages. What was inside? A used surgical face mask with the words 'I am done' written in black Sharpie and some traces of white powder. Frozen physically, I found my mind spinning: American public health expert Dr Tony Fauci had received a similar package, and his words came to mind: 'It had to be one of three things. A hoax. Or anthrax, which meant I'd have to go on Cipro for a month. Or, if it was ricin, I was dead, so bye-bye.' Was this anthrax, a serious bacterial infection that can be treated only with the nasty side-effect-addled medicine ciprofloxacin? Was this ricin, a chemical poison with no antidote? After calling university security to report the incident – which, as it happened, turned out to be a stress-inducing hoax – I walked home reflecting on a surreal eighteen months of working on the response to COVID-19.

My life before COVID-19 was a typical academic one: I spent my days preparing lectures for first-year medical and public health students and trying to figure out how to keep them entertained and awake with bad jokes and pop quizzes. My other responsibilities involved grant-writing to fund new research projects, running the Global Health Governance Programme Team, and writing research papers for medical and public health journals. From time to time I would do a niche media interview on a health topic, or work on a policy report, or brief a UN agency, and I sat on the Board of Save the Children UK as a health expert. Overall, despite a busy work schedule, my life was generally quiet and hidden within the walls of the university.

I'm not sure the exact moment when it hit me that life had changed irrevocably. Was it when a local officer rang my doorbell after an anti-masker social media influencer sent out a request for my address on Facebook, so the police had to mark my home as a high-risk target?

Or was it when hundreds of messages from anti-vaxxers – accusations of hurting children and threats to hurt me as well – started to flood not only my inbox but also those of the university and my line manager? This was driven by their cult leader, a notable conspiracy theorist, anti-vaxxer and AIDS deniers, who had asked his followers to go after me. Why couldn't he have spent his retirement watching Netflix and going for walks instead of leading an anti-science movement?

Or was it when well-known TV presenter Piers Morgan jumped to my defence after a professional motorcyclist saw me on TV and called me 'a bird' who waffles 'sh#t'?* Decades of study and working at top universities to climb the ranks and make it as a professor in my field erased.

On the flip side, I regularly received marriage proposals, flowers, art, books, gifts and cards from across the world. People sent me photos of new puppies they had named 'Devi', which felt slightly odd, and a friend texted me on the 27th of January 2021 that she had just got *Grazia* magazine through the door, which had me first on their 'Chart of Lust', alongside Omar Sy, Nicholas Hoult, Octavia Spencer and Joan Collins. I came to see my public profile as an avatar: both adulation and hate oscillating on a daily basis, reflecting whatever people came to see me as representing, and resulting in an uncomfortable level of public exposure that most academics are not used to.

These are just a handful of the many stories I can tell from two years of working as a scientist, government adviser and media expert through the COVID-19 pandemic. But, in fact, as a professor, I was working on fast-moving health issues, such as Ebola, AIDS and infectious diseases, long before COVID-19, and being connected all the time was emotionally exhausting. To recover, every Christmas break, when the uni was closed, I would take one week in which I

* Piers always had my back although I knew to avoid two topics with him: veganism and Meghan Markle.

turned off my email, checked out from the news and just focused on clearing my mind.

In early January 2020 I opened my email after a week of being offline and saw the memo from WHO about a new pneumonia-like cluster of infections in Wuhan that the Chinese government had notified them about. Emails and WhatsApp messages are how bad news arrives in modern times. The global health security community was buzzing: was this SARS, or MERS, or a novel influenza? What symptoms did the disease cause? How severe were these? How did the infection spread? While our community was fixated on this new pathogen in early January 2020, the rest of the world seemed obsessed with Meghan and Harry leaving the British royal family. Since that notification until now, at the time of finishing this book in August 2021, we haven't stopped working.

Part of the difficulty in knowing how serious these notifications are is the vast number of them received each month. WHO picks up 3,000 signals of potential new outbreaks a month, and, of those, it follows up on 300 and investigates 30. As new information arrives, quick judgements need to be made on the seriousness of the situation and whether it will remain a localized event or spread much further. The past few years have seen Zika virus, plague, dengue, multi-drug resistant TB, polio, cholera, Lassa fever, Nipah virus, yellow fever and numerous other infectious diseases flare up in different countries. These largely have remained national or regional epidemics. But in Wuhan the news only got worse. With just 500 cases, China put Hubei Province into lockdown in mid-January 2020. Almost 60 million people were locked down – that's the population of, say, Italy, or South Africa, or England. This was unprecedented in scale and severity, and it was clear that the spread of the new coronavirus would prove challenging, even for the hammer of the Chinese government. I remember being in a fitness centre reading about the Chinese lockdown on my phone and thinking that this was going to be unlike any other outbreak our research team had tracked. I stared at the people surrounding me on treadmills, spin cycles and elliptical machines, all blissfully unaware about how their lives, and those of the entire world, would change in the coming weeks and months.

In this book I draw upon inside knowledge of being intimately involved with the response to the unfolding pandemic since the start. As a professor at the University of Edinburgh Medical School I have been producing key research for the UK and Scottish governments, and running a large research group providing rapid global COVID-19 analysis and policy advice. I have also served on several Scottish government advisory groups, chaired a working group of the Royal Society DELVE (Data Evaluation and Learning for Viral Epidemics) initiative that feeds into the UK government Scientific Advisory Group for Emergencies (SAGE), attended several SAGE and Cabinet Office advisory meetings, and also served in an informal advisory capacity to the Director-General and Health Emergencies Team of WHO, as well as to governments around the world.

My main job is running a research team, the Global Health Governance Programme, which is largely funded by the Wellcome Trust. It investigates international cooperation in health with a particular emphasis on managing infectious disease outbreaks in low- and middle-income settings. Our team members have spent time in Haiti studying cholera, in Senegal studying malaria, in Bangladesh studying childhood pneumonia and in numerous other countries too. I started my career in medicine at the University of Miami but pivoted to public health quickly once I realized that, while medicine is about treating those who are ill, public health is about preventing people becoming sick in the first place. I did my PhD at Oxford University (Oxford call it a DPhil – and, no, in spite of what some Twitter people think, that does not mean I am a philosopher) and spent several months in India studying infectious diseases and malnutrition in children. And, no, despite the accusation that some in the anti-science community have levelled at me, that doesn't make me a dietician: in poor countries, malnutrition and infectious diseases go hand in hand. After my PhD, I stayed at Oxford as faculty, before leaving in 2012 to join Edinburgh University.

I fell in love with Scotland, with its beautiful green spaces and warmth and friendliness. I see Edinburgh as the 'Miami of the North': international and diverse, with great beaches, easy paddle-boarding, seagulls and a relaxed, artsy vibe. The climate is slightly different, as is the water temperature for swimming. Most people in the UK probably know me from my regular slots on ITV's *Good Morning Britain*, Channel 4 and BBC News, or from my bi-weekly *Guardian* column. Or, if they don't know me by name, they may recognize the wall art in my home office, which can be seen over my shoulder in virtual interviews: the *Superman* painting by Alex Ross, which I sometimes switched for Snoopy and his yellow bird Woodstock modelling responsible behaviour by staying at home, Wookiee Chewbacca on a surfboard waiting for the next wave, or a chimpanzee wearing headphones and not wanting to hear hard truths.

Communication during a crisis is vital, as people search for trusted information and basic scientific understanding. COVID-19 affected every single person in one way or another. But it is challenging to share information during a pandemic, because the data and scientific understanding constantly evolve. On a personal level, it's not easy to stick your neck out and be exposed to the court of public opinion, especially over an issue with as many dimensions and opinions as COVID-19. As a scan of newspaper headlines easily shows, I seemed to get into more trouble by accident than most people do on purpose. For better or worse, my role quickly became a public one, as I tried to explain to people in simple language what was happening, the ongoing scientific developments and the basis of government decision-making, while also advising governments directly on their responses – all while running my research team, who were gathering data on COVID-19 from across the world, drafting this into policy briefs and articles, and liaising with me on where to turn their focus next.

## *COVID-19 in History*

In early 2020 a virus originating in China spread across the world and affected the lives of the 7.8 billion people living on earth. Different countries took drastically different approaches to managing a challenge unprecedented in the era of globalization. What became clear was the critical importance of the role of individuals within or advising governments in shaping each country's specific response. Scientific communities also raced to find solutions to save humanity, with science being seen as the only true exit strategy from the pandemic. For years there had been warnings from the scientific community that the greatest threat to humanity would be a pandemic of an acute respiratory pathogen. These warnings were largely ignored.

It would be hard to overstate the significance that will be attributed to the 2020 crisis by history – on a par with the 1918 flu pandemic as a once-in-a-century event that touched every person's life on this planet. As the science writer Ed Yong said, 'The pandemic is not a hurricane or a wildfire. It is not comparable to Pearl Harbor or 9/11. Such disasters are confined by time and space. The SARS-CoV-2 virus will linger through the year and across the world.' The human race had never before been so interconnected, as people faced a virus that just kept on spreading.

If aliens wanted to run an experiment on earth to understand human behaviour, the COVID-19 pandemic would be the ultimate test and revelation. In a crisis, do humans turn on each other or come together? Where are the fracture lines of society? SARS-CoV-2, the virus responsible for the COVID-19 pandemic, causes absolutely no symptoms in some people and leads to deadly disease in others. It pits the healthy against those with underlying health issues, the young against the old, and essential health care workers against those who want their normal services and lives back. Infectious diseases bind us together: a lesson developing countries that face multiple outbreaks a year know well, and one that richer countries like Britain and the US painfully learnt. What would aliens have thought after eighteen months observing our world?

At a global level this disease resulted in a perverse *Hunger Games*, in which countries in the midst of coping with the disease competed in the league tables of COVID-19 mortality, while also trying not to sink their economies and societies. Countries fought over vaccines, personal protective equipment (PPE) for health workers and treatments; and, within countries, people fought across ideological lines about how to respond to the pandemic. In February and March 2020, weeks after WHO rang the alarm bell on COVID-19, all governments chased down limited PPE stocks, ventilators, oxygen, out-of-stock reagents (the main ingredient for chemical testing) for their labs, and experimental steroids and drugs. These were all needed to prepare health systems for an onslaught of COVID-19 patients in hospitals and clinics. The US stole ventilators from Barbados, PPE from Germany and bought up the rights to remdesivir, seen at the time as a promising treatment for COVID-19, limiting stock to other countries. Despite a World Health Assembly in May 2020, where all member states attending committed to sharing research products and working collectively to address COVID-19, this co-operation broke down when tough decisions had to be made over allocation of scarce resources.

All of this raises the question: where does selfishness begin and end in a pandemic? What is the responsibility of richer countries to poorer countries in the context of limited vaccine supply? In May 2020 richer countries agreed at WHO to share vaccine supply. Months later, by November 2020, European Union countries, the UK and the US had bought up more than 80 per cent of Pfizer/BioNTech vaccine doses. The Director of the African Centres for Disease Control and Prevention noted that he didn't expect African countries to receive any doses of vaccines until late 2021, or even into 2022. This has always been the case in global health: those who pay the most acquire the research products. WHO actively tried to warn against this nationalistic approach, but in the end words and resolutions are just that: as has been shown, it's money and power that count.

This is the story of global health: the massive progress in rich countries over the last century in reducing child mortality, increasing life expectancy and eliminating infectious diseases like polio, smallpox and malaria, set against that of poorer countries, where we still see the rampant spread of preventable diseases and the continual suffering and deaths of children from measles, diarrhoea and pneumonia.

The same questions about selfishness can be asked of our commitment to each other locally. What is our responsibility to our communities? Over the course of the COVID-19 pandemic, families, friends and neighbours have become divided over whether they prioritized the self-interest of enjoying the most from their life and interpreting the rules in ways that met their individual wishes, or agreeing to sacrifice their own desires to go to weddings, funerals or holidays, for the collective good of society. The mixed reaction to taking summer holidays reflected this, as did the return to schools in the UK. A clear example is when entire 'bubbles' of children (school classes kept together to enable contact tracing of positive cases) were sent home from school in September 2020, because one child had been on holiday abroad over the summer and the parents had decided not to abide by the fourteen-day quarantine rules. Children and families in the entire bubble had to pay for the decision of one family to break the rules.

Families have also been fractured over weddings; whether they should take a more cautious approach and delay indoor parties; and whether they should ignore the rules and continue to have dinner parties and sleepovers. And each of us may have re-evaluated the people in our lives. We have become closer to some families who share our thinking and distanced ourselves from others we have seen to have different values. As we compare our pre- and post-COVID-19 selves, perhaps the question is not how much the pandemic has changed us, but rather how it has shown us who we really are as people.

But there are definite moments of selflessness, courage and brightness to be found in 2020 and 2021. We must never forget the personal sacrifice that health care staff have made to treat all patients who showed up needing care, often going to work on wards without having adequate PPE. Thousands of health care workers have died, having contracted the infection at work: they have borne the brunt of this pandemic and cared for all those who needed it, even if those people were COVID-19 deniers, refused life-saving vaccinations or blatantly held parties despite warnings against the risk.

The science community started sprinting in early January 2020, and within weeks had developed PCR (polymerase chain reaction) tests for laboratories; then, soon after, rapid lateral flow (LF) tests for home-testing; and then antibody tests to check for prior infection with COVID-19. Within months they had effectively trialled treatments for COVID-19, such as dexamethasone, which improves survival for the sickest patients; and within a year they had developed multiple safe and effective vaccines. It can't be overstated how incredible these developments are: never before has science worked as collaboratively, effectively and rapidly to find solutions.

But perhaps inequality is the most revealing lesson of COVID-19. Time and time again we have seen that it is one rule for some, another rule for others, whether in early access to testing, compliance with restrictions or creating loopholes in restrictions to allow 'high value' members of society to not abide by quarantine, while ordinary people do.

Wealth was indeed the best shielding strategy, not only from COVID-19 but from the response to it as well: lockdowns around the world exposed the plight of the poor in overcrowded housing compared with the country estates of the rich.

How do we take the open wounds that have been exposed and build a more equal and resilient society? How do we ensure that all people in all parts of the world have the same access to research developments and protection from disease? It starts with government. Abraham Lincoln's words ring clearly: we need 'government of the people, by the people, for the people', not just government for the ruling classes. Perhaps that's the strongest historical legacy of COVID-19, as explored in Chapter 11. And this also gets to the core of why I wrote this book: to show how COVID-19 put into stark relief how global politics shape our health.

## *Warnings Ignored*

For years scientists working within global health and health security, and leaders like Bill Gates, Angela Merkel and Barack Obama, have been warning that the greatest threat to international stability and security would be a global pandemic. In a 2015 TED (Technology, Entertainment and Design) Talk, Gates warned that the biggest potential killer would be not a war but a pandemic. In a speech three years later, he noted, 'The next threat may not be a flu at all. More than likely, it will be an unknown pathogen that we see for the first time during an outbreak, as was the case with SARS, MERS and other recently discovered infectious diseases.'

SARS (or severe acute respiratory syndrome), caused by another coronavirus (SARS-CoV-1), created waves in 2002 and 2003 when it spread largely across East Asia, infecting at least 8,000 people and killing 10 per cent of them. Similar to SARS-CoV-2, SARS is a fast-moving virus with an even higher fatality rate. There was serious concern that, if the virus kept spreading, it could cause millions of deaths across the world.

Fortunately, this did not happen. How did the SARS epidemic end? It neither magically disappeared nor became endemic. Rather, a strong public health response was mounted, orientated around testing those who were unwell, tracing their contacts and isolating any suspected cases. SARS spreads when people are ill (and almost everyone infected gets unwell within two to three days), and so, by ensuring isolation of positive cases and their contacts, it was possible to eliminate the virus from human populations, country by country. SARS needed to jump from one human host to another during the infectious period. If the infectious period passed without any spread occurring, that line of infection died out. As in classic infectious disease management, breaking chains of infection was the way to stop spread and eliminate disease.

Yet this near miss of a pandemic event was not acted upon globally. While East Asian countries bolstered their response mechanisms, having borne the brunt of the SARS epidemic, Western countries remained largely unaffected. No real investment was made in developing a vaccine against SARS, or in devising country preparedness plans for a SARS-like event in most countries, even though SARS-like coronaviruses continued to circulate in bats. There was always a strong likelihood that one of these would jump into humans again and be harder to stop using the traditional infectious disease response.

Even with his repeated warnings about pandemic preparedness, Gates felt he had been ignored, reflecting in December 2020: 'I wish I had done more to call attention to the danger. I feel terrible. The whole point of talking about it was that we could take action and minimize the damage.'

Already in 2014 former US President Obama had highlighted the potential dangers of an airborne pathogen in a speech at the National Institutes of Health. He noted: 'And we were lucky with H1N1 [the 2009/10 swine flu pandemic] – that it did not prove to be more deadly. We can't say we're lucky with Ebola because obviously it's having a devastating effect in West Africa but it is not airborne in its transmission. There may and likely will come a time in which we have both an airborne disease that is deadly.' And, in fact, in 2013 a Worldwide Threat Assessment of the US Intelligence Community noted that 'an easily transmissible, novel respiratory pathogen that kills or incapacitates more than 1 per cent of its victims is among the most disruptive events possible.'

In 2015 German Chancellor Angela Merkel spoke at the opening of the annual World Health Assembly in Geneva, pushing for a better global response system for pandemics. She used Germany's presidency of the G7 that year to ensure pandemic preparedness was a top priority.

Even more recently, in early December 2019, just before COVID-19 emerged, I was asked by the University of Edinburgh magazine about predictions for urgent issues for the year ahead. This is what I said: 'The next deadly disease that will cause a global pandemic is coming. A major priority of my work in 2020 is looking at how governments, international institutions and the private sector can better prepare for and respond to outbreaks. With increased urbanization, movement of people, and closer interaction between animals and humans, it is certain that we will have a rising number of outbreaks of infectious disease.'

Warnings from scientists have been even more specific. In a 2007 research paper in the prestigious journal *Clinical Microbiology Reviews*, the authors warn about wet markets and leaks from biosecurity labs handling virus samples as sources of new infections. Both these settings have been highlighted as potentially linked to the first human infected with SARS-CoV-2. They noted, 'The presence of a large reservoir of SARS-CoV-like viruses in horseshoe bats, together with the culture of eating exotic mammals in southern China, is a time bomb. The possibility of the re-emergence of SARS and other novel viruses from animals or laboratories and therefore the need for preparedness should not be ignored.'

In both a 2018 and 2019 threat assessment, the US intelligence community warned that the US, and the world as a whole, would be vulnerable to the next flu pandemic or large-scale contagious disease outbreak. The 2018 report mentioned that a 'novel strain of a virulent microbe that is easily transmissible between humans continues to be a major threat, with pathogens such as H5N1 and H7N9 influenza and MERS coronavirus having pandemic potential if they were to acquire efficient human-to-human transmissibility.' The 2019 report again noted that we would see more frequent outbreaks of disease due to unplanned urbanization (such as people moving to outskirts of cities and putting up concrete houses on forest land), prolonged humanitarian crises (such as conflict), climate change (such as mosquitos having a wider ranging ground due to increased temperatures), and the expansion and speed of international travel and trade, which make it easier for a virus to spread around the world.

In the Johns Hopkins 2019 *Preparedness for a High-Impact Respiratory Pathogen Pandemic* report, the authors noted it would be difficult to control the spread of a respiratory pathogen if it had a short incubation period (small period of time of being exposed to the virus and being infectious to others) and could spread asymptomatically (someone could feel perfectly well and still infect others). They were basically describing SARS-CoV-2. The authors also highlighted certain families of viruses that would be most likely to cause the next pandemic: they identified both influenza (flu) as well as coronaviruses.

WHO, and its Director, Dr Tedros Adhanom Ghebreyesus, also repeatedly warned countries, noting that the next pandemic was not a question of if, but when: 'The threat of pandemic influenza is ever-present.' In March 2019 WHO released a global plan to fight influenza, which was described as the most comprehensive to date. It included measures to try to protect populations as much as possible from annual outbreaks of seasonal flu, as well as to prepare for pandemic flu. The two main goals were to improve worldwide capacity for surveillance and response, by urging governments to develop national flu plans, as well as better tools to prevent, detect, control and treat flu.

While the focus has been largely on flu off the back of the 2009 swine and 2005 avian flu outbreaks, the 1918 flu pandemic is scarred into the memory of those who have trained in public health because of the tens of millions of people that died from it. In contrast, 1918 is largely forgotten by the general public whom scientists are trying to protect. Jeremy Konyndyk, Director of Foreign Disaster Assistance for the US Agency for International Development, said in 2017, 'At some point a highly fatal, highly contagious virus will emerge – like the 1918 Spanish Flu pandemic.'

The major unknown has been when exactly and which one of the thousands of viruses circulating in the animal kingdom would make a jump to humans, and, once in a human, be able to sustain easy human-to-human transmission. These 'spillover' events have occurred with increasing frequency since 2000. Part of the explanation is having better tracking systems to detect when they occur, but the larger trend is the increasingly closer contact of humans with wild animals, particularly bats, through deforestation and live animal markets (or wet markets), as well as intensive farming of animals in crowded conditions.

But it's not enough for a virus to jump from an animal over to human-to-human transmission for an epidemic to occur. The virus must have an effective way of transmitting, whether through droplets passed through air or bodily fluids like blood or sweat; and it shouldn't kill off the host too quickly, otherwise it won't spread to another human before the original patient dies.

The virus that would cause the most harm in humans would be one that transmits quickly and easily, like the common cold or flu, where most who carry the virus can transmit it to others before falling extremely ill (presymptomatic transmission), and one where the percentage of people dying from the virus (the case fatality rate) is low enough for it to be tolerated by governments and communities as a lurking threat. Add in a virus that causes multi-system syndromes, such as lung scarring, heart inflammation and attacks, blood clotting, and recurrent fatigue, fever and pain, and it's a recipe for a disease of nightmares.

Unfortunately for the world, the virus SARS-CoV-2, and the

disease it causes, COVID-19, fit all those criteria – and more. The virus also can reside in human carriers who don't even know they are infected (asymptomatic) and these people can then continue to transmit and pass it on to others.

## *What Scientists Could and Couldn't Predict*

While the emergence of a new pathogen was entirely foreseen, what became apparent during the first year of the pandemic in 2020 was that the response taken by countries and their citizens was not quite so predictable.

First, few in the scientific community could have predicted that global panic at the start of a pandemic would set in worldwide over . . . toilet paper. Whether in Hong Kong, Germany, Singapore, Japan, Australia, the US or New Zealand, toilet paper became the 'must-need' pandemic buy, leading to fights in grocery stores, mass buying and hoarding of hundreds of rolls, and even people breaking into each other's homes to steal loo roll. #toiletpapergate and #toiletpapercrisis trended on social media. Australia's Chief Medical Officer had to issue a statement to parliament in March 2020: 'We are trying to reassure people that removing all the lavatory paper from the shelves of supermarkets probably isn't a proportionate or sensible thing to do at this time.'

The panic buying of toilet paper was clearly irrational behaviour, probably driven by FOMO – Fear of Missing Out – during a crisis. People thought, well, if their friends were all buying it and media was reporting on this phenomenon, they needed to buy it too. And with a situation spiralling out of control, at least buying toilet paper was one thing that people could do to ensure their comfort in the bathroom during the pandemic. Even during my childhood in Miami, when people would stockpile ahead of hurricanes, they bought petrol, bottled water, torches, batteries and long-lasting tinned food. Toilet paper was never top of that list.

Second, no one could have anticipated that the US and UK, consistently ranked by pandemic preparedness indices as the top-two countries for capacity and readiness, would suffer as badly as they did. I describe how exactly this happened in Chapters 5 and 6. Other countries ranked lower on the list, like Senegal, Ghana, Vietnam and Liberia, reacted quickly and effectively to contain their outbreaks and manage the response. As we will see, failure can be attributed to poor leadership, lack of humility in the face of an infectious disease, not understanding the history of humans' fragile relationship with germs and falsely trying to save the economy rather than human life.

Third, it was surprising that the strength of a health care system would not determine the toll COVID-19 would take. Richer countries made health services their front line instead of recognizing that infectious disease prevention is about all the steps put in place to stop someone becoming infected and arriving at hospital in the first place. Poorer countries knew they couldn't rely on their health services so focused more on preventing infections in the community. In January 2020 estimates out of Wuhan were that 20 per cent of those infected would need hospital care, and a third of those would need an intensive care bed. The patient numbers become astronomical at a population level and far beyond the reach of any health care system.

Fourth, we didn't expect that some leaders would falsely claim the choice was between the economy (affecting millions) and COVID-19 (affecting thousands) instead of realizing that minimizing COVID-19 harm also reduces non-COVID-19 harm and vice versa.

A fifth unpredictable aspect was that the virus would spread across the world through business travellers, luxury holiday-makers on skiing holidays and cruise ships, and affect wealthier and more connected parts of the world first. The prior expectation was that an outbreak would emerge, like the 2014 Ebola outbreak, from poor and fragile settings like rural Guinea or earthquake-torn Haiti.

Mathematical modelling, based on air traffic passenger data, identified key hubs where the virus would travel to first. This didn't predict that in February 2020 Lombardy in northern Italy would be the first place badly hit in Europe, or that Iran would have one of the first major waves outside China, or that touristy Costa Adeje on the southern tip of Tenerife in the Canary Islands would see entire hotels isolated. Travel restrictions and border control would become one of the most important tools governments had to manage the seeding of infection in their populations, but this meant ignoring WHO, which cautioned against using travel and trade restrictions.

Finally, no one could have predicted how difficult it would be to put joined-up strategies in place in Europe and North America, let alone globally. If SARS-CoV-2 killed as many people as SARS or MERS, it would have been easier to have countries all heading towards elimination together. No political leader could have had an 'acceptable level of infection' of a disease that killed 10 per cent or 35 per cent of people, or even a disease that badly affected children or young people. Because of *who* SARS-CoV-2 kills, countries chose various strategies, resulting in a patchwork of mitigation (living with the virus), suppression (control it at low levels) and elimination (Zero COVID), and an absence of global cooperation and strategy. The uncontrolled circulation of the virus allowed variants to emerge that became increasingly transmissible and severe. This was in direct opposition to what social media pundits were sharing with their followers online: they kept arguing that the virus would mutate to a milder form and thus mutation would be beneficial.

In the end, as this book will show, vaccination, testing and treatments were the routes out of the pandemic. Mitigation, suppression and elimination were just holding strategies until science could deliver a solution. Then the race began to vaccinate enough of the adult population to limit both the number of susceptible people and the spread of the virus, as well as to stop rising hospitalizations and severe disease caused by COVID-19.

## *The Toll of the Pandemic*

At the time of writing this book, nearly 5 million people have lost their lives to COVID-19 and millions more have recovered but are living with symptoms that could cause pain and illness for years to come. It is hard to know accurately how many lives were lost to COVID-19, given the lack of testing in most parts of the world, and the lack of vital registration systems in some countries – if someone had died in a remote village, it may have been impossible to take note of this.

But the additional toll of this pandemic is reflected in the economic recession that has caused unemployment figures to rise, more people to go hungry, small businesses to go under, and international travel and the aviation industry to collapse. It is seen in the hundreds of millions of children out of school; the elderly and vulnerable locked up at home and in care homes, unsure whether it is indeed 'safe' to venture out or to hug their grandchildren; and in the mental health anguish of health care workers watching their colleagues becoming ill, and sometimes even dying, all while fighting for adequate PPE and decent kit.

Poorer countries have suffered badly in numerous ways. Working-age adults fell ill and were not able to work for days or weeks, leading to a fall in income and increased poverty. In most low- and middle-income countries, the informal economy (i.e., the parts of the economy that occur off books and that are untaxed and cash-based) makes up a significant share of employment. Owing to poor health and safety conditions, informal workers are most at risk of getting COVID-19, but also most at risk of income loss if they don't show up to work. There is often no paid leave or ability to take time off. Plus the costs of going to hospital can force families into debt, bankruptcy or below the poverty line, because of the need to pay for medical care when it is provided. Many poor people don't have any form of health insurance so have to pay these costs 'out of pocket' especially when no public health services are available.

Changes in consumption behaviour by those avoiding the virus (choosing not to visit markets or to go out and spend money) also lead to job losses. In 2020, Sub-Saharan Africa experienced its first economic recession in twenty-five years, with an economic decline of 2.0 per cent. In Latin America and the Caribbean the economy declined by 6.7 per cent in 2020, the worst economic contraction in the region's history. Unemployment reached an eye-watering 13.5 per cent in the region, up from a typical 5–6 per cent. It would be impossible to understand the pandemic's impact on humans without considering not only the deaths but the economic devastation and societal upheaval it caused.

# The Story of COVID-19

The true story and toll of COVID-19 will become clear only in the years to come. In this book I attempt to take a look at the first eighteen months of the pandemic, and to explain how a simple jump of a virus from a bat or a pangolin to a single human – an incident that took place within seconds – had repercussions for the entire world: a Pandora's Box event that with each passing day became harder and harder to control, with the numbers spiralling from a few dozen cases to 100,000, then a million, then nearly 10 million within just a few months, and then hundreds of millions a year later. And I argue that with the right politics and leadership, much of the suffering and death was largely preventable. Looking ahead, effective leadership, preparation and cooperation can prevent catastrophic pandemics in the future.

This pandemic illustrated that the world is now interconnected as never before, but it also revealed the importance of sovereign states and their protection of their borders. Every person in the world has been affected by this outbreak. How exactly they have been affected has more to do with their government's handling of the virus than anything else. Successful responses in 2020 did not correlate with country wealth but rather with decisive political will and a clear strategy. In this book I compare the responses in the US and UK with those of South Korea, Senegal and China, among others. Everything – whether it was schools reopening, shielding of the elderly and vulnerable individuals, the viability of businesses and families being able to see each other – depended on the ability of a government to contain this virus aggressively and quickly.

We will also explore how global politics shape our health. I use the story of the COVID-19 pandemic to explore deeper issues within global health politics, and what these tell us about major gaps in our systems of response as well as about the way to best prepare for the next crisis. I start by tracing the evolution of the virus from China and East Asia to Europe, North America and the Western Pacific, and finally to Sub-Saharan Africa, South Asia and Latin America. Combining science, politics, ethics and economics, I dissect the global structures that determine our fate, and the deep-seated economic and social inequalities at their heart. Highlighting lessons learnt from past and present, the book ends by setting out a vision for how we can better protect ourselves from the inevitable health crises to come. In the following chapters I describe pivotal developments in the outbreak: the arrest of the Chinese doctor in Wuhan who was trying to warn about a new respiratory pathogen and later died of COVID-19; the horror of the Chinese virologists who identified SARS-CoV-2 and realized this could trigger a global pandemic against which humankind had no defences; the inside deliberations in the WHO headquarters in Geneva between Dr Tedros Adhanom Ghebreyesus (the Director-General) and Dr Mike Ryan (Executive Director of the Health Emergencies Programme) over how to handle the increasing geopolitical tension between the US and China and keep all governments sharing data and at the table; the boredom of passengers stuck on the *Diamond Princess* cruise ship as Japanese authorities panicked over how the virus was spreading and what to do with the people on board; the frustration and anger of President Donald Trump, who came to realize that COVID-19 would be his main opponent going into the 2020 presidential election; the blind optimism of UK Prime Minister Boris Johnson that 'taking it on the chin' would be the best way to get through the pandemic; the fatalism of Swedish health authorities that the virus was unstoppable, so normal life must continue even as the death toll mounted; the massive logistical operation that South Korea put in place to avoid lockdown and contain the virus; the anxiety of health workers being sent on to COVID-19 wards without appropriate protection while hearing the clapping of communities outside of hospitals recognizing their sacrifice and martyrdom; and the tragedy in several low-resource countries like India, Brazil and South Africa of seeing a tidal wave coming but knowing that there would be no way to cope. While telling these stories, I also unpick the major themes that ran throughout the pandemic: the accentuation of underlying inequalities within society, such as discrimination, racism and poverty that became impossible to ignore; the relationship between governments and their scientific experts; the debate over whether lockdown would do more harm than COVID-19 itself, through rising unemployment, poverty and starvation; achieving a balance between engaging the research community in accelerated treatment and vaccine development and containing Big Pharma's drive to cash in for their shareholders' benefit; the rising anti-science movement that contributed to the US's being blown off course; and how the populist leadership styles from the US, UK and Brazil struggled to cope with the need during a pandemic for solid and cautious governance.

## *How It Began*

The story of COVID-19 will be told in myriad ways; it already has been. And we all have individual stories of how our own lives and families were changed from January 2020 onwards. Almost a sliding-doors moment of how life would have looked if COVID-19 hadn't happened, versus how it turned out instead, for better or worse. This is my attempt to make sense of how it started and evolved, and how not only this pandemic but also future outbreaks may be brought to an end.

I hope you enjoy reading the analysis, tolerate my bad sense of humour and perhaps even see the world in a slightly different way afterwards. I hope reading this makes you smile and tear up and reflect on who we are as people, as communities and as members of humanity. The entire pandemic and the devastation it brought to lives and livelihoods, as we will see in the next chapter, all started from a single case. And from there it spiralled out to all corners of the world, from the coldest parts of Antarctica to the most remote tribal communities in Brazil. But the story of COVID-19 begins in China.

# 1. Spillover

It is likely impossible to identify the exact moment that SARS-CoV-2 jumped from its host reservoir in bats, into some kind of intermediary animal host, and finally into humans, as WHO suggests. This random and seemingly trivial act in the universe was one of hundreds that occur across the world each day. But in this instance coronavirus managed not only to infect a human but to sustain human-to-human transmission. It also managed to sustain this in an easily transmissible way: through people breathing close to each other. And, even more perniciously, it was able to spread from people who felt completely well.

This was just bad luck for us humans. At each step in the chain, from contact through to human-to-human transmission, the odds stack ever more greatly against the virus. But, once a virus can spread through breathing, it's a recipe for a very difficult pathogen to stop. In this chapter we take a closer look at the concerns over diseases at the animal–human nexus and why spillover events have been happening more and more often. And we examine the various theories on the origin of SARS-CoV-2 to address whether there was ever an opportunity to stop its introduction into humans. Where did SARS-CoV-2 come from? How was it first detected, and how long was the delay in recognizing there was a new virus spreading? When did scientists realize the world would never be the same and that SARS-CoV-2 could indeed be the 'Disease X' we had all been preparing for? And how much is the Chinese government to blame for the COVID-19 pandemic? At an early stage could the pandemic have been prevented? Simple questions with important and often not-so-easy answers.

## *No, It's Not Witchcraft. It's Science*

Scientists have repeatedly warned that we would face another pandemic in the future. Almost all agreed that the question was when, not if. I was often asked in 2020 how I was able to 'predict' SARS-CoV-2. This stemmed from a *Daily Mail* article titled 'Leading UK Professor warned of coronavirus type outbreak two years ago'. The article pulled out a video clip from a 2018 Hay Festival talk where I spoke about a fictional scenario of an animal passing on infection to a farmer in China, and then the farmer infecting those in their community before getting on a plane to the UK.

I chose China as the location for this spillover event to illustrate the point that our health threats are interconnected across the world, and to show that countries must cooperate in order to manage threats to their people. In my fictional scenario, for the UK government to worry only about what was happening within its own borders didn't make sense in a globally interconnected world. This vulnerability was clearly revealed in January and February 2020, when the UK government was more concerned with Brexit and internal politics than about an emerging health threat in China, as we will see in Chapter 5.

Rich countries like the US and UK are often reluctant to sit at the same table as small and poor countries like Malawi or Haiti. This is clearly apparent at the World Health Assembly, the main decision-making body within WHO, where all countries, regardless of wealth, population size or power, are given the same speaking time and the same voting power.

This 'one country, one vote' dynamic has frustrated richer countries that have shifted their financial resources into newer, multilateral agencies, like the Global Fund to Fight HIV/AIDS, TB and Malaria in order to have more control over decision-making and priorities. The US is the largest funder of the Global Fund, and minutes reveal that the board of the Global Fund has never disagreed with a US position on any issue. This contrasts with WHO, which often pressurizes the US into international agreements that are driven by other countries. For example, after strong opposition, the US finally capitulated and agreed to the resolution adopted at the May 2020 annual World Health Assembly: that the research and development of COVID-19 products, such as vaccines, therapeutic drugs and testing technology, would be shared globally.

Returning to the Hay Festival: the truth is that I didn't predict a particular coronavirus outbreak but nor were my comments random guesses at the future. I was just outlining what many health experts and leaders – whether Bill Gates, Barack Obama, Angela Merkel or Dr Tedros – had been concerned about for years, as mentioned in the book's Introduction. The 2014 Ebola outbreak in West Africa revealed our global vulnerability to pathogens, albeit a virus that does not transmit very easily. In fact, the May 2017 *Time* magazine cover warned, 'We are not ready for the next pandemic.' In September 2019 WHO published a report warning about a fast-moving respiratory pathogen and urged countries to prepare. And this pathogen could easily be a novel influenza or a SARS- or MERS-like coronavirus, or even a new bacterial strain resistant to our current crop of antibiotics.

The history of global and international health is the history of infectious disease control. In 1851 the first International Sanitary Conference, hosted by the French government in Paris, was called because states were concerned about the spread of infectious diseases along international trade routes. Twelve countries participated to discuss whether cholera should be subject to quarantine regulations. Thirteen more conferences followed, with an increasing number of governments and an expanded list of diseases, including yellow fever, bubonic plague, smallpox, typhus and cholera. Aside from smallpox, which has been eradicated completely, all these diseases are still challenges, as shown by outbreaks in the past few years of yellow fever in Brazil, plague in Madagascar and cholera in Haiti.

These viruses and diseases have been around for centuries, circulating and afflicting humans. But where did they come from? Let's look at COVID-19 to find out.

## *Where Did COVID-19 Come From?*

As I write this in August 2021, the exact origin of SARS-CoV-2 is still contested, with various scientists suggesting that it may not have had a fully natural origin: for example, they say that the virus accidentally escaped from a virology laboratory or during a field experiment when researchers were visiting bat caves.

Initially the outbreak (known as the 'Wuhan pneumonia cluster') was linked to a wet market where live animals were kept in caged conditions. It was believed that several individuals might have been infected by the same animal there. Wet markets are always a risk because of poor ventilation, poor drainage, close proximity of wild animals to one another in unhygienic conditions and thousands of daily visitors packed together.

Phylogenetic evidence, that is, evidence that looks at the evolutionary relationship among various genetic samples of virus, shows that a transmission cluster developed in the market, because the genomic sequencing was almost identical among all the cases in the market. Genomic sequencing is a tool that allows us to track how a virus changes and mutates. Each time the virus replicates itself inside a new host, mutations arise randomly from mistakes made in copying, and these can now be tracked in real-time to provide almost a 'fingerprint' of how the virus once looked within a particular human at a particular time. Sequencing has been crucial throughout the COVID-19 pandemic, whether in enabling us to create the first test-kits in mid-January 2020 or in tracking the spread of variants in early 2021 (variants are slight changes to the original SARS-CoV-2 that lead to different characteristics, and we'll hear more about these later).

Once the 'Wuhan pneumonia cluster' was identified in December 2019, the Chinese government immediately shut the market to undertake further investigation. Although the wildlife produce at the market was considered potentially susceptible to SARS-CoV-2, none of the animal products, live or dead, tested positive. In addition, it became clear that other COVID-19 cases had not been to the market, including the man who had been diagnosed. WHO Mission Team member Dominic Dwyer commented, 'The market in Wuhan, in the end, was more of an amplifying event rather than necessarily a true ground zero. So we need to look elsewhere for the viral origins.'

The Chinese government has been difficult and cagey in sharing information on the earliest cluster of cases; based on finding traces in waste-water and reports of China purchasing a number of testing machines earlier in 2019, some have suggested that the virus was circulating in China earlier in 2019. The Chinese government also won't say when exactly high-level authorities were aware that they were facing a novel pathogen of possible pandemic potential. While it notified WHO on the 30th of December under its obligations under international law, it hasn't made it easy for any external group to come into the country and assess the origin.

The Chinese government has put forward a theory called 'the cold-chain hypothesis': that the virus originated outside China, potentially in Italy, and was brought into China in late 2019. This theory postulates that bits of the virus could have been frozen into food in Place A (e.g., Italy) and sent to Place B (Wuhan). Chinese workers in these factories would have been exposed to the viral bits and become infected. Therefore, the government has suggested that the virus could have been imported from another part of the world and therefore it is not to 'blame' for the pandemic. It is hard to investigate this theory thoroughly: one would have to test lots of frozen products and identify the origins for any positive samples. But it seems highly unlikely that the virus was circulating earlier in 2019 (either in China or elsewhere), in light of the fact that no exposure to the virus was detected anywhere else in the world pre-November 2019.

The caginess of the Chinese government, and its lack of transparency about the origin of the virus, has led to outlandish conspiracy theories by several individuals from the US, such as the virus being bioengineered as a weapon and released intentionally, as well as to somewhat more reasonable allegations that SARS-CoV-2, while having a natural origin, may have been leaked from a lab by accident.

Along with US former Secretary of State Mike Pompeo and former head of MI6 Sir Richard Dearlove, former President Donald Trump often said that he believed that China knows more than it's sharing. All three have pointed to the high-security biosecurity lab in Wuhan, a key research hub for coronaviruses, and suggested that experiments within the lab created this virus, which then escaped into the local community. This hypothesis has gained traction as a result of China's draconian response to any spread of the virus within their country, and the harsh lockdown in Wuhan at just over 500 cases in order to prevent further spread throughout the country. People who believe this wonder whether China would have responded so swiftly had it not originated from the lab.

A second theory of lab escape does not assume that the virus was designed as a biological weapon. Instead, it notes that it originated zoonotically (in bats) but lab modifications produced greater transmissibility. The benign intent might have been to use virus samples to help to build a vaccine to protect whole populations against future coronaviruses. Richard Ebright a molecular biologist, has said, 'There are indeed many unexplained features of this virus that are hard, if not impossible, to explain based on a completely natural origin.'

In 2017, when the Wuhan Institute of Virology was launched, Ebright raised biosecurity concerns about the procedures for handling samples, and entry and exit points to the lab not being secure enough. When he heard about the Wuhan outbreak, he noted, 'The news of a novel coronavirus in Wuhan screamed lab release.' Similarly Jamie Metzl, a geopolitical expert, said that lab escape could indeed be plausible, considering the close proximity of the Wuhan Institute of Virology (the only Level 4 lab in China, and the world's research centre for coronaviruses) to the wet market associated with the first cluster of cases. (Levels 1–4 are used to describe the biosafety level for various biological agents, from Level 1 labs, which work with agents that do not cause disease in healthy humans, to Level 4 labs, which work with samples that can cause severe to fatal disease for which there is no vaccine or treatment.) Metzl was also frustrated by the failure of the Chinese government to authorize an independent investigation into the natural origin.

Metzl often cites a paper in *Nature*, the leading scientific journal, to support his scientific claims. The paper shows that the RaTG13 coronavirus (a viral relative found in horseshoe bats) sampled in Yunnan Province in 2013 shared 96.2 per cent of the genomic sequence, across the 29,891 genomic base-pairs, with SARS-CoV-2. In other words, it was pretty similar across the entire genome. However, in the region encoding the reception-binder domain of the spike protein, the sequence identity dropped to 85 per cent. A specific part of the genome, therefore, was different, and it is this part that reveals how the virus infects cells. Since the spike protein is critical to the virus's behaviour, Metzl thinks it is puzzling that this particular region does not match the rest of the genomic similarity.

Studies re-engineering viruses have been funded around the world, even by the US National Institutes of Health (NIH), 'because they help define the fundamental nature of human-pathogen interactions, enable the assessment of the pandemic potential of emerging infectious agents, and inform public health and preparedness efforts', as US NIH Director Dr Francis Collins has said.

In 2015 *Nature Medicine* published a study on an infectious disease team's efforts to engineer a new virus that combined elements of existing ones. They joined the SHC014 coronavirus surface protein (like those found in Chinese horseshoe bats) to the entire genomic sequence of a virus that causes human-like SARS in mice. The study found that this virus could infect human airway cells in samples and cause disease in mice.

However, these studies were stopped in the US after the NIH ceased federal funding for gain-of-function studies (studies that intentionally modify pathogens for certain characteristics) for influenza, SARS and MERS – the reason being the biosafety and biosecurity risks of viruses escaping. Even with clear procedures in place, the US has suffered from numerous escapes of pathogens from labs across the country. For example, in May 2015, the US Department of Defense sent live anthrax samples (instead of dead ones) to eighteen labs across the US and to a military base in South Korea. By contrast, China has continued these kinds of gain-of-function studies leading Dr Philip Murphy, an NIH scientist, to note, 'All possibilities should be on the table, including a lab leak.'

However, several senior scientists, including virologists, have firmly rejected the lab-modification hypothesis. They point to several clues that lead to natural origin. For example, the genomic sequencing evidence indicates that, instead of signs of inserted gene sequences, the positioning of the points of variation (from the bat sequences) are scattered randomly, as though it had evolved naturally. SARS-CoV-2 has a high affinity for binding to ACE2 receptors. But why this receptor, especially given it is so abundant in humans? Computation analysis suggests that the high affinity binding for SARS-CoV-2 to ACE2 is not ideal. Therefore, this characteristic probably arose from the virus fastening itself on to a human or human-like ACE2, i.e., through a process of natural selection. Virologists have pointed to this finding as 'strong evidence that SARS-CoV-2 is not the product of purposeful manipulation'.

Similarly, a *Nature Medicine* correspondence in 2020 noted,

> It is improbable that SARS-CoV-2 emerged through laboratory manipulation of a related SARS-CoV-like coronavirus. The RBD [receptor binding domain] of SARS-CoV-2 is optimized for binding to human ACE2 with an efficient solution different from those previously predicted. Furthermore, if genetic manipulation had been performed, one of the several reverse-genetic systems available for betacoronaviruses would probably have been used. The genetic data irrefutably shows that SARS-CoV-2 is not derived from any previously used virus backbone.

This argument against lab escape was reinforced by Dr Peter Embarek, Mission Leader of the WHO team sent to Wuhan, who noted in February 2021, 'I think that's in line with what other laboratories around the world have said as well, that this virus has not been worked with knowingly in any labs around the world working with coronaviruses.' He continued that the Mission Team had visited labs in Wuhan, where Chinese scientists firmly refuted that they had been working with or had the SARS-CoV-2 virus in their collections or in their laboratories. Embarek later did a TV interview in August 2021 in which he acknowledged the tight scrutiny of the Chinese government on the WHO Mission Team, and shared his hypothesis that the first person infected with SARS-CoV-2 worked at the Wuhan Institute of Virology.

Given the heavy-handed pressure of the Chinese government, and its repeated attempts to say the virus came from abroad, it is not enough just to take what is said at face value. More evidence is needed to firmly rule out the lab escape hypothesis than just Chinese scientists (under potential pressure from the Chinese government) saying it's not true. The US government, even under a Biden presidency, has been hesitant to accept the Chinese government's version of events. It has continued to push for an independent team of scientists to have full access without supervision by authorities. Tedros supported an independent investigation including a lab audit in March 2021, which led to the Chinese government growing increasingly dissatisfied with WHO and even hinting that it would run its own candidate to be the Director-General against him if he didn't agree to close the investigation. This shows the difficulty in conducting science independently under political pressure and governments' own agendas.

Where else could COVID-19 have come from? Another hypothesis is that the virus is of natural origin and may have been circulating in bats for many years before it spilt over into humans. Several virologists have expressed confidence that the virus had a natural origin based on genomic sequencing and phylogenetic (evolutionary relationship among genes) analysis. A *Nature* study suggested that the lineage giving rise to SARS-CoV-2 had been circulating unnoticed in bats for decades. Collins, the US National Institutes of Health Director, has said that he thinks the virus emerged 'as a result of gradual evolutionary changes over the years or perhaps decades, the virus eventually gained the ability to spread from human to human and cause serious, often life-threatening disease.'

But tracing the natural origin of coronaviruses is challenging. It was found that 9 per cent of 12,333 bats from Latin America, Africa and Asia carried at least one of ninety-one distinct coronaviruses. Bats have been known to harbour rich gene pools of SARS-related coronaviruses. A 2017 five-year study tracking bats and viral emergence concluded with 'this work . . . highlights the necessity of preparedness for future emergence of SARS-like diseases' and proposed 'examination . . . to determine if spillover is already occurring . . . and to design intervention strategies to avoid future disease emergence'. To establish from which bat population the virus emerged and whether there was an intermediary animal involved, researchers need to find evidence of the exact same virus in that species. This can be like looking for a needle in a haystack. As Dr Cui Jie, a virologist, said, 'the most challenging work is to locate the caves, which usually are in remote areas.'

Virologists who have looked at the genetic sequencing have noted that there is enough evidence to conclude that COVID-19 originated in bats, in a hotspot of viral evolution along the border of Yunnan Province in China, Myanmar, Laos and Vietnam. In fact, there are dozens of other coronaviruses circulating within bats in southern China that are closely related to SARS-CoV-2, indicating that it was likely that a virus of this kind would evolve naturally. It may have been circulating among bats for years without causing any disease for them, and at some point, a bat infected a human directly.

Alternatively, the virus could have infected an intermediary mammal, spread and evolved within that species, and then jumped into humans when they had close contact. Dr Stanley Perlman, a microbiology and immunology professor, said 'going after bats will only give you partial information – the viruses you are looking at may or may not get the additional mutations they need to be transmissible among humans . . . There has almost always been an intermediary involved, and without knowing what that is and what changes the virus would have to undergo, it is very hard to make any kind of predictions.'

Pangolins are suspected: a study of Malayan pangolins in a wildlife rescue centre in Guangdong noted that 17 out of 25 carried coronaviruses in their lungs. There was 81.6 per cent similarity between SARS-CoV-2 and the coronaviruses in the pangolins.

Chinese scientists at the Wuhan Institute for Virology have started investigations into the natural origin and found that horseshoe bats at the China–Myanmar border carried coronaviruses that shared roughly 96 per cent genomic similarity with overall SARS-CoV-2, except for the spike protein, which strangely had less similarity. Therefore, the virologists looked for the origin in other animals and found that Malayan pangolins had been infected by a virus with a similar spike protein, and this finding was confirmed by research teams in Hong Kong as well as in Guangzhou. This led to some speculation that the genomic sequence of SARS-CoV-2 was a combination of the bat and pangolin viruses. However, pangolins became ill from their coronavirus infection, suggesting they are not a natural host for this virus, as viruses typically find animal hosts that can carry them without becoming ill, like bats; plus the distance between Wuhan and the Yunnan Province is roughly 1,500 km.

Some virologists, like Dr Arinjay Banerjee of the University of Saskatchewan, say the similarity is not close enough to confirm the involvement of pangolins. Other animals have been considered as potential intermediaries, including civets, camels, pigs and snakes. To assess which could be the host, similar methods to those applied to SARS were used to determine if civet cats could be the intermediary. Thousands of different animal populations were tested for past or current infection by looking for serum antibodies, i.e., evidence of prior infection with that virus. Many studies in China have looked for antibodies for SARS-CoV-2 among as many as 30,000 wild, farmed and domestic animals, but no clear signal of infection has been found.

Dr Christian Drosten, a virologist from Germany who has studied coronaviruses for decades, noted in an interview with the *Guardian* that if he were to be given a large research grant and free access in China, he would look for the virus in places where raccoon dogs are bred. Raccoon dogs are kept on crowded farms in China and used for their fur. And, while civet cats are often pointed to as the source of the original SARS, it was also found in raccoon dogs.

Another theory is that bats infected a human with a precursor to SARS-CoV-2, and then the virus developed within a natural selection process among humans. The virus may have then changed and mutated into a more severe form after spreading undetected for quite a while among humans. However, there is no evidence for this theory.

Given all the information we have so far at the time of writing this book in August 2021, the lab leak hypothesis seems to be as likely an explanation as natural spillover and should be pursued until evidence emerges to the contrary. My view is that this was unlikely to be an intentional leak or a man-made virus but rather someone working at the lab becoming infected with SARS-CoV-2 during the course of their job, and then going on to infect others in the community. It just seems a remarkable coincidence that a new coronavirus would emerge within kilometres of one of the only Level 4 research labs working on coronaviruses in the world. Moreover, no exact matches of SARS-CoV-2 have been found in any intermediary animals like civet cats or pangolins, indicating a natural host reservoir. The caginess of the Chinese government and its reluctance to allow any kind of independent analysis or lab audit suggest it might not want the full picture to be known. While the Chinese government is worried about being blamed for causing COVID-19, its lack of transparency and secrecy add to concerns. It would be better served reputation-wise if it were to open its books and show the international community that it is willing to cooperate and identify the origin.

## *The WHO Mission to China*

To move from theories and debated hypotheses to confirmation requires investigation and hard evidence. After months of negotiation with the Chinese government, which had resisted an independent mission, the World Health Organization finally got permission to let an international team (approved by the Chinese government) spend four weeks in China, from January to February 2021.

Embarek, the Mission Team leader, noted:

> It's important to understand the origin of the virus for three reasons. One is if we find the source and if it's still out there, we can prevent future reintroduction of the same virus into the human population. Second, if we understand how this one jumped from bat origin into humans, we can perhaps prevent similar events in the future. Third, if

> we can find the virus, what it looked like before it jumped to the human population, we could potentially be in a better position to develop more efficient treatments and vaccines for this disease.

However, various team members expressed concern over whether they could really identify the origin in just a few weeks, and the final report was delayed several times because of tussling between the US and Chinese governments over the independence and content of the report – another example of politics influencing the scientific research taking place.

The team spent the first two weeks in online discussions with Chinese scientists, epidemiologists and doctors, and the final two weeks visiting sites in person. The working day was long – up to fifteen hours – so as to better understand the situation and ensure enough face-to-face contact to build respect and trust. The team also met and questioned people such as Wuhan doctors, the relatives of deceased health workers and the first confirmed COVID-19 case (a man, since recovered, and first identified on the 8th of December 2019). All this was organized and chaperoned by the Chinese government.

Some have been sceptical about whether we can use this type of mission to answer such a complex question. Dr David Heymann, a former WHO Assistant Director-General, noted, 'it is unlikely that the WHO team working with Chinese investigators will be able to determine the origin of the pandemic – it is very difficult to do this retrospectively by identifying early cases and then proceeding with case-control studies to identify risk factors for infection. And what matters now is controlling the current outbreak and understanding how to better prevent such pandemics in the future.'

Looking forward, Peter Daszack, another WHO Mission Team member, agreed that we must learn from the delays in getting an international team on site to investigate origins. Instead he thinks we should prepare international stand-by investigation teams, which would be ready to fly out and conduct on-the-ground investigations within hours of spillover events being confirmed. This of course requires governments to cooperate and agree to have these missions enter the country and have access to the required data.

## *Dr Li Wenliang*

Part of the difficulty in organizing these kinds of investigations is the transparency required from governments. The Chinese government tried for weeks to underplay and hide what was happening in Wuhan. One doctor's story, Dr Li, should never be forgotten.

Dr Li, born in 1986, was an ophthalmologist working at the Wuhan Central Hospital. He went into medicine to serve his patients and was also a father to two children (aged four and ten months) with a third on the way, due in June 2020. In late December 2019 he became aware of a potential new disease outbreak after seeing seven patients with SARS-like symptoms and medical reports showing high confidence of SARS-like infection.

Using the WeChat (the largest social media platform in China) group 'Wuhan University Clinical Medicine 2004', Li wrote, 'There are seven confirmed cases of SARS at Huanan Seafood Market.' He then shared a picture of the diagnosis report and a video of CT scan results. He continued, 'They are being isolated in the emergency department of our hospital's Houhu Hospital District.'

The response to his messages was discomfort and, largely, silence. Another group member warned, 'Be careful, or else our chat group might be dismissed.' Li ignored this warning and texted back, 'The latest news is, it has been confirmed that they are coronaviruses infections, but the exact virus is being subtyped' and 'Don't circulate the information outside of this group, tell your family and loved ones to [take] precautions.' A few hours after sending the message, he was tracked down by Wuhan City health officials who questioned where this information was gathered, which led to the closure of the suspected source, the Huanan Seafood Market, and the public announcement of the outbreak.

Days after sending these messages, Li was censured by the hospital leadership for sharing information about the cases and was forced to sign a declaration at the Public Security Bureau in Wuhan, acknowledging that he had made false statements that disturbed the public order. The Bureau noted, 'We now warn and admonish you about the violation of the law that you committed when you published untrue information on the internet. It is illegal conduct. The Public Security Department hopes that you actively cooperate, follow the advice of the people's police and stop your illegal behaviour. Can you do it?'

Li agreed to this and faced public humiliation. China Central Television broadcast to the nation that the Wuhan Bureau of Public Security had investigated and punished Li and seven others for disseminating misinformation online about the virus. While Li agreed to comply with the government, he expressed frustration to journalists abroad. He noted, 'The police believed this virus was not confirmed to be SARS. They believed I was spreading rumours. They asked me to acknowledge that I was at fault. I felt I was being wronged, but I had to accept it. Obviously I had been acting out of goodwill. I felt very sad seeing so many people losing their loved ones.'

On the 10th of January 2020 Li started coughing and became severely ill. Wuhan Central Hospital released a statement reporting that he was infected with SARS-CoV-2 and in critical condition. He was placed on oxygen and initially declared dead at 9.30 p.m. on the 6th of February. The public responded in anger, wanting more details, and the Chinese government then declared him dead again, age thirty-three at 2.58 a.m. on the 7th of February. The circumstances around how exactly and when he died remained suspicious and clouded by ambiguity. He was the second medical worker to die of COVID-19, after 62-year-old Liang Wudong.

Whether he died because of COVID-19 itself, or because of mistreatment by the government, Li paid with his life for his commitment to whistleblowing and transparency. In an interview with the *LA Times* he said, 'It's not so important to me if I'm vindicated or not. What's more important is that everyone knows the truth.' Public outrage followed his death with the hashtag '#WeWantFreedomofSpeech' trending on Sina Weibo, a Chinese microblogging website, until it was deleted by the government. It also generated outrage across the world, with the Chinese government then forced to award Li the title of 'martyr', the highest honour the government can bestow on a citizen who dies while working to serve the state.

Li's story also raises the question of how whistle-blowers can be protected in a system in which governments have clear incentives to hide outbreaks and novel pathogens given the economic and reputational damage associated with them. Professor Tom Inglesby, who runs a large health security centre at Johns Hopkins University, noted, 'One of the world's most important warning systems for a deadly new outbreak is a doctor's or nurse's recognition that some new disease is emerging and then sounding the alarm.' This points to the need for international rules explicitly to recognize the importance of transparency by governments and to create incentives for the early reporting of outbreaks.

## *The International Health Regulations*

The key global mechanisms for preventing and responding to outbreaks are the International Health Regulations adopted by all member states of WHO in 2005. These require that governments develop, maintain and report on core public health capacities that enable the detection of, and response to, public health threats with the potential for international impact. WHO, in turn, assumes responsibility for leading the international coordination of these strategies. The WHO Health Emergencies team detects hundreds of signals of new outbreaks each month from across the world, and from these has to distinguish which are the roughly thirty that require further investigation and investment of resources, and, of those, which are the one or two that require WHO operational assistance.

In the next chapter we get more detail on the IHR and COVID-19, but, quite simply, the IHR state that notifications to WHO of potential Public Health Emergencies of International Concern (or PHEIC) must be made within twenty-four hours after assessment by public health authorities, providing they satisfy more than two of the following four criteria. Each potential outbreak event occurring within a country's territories should be assessed using the following criteria within forty-eight hours:

1. Is the public health impact of the event serious?
2. Is the event unusual or unexpected?
3. Is there a significant risk of international spread?
4. Is there a significant risk of international trade or travel restrictions?

Several infectious diseases immediately bypass this assessment, as they are always notifiable to WHO. These include: smallpox, polio, and human influenza caused by a new subtype or SARS. While all governments have agreed to the IHR, there are no official penalties for non-compliance, and no hard mechanisms for WHO to punish countries for not complying with their IHR notification obligations. If countries are in doubt as to whether an outbreak is indeed notifiable, governments can always informally consult their WHO IHR Contact Point.

On the 30th of December 2019, under its IHR obligations, China notified WHO of a new cluster of pneumonia-like cases. Early evidence suggests that Wuhan authorities had held this data for several weeks before notifying WHO, and did so only when it became clear that the government was worried that this outbreak was not going to be easy to contain or manage. On the 4th of January 2020 WHO sent a memo to the world. This sent South Korea and other East Asian countries running but was largely ignored by Western countries.

Reflecting on the difference between Taiwan (which reacted early and aggressively) and Western countries, the Taiwanese Health Minister, Chen Shih-chung, said, 'Taiwan dealt with SARS so we knew that this kind of epidemic is serious . . . I think Western countries, because they felt they had comprehensive medical, social welfare and insurance systems, neglected the epidemic in the early stage . . . they were not humble enough. But that over-confidence led them to take action late.' We will look closely at the slow responses in Europe, North America and Britain later.

Could the pandemic have been avoided, and if so, at what moment? Dr Shih-chung is agnostic about this. He notes that, 'It's quite hard to avoid this kind of situation but obviously I think a lot of people have not shown a degree of respect, that human beings are not humble enough in facing this kind of pandemic. Some thought we were able to cope because of our scientific developments, but the reality is that we, as human beings, are not strong enough.'

## *Spillover Events*

Humans need to have humility in the face of viruses and other pathogens that have plagued us for centuries. Aside from climate change, the largest risk to our survival as a race is a new virus spilling over from animals into humans (also affected by climate change); one has only to look at what happened not only with COVID-19 but also with SARS, MERS, avian and swine flu. It has been estimated that there are 1.7 million undiscovered viruses circulating in the animal kingdom, especially in emerging disease hotspots such as rural South-East Asia, as well as rural West Africa.

Rather than being once-in-a-lifetime events, spillovers have been occurring with increasing frequency. It has been found that 60.3 per cent of emerging infectious disease events are by zoonoses, of which 71.8 per cent originate in wildlife (the rest from domesticated animals) and are increasing significantly over time. The authors of a global comparative study found that hotspots can be mapped based on socio-economic, environmental and ecological factors.

The prestigious medical journal the *Lancet* published a series on zoonoses that proposed pandemic potential in three stages. Zoonosis is the process whereby animal diseases jump or shift from different host species, particularly from other mammalian species to humans. The first is no human infection (Stage 1); the second is localized human infection (spillover, Stage 2); and widespread transmission and global dissemination (Stage 3).

A Stage 2 (localized) emergency is the initial spillover of a wildlife or livestock pathogen to people. This could be anything from the handling of slaughtered animals to exposure to infected small particles in markets or farms, or in the wild. It can result in either a handful of cases or a large outbreak, and either limited human-to-human transmission (like Ebola) or some without (Hendra virus, which causes a severe disease in humans when exposed to an infected horse but cannot spread from human to human).

Only a few outbreaks reach the point of Stage 3, but when they do it is catastrophic, as we have seen with COVID-19. As Professor Christine Johnson, University of California, Davis, said, 'We need to be really attentive to how we interact with wildlife and the activities that bring humans and wildlife together. We obviously don't want pandemics of this scale. We need to find ways to coexist safely with wildlife, as they have no shortages of viruses to give us.'

The challenge for any virus is not only to infect a human but also to start a transmission chain. It happens more often than most people would think. Just in the past half century, we have seen HIV-1 emerge from chimpanzees, SARS-CoV emerge from bats and civet cats, Nipah virus from bats and pigs, and influenza from pigs and birds. MERS moved from camels to humans.

As I noted earlier, wet markets, where many scientists think SARS-CoV-2 moved into humans, are ideal environments for spillover events. The markets bring together a broad range of animal species from all across the world into crowded and small cages. In these cages they are kept in poor sanitary conditions with excrement shared between cages, and animals being fed with random assortments of food that is rarely their natural diet. Once selected by a customer, the animals are killed at the market, with blood circulating freely in conditions of poor sanitation. This is usually when a spillover event occurs.

While there is limited data on exact level of risk, we can generally say that wet markets are a hazardous setting for the transmission of zoonotic diseases, as they enable the emergence, amplification and dissemination of new infectious diseases. Coronaviruses are of particular concern because their unique virological properties of replication enable genetic recombination between different animals. Coronaviruses have large RNA genomes, which result in their having higher rates of mutation. These pathogens have greater 'plasticity' in infecting new hosts and finding new ecological niches.

Various transmission routes are involved in marketplaces, including the faecal-oral route (although not officially recognized for

SARS-CoV-2). Many markets do not have toilets or handwashing stations, which is why outbreaks of other faecal-oral pathogens like *E. coli* and salmonella are a common occurrence. Quite simply, as Professor Maya Nadimpalli of Tufts University and Professor Amy Pickering of Berkeley wrote in an article in the journal *Lancet Planetary Medicine* in 2020, 'faecal bacteria from animals can contaminate meat and nearby produce during slaughtering or through cross-contamination of surfaces.'

Bats, reservoirs for many viruses that can infect humans, are often stored and sold in wet markets and may be transported next to exotic animals like civet cats, pangolins and beavers. These processes of storage and transportation aid the spread of microbes between bats and other animals; they're also kept in stressful situations (in small cages and uncomfortable temperatures, not fed properly), which means they're particularly susceptible to infection.

Farming practices that keep animals in crowded and unhygienic conditions have triggered viruses such as avian and swine flu. Stressed animals are also more prone to become ill. Animal stress can arise from unsustainable livestock industries, wildlife trade and artificial co-housing of different animal species (like at wet markets) – all of which provide opportunities to find new hosts, usually those that are unlikely to occur in a natural setting. Animal influenza (flu) viruses rarely infect humans, and, when they do, usually are not transmitted between humans, but certain mutations can cause them to start infection chains, as seen in the H1N1 (swine flu) pandemic in 2009 and the H5N1 (bird flu) outbreak in Asia in 2007.

Because of the risks inherent with wet markets, and the initial suspicions that SARS-CoV-2 emerged from a wet market, calls have been made from both high-profile politicians and public health experts to shut them down fully. As US infectious disease expert Dr Anthony Fauci said, 'It boggles my mind how when we have so many diseases that emanate out of that unusual human–animal interface, that we don't just shut [wet markets] down. I don't know what else has to happen to get us to appreciate that.'

US Republican Senator Lindsey Graham was clear on this. He said, 'I'm going to write a letter to the World Health Organization and to the Chinese ambassador asking them to close the Chinese wet markets. These are open-air markets where they sell monkey, they sell bat. We think this whole thing started from the transmission from a bat to a human.' The President of China himself, Xi Jinping, said, 'We must resolutely close and crack down on illegal wild animal markets and trade. The bad habit of eating wildlife without limits must be abandoned.'

However, the problem is more complex than it first appears. These types of markets provide a source of protein to low-income populations in a way that is affordable and easily accessible. Millions of people, particularly in low-income rural areas, rely on wild animals to sustain their livelihood. Unless these communities have alternative income and food sources, banning wet markets could indirectly open up the illegal trade in wild animals, which could bring these species to the brink of extinction as well as create unregulated opportunities for viruses to spill over, with even less control over infection prevention and management. This is particularly true across Asia, Africa and Latin America.

Would it be better instead to have well-regulated wet markets? This would involve installing handwashing facilities and toilets, requiring adequate drainage, separating live animals from meat and produce, and implementing protocols for cleaning food and slaughtering animals. A study examining live poultry markets in Hong Kong after the H5N1 outbreak concluded that enhanced disinfection and regular closure of live poultry markets could decrease the infection risk of poultry in the entire region.

Aside from the economic factors, cultural attitudes around wet markets make them hard to shut down completely. The culture around food in China, for example, is that freshly prepared food from live animals is the most nutritious and tasty, compared with frozen meat, which is considered inferior. For daily ease of access to freshly slaughtered meat, wet markets are often located close to residential areas across most parts of China.

In addition, advocates of traditional Chinese medicine believe eating wild animals and hard animal tissues like bones and horns strengthens the body and cures diseases. Regularly consumed animals in the Guangdong Province include pangolin, monitor lizard, giant salamander, wild snakes, owls and yellow-breasted bunting. People in the southern Chinese provinces, especially Guangdong, eat exotic animals because it is believed they enhance health and sexual function.

In China the close contact between humans and live animals has resulted in quite a few pathogens emerging, including strains of antibiotic-resistant infections. For example, in 2015 colistin-resistant *E. coli* bacteria was detected in pigs in China. Colistin is one of our antibiotics of last resort and used by doctors when other antibiotics are ineffective against an infection. The pigs had been fed colistin in order to promote faster growth as well as prophylactically to prevent infection on crowded farms; this overuse of antibiotics leads to selection for those bacterial strains that are resistant.

Local researchers then conducted a wider sweep for colistin-resistant *E. coli* in not just animals but also patients in local hospitals. They detected the gene MCR-1, which directs colistin resistance, in 1 per cent of hospital patients with infections. MCR-1 is contained on a plasmid, a small piece of DNA that can move between bacteria freely, and is thus the perfect vector to spread colistin-resistance not just in *E. coli* but in other bacteria as well.

We rely on antibiotics for safe surgeries and C-sections, supporting chemotherapy and organ transplants and even for daily issues like urinary tract infections and ear infections. Routine hospital appointments would suddenly become life-threatening if bacteria were able to develop 'pan-drug resistance', meaning infections becoming untreatable by any antibiotic.

This is what a post-antibiotic world looks like, when drugs just don't work any more. And all it takes is for pigs in China to be fed antibiotics, and then for an antibiotic-resistant strain to spill over from pigs and infect a farmer, and for that farmer to infect others in their community, who then travel across the world.

These types of spillover events also happen when humans take over wild animals' habitats, as was the case with Ebola and HIV/AIDS. Disease ecologists have been warning about how deforestation can be a strong driver of infectious disease transmission. The expert Peter Daszack has noted that nearly one in three outbreaks of new and emerging diseases is linked to land-use change like deforestation. At some point it is a pure numbers game: the fewer forest habitats there are for wild animals to live, the more overlap there will be with humans' living situations.

For example, reduced space for wild primates means they have less forest to forage in and so must risk encounters with humans, raising the chance of an exchange of pathogens. Another example is Lyme disease, a painful and debilitating illness, which spreads from ticks to humans particularly when urban populations move into areas where deer and other tick-carrying animals live. This has been a particular problem in suburbs in the US.

Accumulated deforestation, leading to major landscape change, was found to be an important driver of malaria incidence in the south-western Brazilian Amazon. A study estimated that a 10 per cent increase in deforestation leads to a 3.3 per cent increase in malaria incidence.

There is also the 'spill-back' phenomenon, where pathogens are transferred back to humans from animals, after jumping between different animal species. This occurred with COVID-19 when the virus jumped into a large mink population in Denmark, mutated into a new strain, then jumped back into humans living in that area of Denmark in November 2020. Through rapid action, including the culling of almost 20 million mink, the Danish government was able to contain the spread of this new strain and ensure it was eliminated. However, COVID-19 has continued to be found in all kinds of animals, including tigers, lions, cats, dogs, mink and even gorillas, leading San Diego Zoo to start vaccinating its primates against the virus in March 2021. This was before over a hundred countries had started to vaccinate humans.

All this points to the need for government action to ban wet markets, regulate deforestation, protect wildlife reserves and habitats, and reduce the interactions between wild animals and humans. The place to stop new outbreaks is at the moment of spillover. It is impossible to stop viruses circulating within the animal population. There are estimated to be over a million different viruses infecting animals, and trying to chase each one would be a losing battle, and not the right focus for public health investment. Instead we must look at how to reduce opportunities for viruses to make the jump into humans.

## *Could Spillover Have Been Avoided?*

The current way we think about health is quite narrow: WHO focuses on humans. Medical doctors focus on humans. Even public health departments have a clear focus on human health. But just focusing on human health leaves us vulnerable to what is happening in the animal kingdom, and how perhaps the greatest threats to human health are coming from just beyond this restricted gaze. The challenge is how to bridge the wellbeing of animals, usually the purview of vets, with human health, as well as with concerns about the environment and climate change. This approach is often referred to as 'One Health'.

As climate change becomes a growing challenge in the years to come, environmental shifts will have major ramifications for where animals live, roam, feed and are handled, which will in turn impact on humans. This impact will feature most prominently in increasing the chances for viruses to make the jump into humans.

If we could go back in time with knowledge of exactly when and how SARS-CoV-2 emerged in Wuhan, we could perhaps have stopped the pandemic. We could have closed or better managed the wet market where it is likely a person was infected from an animal host, or better regulated the lab from which SARS-CoV-2 may have escaped. When it was alerted to the fact that spillover of a virus had occurred, perhaps the Chinese government could have moved faster to bring in an investigation team to do a broad sweep of all facilities, labs and wet markets as well as nearby bat caves and farms.

SARS-CoV-2 has the perfect virus recipe for respiratory transmission, making it hard to control once it is seeded within humans. At this point the Chinese government would have had to stop all flights out of China and locked down the entire country to try to contain it. Early, aggressive measures by the Chinese government could have potentially stopped further spread and elimination from human populations, as occurred with SARS and MERS. Yet even these strong steps might not have stopped planes taking off across the world in the weeks prior to SARS-CoV-2 being identified, with people unsuspectingly carrying a new virus with them. Global air travel means that bugs travel far and fast to all corners of the planet. And, unlike SARS and MERS, people can feel fine with SARS-CoV-2 for weeks and remain completely unaware that they are infectious to others.

Once spillover occurred, SARS-CoV-2 spread rapidly in Wuhan, sounding alarm bells among medical professionals. Unlike low-income countries, which lack laboratories and the capacity to diagnose new infections before substantial spread, China can boast about the quality of its universities and its research; in addition, it possesses a well-functioning health system in urban settings like Wuhan, and good administrative control to manage policy challenges in a forceful and controlled way. The real question was whether China, even with such a powerful state, could contain coronavirus and stop its spread, within its borders and beyond. We will look at China's remarkable and unprecedented response in the next chapter.

# 2. China's Hammer

China initially tried to downplay the extent of the spread of the virus, but, when the threat it posed became clear, the government reacted with an authoritarian and harsh response to ensure containment within Wuhan and Hubei Province as well as throughout China. Its measures, which put the health of the entire Chinese population above the human rights of individuals, were seen as brutal and extreme by those watching from afar. They involved sealing people into their homes, forcibly moving positive cases into isolation facilities and stopping citizens from blogging or sharing what was happening with people outside China.

China's response hinted at the difficult future political decisions to be taken around the world balancing the rights of the collective versus the individual, and balancing the state's responsibility to offer safety at the cost of infringing personal freedoms. It also indicated the incentive for countries to downplay their outbreaks, either by controlling the narrative or by testing fewer people to ensure their COVID-19 rates stayed artificially low. Many of the early decisions by the Chinese government were a glimpse into what every leader in every country would soon have to face.

By the 22nd of January 2020 China had 571 cases and a death toll of 17. Infectious disease modellers sounded the alarm, noting that China could experience 100,000 new infections a day, with potentially hundreds of millions of people becoming infected within weeks. The most pessimistic predictions from modellers outlined that 80 per cent of China's 1.3 billion people would get this virus.

On the 23rd of January the central government of China imposed a lockdown in Wuhan and other cities in Hubei, affecting 57 million people. A lockdown of this kind had never been done in human history, and, while only a trickle of information flowed out of the country, it became clear that this was not, as some at the time said, a bad flu. However, something unexpected occurred, and the exponential rise of cases in China started to flatten. In fact, China managed to keep its total confirmed cases under 100,000, with daily new cases coming down as of the time of writing of this book.

## *Lockdown*

The way China set about eliminating SARS-CoV-2 could be described as draconian. It undertook house-to-house testing and removed individuals to quarantine facilities if they tested positive (sometimes against their will); it used tracking technology to trace 99–100 per cent of those who had had contact with the infected; it locked down entire buildings so individuals could not leave their flats or have free movement; and it constructed completely new hospitals within days.

Wuhan's lockdown was carefully executed and planned to use controls on movement, privacy and extensive surveillance to find all cases of SARS-CoV-2 and to ensure they didn't spread further into China. First, Wuhan officials restricted all transport in and out of the city. Planes and trains leaving the city were cancelled, and buses, subways and ferries were suspended. The rest of China also fell under traffic restrictions (over 14,000 health checkpoints at road junctions to stop travel) to reduce the number of people travelling outside and to encourage more of the population to self-isolate at home. Chinese residents were instructed to stop their inter-city travel and intra-city activities. People were ordered to stay home unless for groceries or medical care. Schools, offices and factories were closed. Private vehicles were banned from city streets.

The Chinese government understood well that the virus moves when people move. So it stopped people moving internally. In fact, the word *quarantine* comes from the Italian term *quaranta giorni*, which means 'forty days' and dates from the fourteenth century. Between 1345 and 1360, the Great Plague, also known as the Black Death, killed roughly half of Europe's population, to the point that records noted that there were not enough people left living to bury the dead. To protect the city from the disease spreading across Europe, in 1377 a law was passed that all ships and people entering Dubrovnik had to quarantine for forty days so they could be screened for infection. During this time the city was under Venetian rule. This order understood that sailors could feel well but still be incubating the virus, and display symptoms only later. Even in the times of medieval medicine, quarantine recognized that isolating people during incubation periods was important to stop further spread. This vital lesson of controlling the movement of people in order to control the spread of the virus would not be learnt by many countries during COVID-19, nearly 650 years later, until it had spread widely through their populations.

Returning to China, the government did not apply this same principle of controlling the movement of people to outward flights from Wuhan to the rest of the world. While it was shutting down internally to eliminate the virus, flights kept taking off from a major airport hub to a multitude of international destinations. This decision seems inexplicable except in terms of domestic selfish interest. These passengers would enter most countries with no checks or screening, setting off chains of infection that would soon become impossible to break without strict lockdown measures. Two months later China shut down travel coming into the country to protect its progress and stop the importation of cases, at a point when the rest of the world was struggling to contain this new virus. If only it had been more concerned about export of the virus to the world to begin with.

With Wuhan's lockdown, China had a logistical upper hand, given how housing is organized. The middle-class cultural desire for both security and exclusivity in living has resulted in estates being the predominant form of housing. Surrounded by walls or security fencing, and with security-staffed entrances, these estates are not just for the wealthy; they are quite universal across urban China. The 'gated' nature of housing is said to have helped the implementation of lockdown, rapidly facilitating enforced containment of infectious people in their homes. Entrances were controlled (usually only one open) and strictly guarded.

In addition, China's governance has *shequ*, the city's most localized form of urban government, which resembles self-governing community organizations. At different phases the Wuhan City Novel Coronavirus Prevention and Control Command Centre would announce targets, and the *shequ* would implement them at a local level. It was able to mobilize resources and services, do basic public health screening and support those isolating on a personal basis. Many people volunteered to help enforce public health measures such as controlling population movement and securing infectious individuals within their homes.

The efforts to contain the spread within Wuhan were effective and focused on reducing the R number. R became a household reference, as it indicated how many people were likely to be infected by an index case. $R_0$, pronounced 'R nought', shows how many people on average will contract a disease from one person with that disease (in a completely susceptible population). It's a good measure of how infectious a virus is, and can range from the high end, for measles (12–18), to the lower end, for Ebola (1.51–2.53). If $R_0$ is less than 1, the disease will decline and die out over time. If it's over 1, exponential growth occurs. If it's at 1, it is a stable outbreak.

Even with Chinese New Year on the horizon, travel reduced significantly. Estimates suggest a 99 per cent reduction in people travelling from Wuhan to other places, with R reducing from 2.35 on the 16th of January to 1.05 on the 30th of January, and infection spread to other cities was slowed by almost three days. In total roughly 11 million people were locked down in Wuhan. Enhanced restrictions in Hubei Province (where Wuhan is located) affected over 66 million people.

These measures to contain spread worked. A study exploring antibody rates found that seroprevalence (population-level surveys of antibodies) was much higher in Wuhan (1.68 per cent) than in the surrounding area of Hubei Province (0.59 per cent) and the rest of China (0.38 per cent), suggesting effective initial containment. By the 1st of March 2020 the overall number of people with confirmed COVID-19 in China was only 80,174, with a total of 2,915 deaths. This was far below the scientific modelling projections of 80 per cent of the Chinese population having COVID-19. In fact, this ended up being 0.00006 per cent of the Chinese population, showing that containment strategies (however draconian) could be effective at stopping this respiratory pathogen.

## *Individuals versus Populations*

China was criticized heavily for what many in the international community saw as its huge infringement of civil liberties. Thomas Bollyky, Director of the Global Health Program at the Council on Foreign Relations, said, 'No other nation (Western or otherwise) can or should seek to replicate China's actions . . . The disregard for civil liberties and human rights that the government has demonstrated in its quarantine and censorship activities are inseparable from the policies and actions of the government that contributed to the outbreak in the first place.'

For example, Beijing keeps track of its 1.4 billion citizens using face-scanning cameras, household registrations as well as neighbourhood informants. These informants, often local volunteers, circulated lists of potential infected people in Wuhan with identifying information including addresses, dates of birth, national ID numbers, phone numbers and occupations. This type of system has been traditionally used for state control over political dissidents.

In addition, those infected with COVID-19 were treated as akin to criminals who had done something wrong and needed to be imprisoned. Meron Mei, a student at Wuhan University, returned to his home village in Xishui (two hours away) and went to hospital with a cough. Soon after, five policemen showed up at his house and posted a warning on the door: 'Do not approach – patient with suspected pneumonia.' This resulted in constant surveillance such as his phone being checked, his camera being disabled, all his photos deleted and forced temperature checks three times a day by doctors in full PPE. Mei complained, 'I am in prison. I'm so angry. I feel physically and mentally exhausted.' His story isn't unique. Hundreds of people shared similar stories on social media and via the international press.

## *Turbo-charged Public Health*

But, while international media picked up on the harsh movement and quarantine policies, China also turbo-charged its public health response to an infectious disease outbreak. National authorities using both PCR testing and symptom-based diagnosis conducted active case finding in all provinces. They traced over 99 per cent of contacts using their surveillance systems and ensured they tracked these within and outside Wuhan city. They did retroactive case finding in medical institutions and hospitals in Wuhan City. They closed the Huanan Seafood Wholesale Market in Wuhan for environmental sanitation and disinfection, as well as inspecting other markets for hygiene standards. And they broadcast clear public messaging on how to avoid acquiring the virus, such as staying at home, avoiding crowded spaces and wearing face masks, and suggested hygiene measures like disinfection of surfaces. In an atmosphere of fear and shock about a new serious virus, there was a very high rate of compliance.

After a period in which there were serious infringements of civil liberties, the lockdown was lifted after seventy-six days and largely people could return to their lives pre-lockdown, including travel to visit loved ones. For example, Zhang Kaizhong, who hadn't seen his wife who lives in Jiangsu for the entire time, said 'It's like being liberated.' Wuhan city officials even put on a light show and special events to mark the reopening of travel hubs such as airports and bus and train stations, and lit up buildings with 'Hello, Wuhan' to encourage people to return to their pre-COVID-19 lives.

But it wasn't fully back to normal. Masks were still used, there were limits on large gatherings, and primary and secondary schools, nurseries and universities stayed shut. Even still, social media users posted, 'Wuhan lockdown lifted – seeing those words almost makes me want to cry' and 'After so much struggle, our Wuhan will return.'

## *The WHO Mission to China*

The Chinese government tightly controlled any information coming out of Wuhan. The international community was desperate for more intelligence, which arrived when WHO sent in a mission to investigate the response. Dr Bruce Aylward, the Senior Adviser to the Director-General on Organizational Change, led the Mission Team, which was jointly coordinated with the Chinese government and involved 25 people – 13 international experts and 12 local experts.

Aylward was a smart pick to handle such a tricky diplomatic task and gain access to a situation about which China was proving to be reticent. He had trained all his life for this role. A Canadian doctor, Aylward had already spent twenty-eight years with WHO, starting in communicable disease control, then moving into leading the Global Polio Eradication Initiative, before becoming the Assistant Director-General of the WHO Polio and Emergencies Cluster. In 2014 he was appointed the Special Representative of the Director-General for the Ebola Response and then was quickly appointed co-lead of the WHO–China Joint Mission on COVID-19. His main message returning from China was 'isolating, contact tracing and testing', which was picked up by some countries like New Zealand, but not by others such as Britain, as I discuss in Chapter 5.

Aylward started the team's mission in Beijing, with meetings with national institutions and the municipality to understand the Chinese response, particularly how they were finding cases, how they were tracing them, and what was working in their response and not working. As soon as he got off the plane in Geneva, he went straight into a press briefing on the 24th of February 2020 to share with the world what their team had learnt. The countries that paid attention would gain valuable intel; the ones that didn't would pay dearly for their neglect.

Aylward, looking jet-lagged and tired, started outlining what the world could learn from China. He said, 'They have taken very standard, and what some people think of as old-fashioned, public health tools . . . and applied these with a rigour and innovation of approach on a scale that we've never seen in history.' He asked for governments to understand the severity of this crisis and pushed for countries to move from preparedness mindsets to 'readiness, rapid response thinking', because it would soon be arriving and kicking off in their countries.

He picked out five key steps that China systematically followed. First, the government had a differentiated approach based on the outbreak situation that obtained locally: zero cases, sporadic cases, clusters or community transmission. It was then in a position to tailor its response and stop people travelling. Lockdown in Wuhan was necessary, because it was in the highest risk category, with community transmission.

Second, it mobilized collective action and cooperation with the people in China to change their behaviour towards managing the spread of the virus. But he did note that this was a specific cultural context. 'It's never easy to get the kind of passion, interest, commitment, and an individual sense of duty [to have such a stringent lockdown].'

Third, it repurposed the machinery of government to be able to handle the logistics of such a large response, including building large hospitals within days and managing public health checkpoints over travel.

Fourth, it used technology to assist in the surveillance of hotspots and could access real-time data on clusters and collectively work out the appropriate response.

Finally, it created quick national clinical guidelines based on emerging data from hospitals, so that best practice for management of severe COVID-19 could be developed. This was all built off its SARS preparedness plans, which had focused on surveillance, case finding and contact tracing on a large scale.

The evidence in February 2020 showed that containment was successful. The key, Aylward pointed out, was a mindset that 'this virus is gonna come, it's gonna show up in our country, we're gonna find it within the first week, we're gonna find every case, we're gonna go after every contact, we are gonna make sure we can isolate them and keep these people alive.' In addition to containment, he encouraged governments to plan hospital beds, oxygen supplies, lab capacity for testing and adequate ventilators. He advised other countries to 'access the expertise of China. They've done this at scale: they know what they're doing. And they're really, really good at it and they're keen to help.'

Aside from the policy response, Aylward shared the early findings from China on who was being affected by the virus and its clinical outcomes. At this press briefing we learnt that the virus transmitted through respiratory mechanisms (droplets) shared through sneezing, coughing, speaking and touching infected surfaces; that children did not seem to be affected severely by the virus, with no deaths reported in children under the age of ten; that most health workers were more likely to be infected at home and not at work, showing that strict infection control standards were effective in hospital settings; that mortality increased with age, that comorbidities, such as heart disease, respiratory conditions and cancer, increased the risk of hospitalization; and that this virus was not like SARS (which had been eliminated before) or seasonal flu (which was much less serious) but rather needed to be treated with caution and full attention. It would take countries in Europe and the US another month or two to realize these facts.

In the short Q&A after his briefing Aylward made two important observations. The first was on vaccines: he warned that the scientific community hadn't been able to produce vaccines for SARS or MERS, so the possibility of developing a coronavirus vaccine within a short- to medium-term timeframe was slim. He didn't specify what this timeframe would be, but vaccines usually take years, if not decades, to develop.

He suggested that the established influenza plan of pushing quickly for vaccine deployment would not be optimal, and that instead there was a need to get on top of cases quickly and aim to eliminate through case finding and isolation. This prediction of how scientifically difficult an effective coronavirus vaccine would be turned out to be too pessimistic; but, to be fair, it reflected scientific consensus at the time, including from American public health expert Tony Fauci in the US. The pessimism about a vaccine being quickly available influenced how Britain, Sweden and New Zealand would approach their management. It pushed countries either to eliminate with strict border control or to try to 'live with the virus' and manage its perceived inevitable spread through the population.

The second important observation was about children. Even with advanced surveillance, effective contact tracing and big-data analysis, the Chinese government had never found any instance in which a child was an index case, that is the initial case identified that leads to a cluster. It seemed like children not play an important role in spreading the virus, which was unlike their role in spreading other respiratory viruses like common colds. Aylward contrasted this with the flu, noting, 'One of the big problems with flu is that all the kids get sick and then all the parents have to stay home, and then you lose billions of dollars.'

The early findings from China indicated that children were less likely to be infected by adults, less likely to transmit to others as well as less likely to be symptomatic and fall ill. Even with this background knowledge, major economies, one after another, were forced into lockdowns and school closures across the globe. While children themselves weren't pointed to as vectors of infection, the fear was that schools exposed teachers to the virus (even from each other as colleagues) and increased mixing through the use of public transport and parents mingling at drop-off and pick-up. Children older than twelve (secondary pupils) were seen as a riskier group than primary children. Most likely this difference was due to older children being biologically more similar to adults than younger children. This, again, would shift with the Alpha variant that emerged in England, which would increase transmissibility within children (and adults) and make keeping schools open an even larger battle throughout the world. I discuss variants and schools in Chapter 8.

## *Getting the Sequencing out of China*

In parallel with these political developments, the first PCR test-kits were being created to test and identify carriers. This was reliant on getting viral samples and then sequencing them.

Kits were created first by recruiting a small number of in-patients from Wuhan hospitals and collecting fluid samples of their lungs, growing any virus within them and isolating the RNA for analysis. Next, generation sequencing was applied to these samples, which is a large-scale DNA sequencing technique used to map entire genomes.

Phylogenetic analysis, like constructing a family tree but of viral evolutionary relationships, of the genomes and those of other coronaviruses was used to track the virus's evolutionary history and to help determine its probable origin. Homology modelling (like jigsaw pieces between a known template protein and an unknown target and working out which fit together) was also run to ascertain various properties of the virus.

Essential to this sequencing was Professor Eddie Holmes, a beach-loving, guitar-playing and straight-talking Brit based at the University of Sydney. Holmes is an evolutionary biologist who revealed the origin, evolution and molecular epidemiology of Hepatitis C, influenza, HIV and dengue. He worked closely with Professor Yong-Zhen Zhang of Fudan University, who has discovered multiple RNA viruses and created a network of labs in China dedicated to monitoring and documenting new viruses. What exactly did Zhang and Holmes do?

On the 3rd of January 2020 Zhang's team at Fudan received swabs from early patients in the Wuhan cluster. By 2 a.m., on the 5th of January, Zhang submitted his data on the genome to the US National Center for Biotechnology Information, noting the similarities to SARS and that it was spreading through respiratory transmission. Holmes looked at the genome and said, 'We just thought, oh no. It's SARS back again . . . I didn't sleep . . . It was weighing on my conscience.'

Zhang urged the Chinese Ministry of Health to move quickly to contain this virus and said China needed 'emergency public measures to protect against this disease'. Unfortunately, as Zhang said himself, 'Nobody listened to us, and that's really tragic.'

Holmes, who, as an Honorary Visiting Professor at Fudan University, has a long history of collaborations in China, contacted Zhang, asking for permission to publish the genome open-access on virological.org, a discussion forum for the analysis of virus genomes. Zhang, who was on a plane at the time, said he needed to think about it, because of Chinese pressure not to release too much information about the outbreak. A minute later he called Holmes back and gave permission to publish the data.

A researcher working at Zhang's lab emailed the sequence to a colleague at the University of Edinburgh, and, after fifty-two minutes, Holmes shared it online, and later reflected, 'If we didn't collaborate with colleagues in China – if I wasn't working with and talking to Zhang – that sequence wouldn't have gone online as early as it did.'

This did not go unnoticed by the Chinese government. Zhang was punished by the Shanghai Health Commission for breaking Chinese regulations on sharing information on the new coronavirus. His lab was closed temporarily on the 12th of January 2020; the official reason given was that changes needed to be made to improve biosafety protocols. It reopened on the 24th of January with the ability to test thousands of viral samples. Holmes has insisted that all the evidence points to an origin in animal species and that 'There is no evidence that SARS-CoV-2 . . . originated in a lab in Wuhan, China.'

While the Chinese government was unhappy with Zhang, the rest of the world celebrated receiving this important scientific information. *Time* magazine chose Zhang as one of their most influential people of 2020 for his work in publishing (with Holmes) the first SARS-CoV-2 genome within days of its detection. *Time* noted, 'That data allowed scientists around the world to begin developing tests for detecting the virus as early as January', and enabled the mRNA vaccine creators (such as Moderna and BioNTech) and the Oxford and Harvard vaccine research teams to design their vaccines off the sequencing. Zhang and Holmes also won the 2021 General Symbiont Prize for their excellence in sharing data.

## *Testing, Testing, Testing*

As scientists across the world started to work to build diagnostic capacity based on the sequencing, WHO was concerned about the poorest parts of the world and their ability to access lab-based diagnoses. Dr Mariângela Simão, the WHO Assistant Director-General for Medicines and Health Products, stated it clearly: 'Facilitating access to accurate tests is essential for countries to address the pandemic with the best tools possible', while Dr Tedros told countries that they must prepare for 'testing, testing, testing'.

Prior to COVID-19, the whole of Sub-Saharan Africa had only two laboratories able to do testing. WHO's efforts in early 2020 resulted in fifty-eight testing sites being set up in the region. In addition countries like Senegal reached out to diagnostic manufacturers such as Mologic in the UK in January 2020 to see whether they could develop a partnership for a rapid COVID-19 diagnostic. Senegalese scientists knew that testing capacity would be a crucial factor in containing the outbreak, and just relying on PCR lab-based methods would not be enough.

However, rapid diagnostics were difficult to develop, and it took until April 2020 for WHO to list two diagnostic 'quality-assured, accurate' tests for emergency use: a Roche Qualitative Assay on the 3rd of April, followed by Primerdesign on the 7th. Both tests use real-time PCR. One of WHO's roles became listing specific products for emergency use based on quality assurance. As Simão said, 'The emergency use listing of these products will enable countries to increase testing with quality-assured diagnostics.'

The Emergency Use Listing (EUL) process of WHO was set up in 2018 to help speed up the availability of tests in emergency situations, after a trial run during the 2014 Ebola crisis. It consists of a set of procedures to evaluate three product categories (diagnostics, treatments and vaccines) for acceptable performance, quality and safety and to accelerate use of these tools during public health emergency situations. The goal is to help procurement agencies in countries navigate the large number and different types of devices across the markets, assess them and provide assurance of the products' quality and performance. It is specifically geared to supporting countries that might lack the capacity to do this themselves on emergency timescales.

## *A Public Health Emergency of International Concern*

Aside from its role in supporting countries in diagnostics, WHO also plays a crucial role in alerting the world to potential new global threats. As discussed in the last chapter, the International Health Regulations are the current international-level framework to prepare and respond to infectious disease outbreaks. Being an artefact of international law, the direct subjects of the IHR (2005) are state parties (governments). They are binding on all 196 member states and require them to develop, maintain and report on core public health capacities that enable the detection of, and response to, public health threats with potential international impact. WHO is responsible for international coordination supported by the Director-General and the Secretariat in Geneva, Switzerland.

The IHR require states to notify WHO of potential threats. The Director-General then has the power to declare a Public Health Emergency of International Concern (PHEIC) – the loudest alarm bell the world has to alert governments to an incoming crisis. PHEIC is pronounced like the word 'fake' and, although an acronym, it is never spoken by its individual letters. An emergency is defined as an 'extraordinary event' that constitutes a public health risk to other states through the international spread of disease and also potentially requires a coordinated response. Declaring a PHEIC is a binary decision: yes or no. Alongside this decision, the Director-General issues non-binding Temporary Recommendations (which automatically terminate after three months) for measures to prevent or reduce the international spread of disease.

To do this, the Director-General relies on a group of experts called the Emergency Committee, composed of unpaid specialists, largely academics in epidemiology, medicine, public health and virology, drawn from WHO's Roster of Experts. The Emergency Committee is convened by the Director-General in order to provide advice on whether the situation meets the criteria for an emergency and on what the recommendations should be to governments.

As of February 2020 eight Emergency Committees have been convened by two directors-general (Dr Margaret Chan and Dr Tedros) since the mechanism began in 2007. Six have recommended a PHEIC declaration for H1N1, in 2009; polio, in 2014 and ongoing; Ebola, in 2014/15; Zika virus, in 2016; Ebola, in 2018 and ongoing; COVID-19, in 2020 and ongoing. No PHEIC was declared for MERS (2013–15) and yellow fever (2016). In every instance, the Director-General has followed the committee's advice and issued their recommendations.

Given how many disease events and outbreaks occur, often in the hundreds per month, the PHEIC mechanism has been used relatively sparingly, leading to frequent criticism of WHO's reluctance to make the declaration, or for the Director-General even to convene the Emergency Committee. The fear of overreacting and causing panic seems to drive this wariness. The first Independent Review Committee, for example, highlighted two other events it judged to have met the criteria for a PHEIC that members felt were neglected by WHO: Nipah virus in Bangladesh, in 2009, and cholera in Haiti, in 2010. More recently, although the Ebola outbreak in DRC was declared a PHEIC in July 2019, the Emergency Committee had declined to recommend a PHEIC in three previous meetings, prompting repeated criticisms from both academic and partner organizations.

One of the main points of contention is the binary nature of PHEIC: something is or isn't an emergency, when there is no graded alarm system. This is a challenging decision to make with limited information, whereas a graded system of 1 to 4 allows some nuance in decision-making. Notably, two Emergency Committees have questioned the flexibility of the PHEIC mechanism: the Polio Committee and the COVID-19 Committee. This fuels the double-edged sword of PHEIC publicity. Although publicity is vital for rapid and widespread risk communication, the declaration of a PHEIC often invites speculation on the financial damage to countries involved caused by the loss of trade and tourism income; but to fail to declare a PHEIC invites criticisms of the WHO's global health leadership status. The decision to declare a PHEIC comprises a tense balancing of political and economic calculations against the expected benefit of announcing Temporary Recommendations.

Even when Temporary Recommendations are advised, there is no way to enforce these, leading to widespread non-compliance. For example, no Emergency Committee has recommended a total travel ban, and only the Polio Committee has recommended bans conditional on vaccination status. Despite this, as was clear during COVID-19, governments have disregarded this advice and put in place their own travel measures. Countries like Vietnam and New Zealand have tended to outperform those that followed early WHO guidance. I discuss travel restrictions and WHO in Chapter 7.

On the 22nd of January 2020 the Emergency Committee convened to decide on whether COVID-19 constituted a PHEIC. Journalists and academics waited for the press briefing to begin; it was several hours late. The briefing, led by Professor Didier Houssin, the Chair of the Committee, and Tedros, was singular in its lack of a decision and postponement to the next day. Tedros said, 'To proceed, we need more information. For that reason, I have decided to ask the Emergency Committee to meet again tomorrow to continue their discussion.' Houssin explained that the committee was split on what to advise the Director-General.

When pushed by a journalist about why the committee was so hesitant, Houssin said that they didn't have enough information on severity and transmissibility, based on the data provided by the Chinese authorities. The severity became referred to as the IFR (infection fatality rate), which is the percentage of people who die out of all those who become infected. This contrasts with the CFR, which is the number of deaths divided by the number of confirmed (by testing) cases of disease. The IFR would become a major discussion point in the scientific community when comparisons to the seasonal flu IFR began to be made. Seasonal flu kills 0.1–0.3 per cent of all those exposed; it was unknown what the fatality rate for COVID-19 was, as there was no clear data on how many people were infected overall. Similarly the transmissibility (or R) of the virus was still unknown, i.e., how many people were likely to be infected by an index case.

Dr Maria Van Kerkhove was also present that day. A soft-spoken, highly qualified American, Van Kerkhove had previous experience in MERS and was appointed as the COVID-19 Response Technical Lead. Her workload increased dramatically overnight, all while she juggled two young kids (9 years old and 18 months) at home, often not seeing them for days on end. She recounted in an interview how she had to quarantine away from them in a room in their home after her return from the China Mission and talk to her children through windows: 'It was awful. Awful!' Her older child was convinced she was going to die of COVID-19 when she went to China, and her young toddler thought she was playing hide and seek and tried to chase after her. The personal toll was hard. 'I would laugh in front of him and then come into the bedroom and cry because it was just a horrible, horrible thing.'

As the pandemic progressed, Van Kerkhove became a focus of attack, not only by anti-science activists but also by academics who felt WHO was underplaying asymptomatic and aerosol transmission. She found it difficult: 'I was talking with my husband and I was saying I'm struggling at the moment with the pushback and the second-guessing and the challenging. I'm trying to get the information out to clarify WHO's position, which is to help people, which is to suppress transmission and to save lives.'

At this press briefing Van Kerkhove shared what steps governments could already start to take: 'The primary issues are to limit human-to-human transmission; reduce secondary infections, especially amongst close contacts and in health-care environments; prevent transmission through amplification and super-spreading events and prevent further international spread; reduce zoonotic spread, by identifying the animal source and limiting animal exposure.' She pointed to case finding and testing as essential to controlling the spread of the virus. In her words, governments had the playbook of what to do for the coming weeks. Some governments listened. And other governments weren't paying attention at all.

With the problem being limited information, journalists pushed Tedros on whether the Chinese government's lack of transparency was creating obstacles for the committee. Tedros lavished praise on the Chinese President and Health Minister for their clear guidance, the immediate action taken on the Huanan market, and their swift work in identifying and sequencing the new pathogen. He also supported the Wuhan lockdown, noting that strong action was necessary to contain the outbreak within China as well as international spread. This early defence of China would come back to haunt WHO, as former US President Donald Trump would use this as an example of WHO being China-focused, and not serving the interests of other countries or the US. Soon after, Trump withdrew the US from WHO, which triggered a one-year grace period. This decision was reversed by the new US President, Joe Biden, when he came to office.

On the 23rd of January 2020 the Emergency Committee met again as a continuation of the first day. At the time there were only 585 cases and 17 deaths reported to WHO across China, Japan, South Korea, Singapore, Thailand, the US and Vietnam, and Houssin subsequently concluded that 'It's a bit too early to consider that this is a public health emergency of international concern.' He noted that the committee was still split: arguments that were used in favour of a PHEIC were evolution of the epidemic, the rapid increase in the number of cases and the severity of the disease; arguments against were that there were still limited cases abroad outside China, and that the rapid action of the Chinese government would contain the disease.

Tedros summed it up: 'Make no mistake. This is an emergency in China, but it has not yet become a global health emergency. It may yet become one . . . I wish to reiterate that the fact that I am not declaring a PHEIC today should not be taken as a sign that WHO does not think the situation is serious . . . Nothing could be further from the truth.'

Once again Tedros praised the Chinese government 'for its cooperation and transparency', especially in isolating the sequencing of the virus quickly and sharing data with the international community. When asked what additional information they had received from China since the previous day, Houssin alluded to 'very precise information about the evolution of the epidemic' from Chinese health authorities. This must have reassured the committee that the situation was under control there. In retrospect the committee was too focused on containment in China, which would be successful, and not enough on the spread that had already occurred to other countries, including those without the political, cultural, financial or geographic ability that China had to contain and eliminate.

As the numbers continued to increase rapidly in China, Tedros convened the Emergency Committee one week later, on the 30th of January 2020, to reconsider the situation. At this time there were 7,711 confirmed and 12,167 suspected cases, 1,370 of which were severe, and 170 deaths; 124 had recovered and been discharged from hospital. There were also 83 cases in 18 countries. Of these, 7 had no history of travel in China. It was now clear that this virus could sustain human-to-human transmission in multiple settings. Based on this significant increase in numbers of cases and additional countries reporting confirmed cases, they recommended to Tedros that a PHEIC should be declared.

## *This is Not a Drill*

On the 30th of January 2020 Dr Tedros walked slowly into the press room in Geneva, aware that health ministries from across the world were hanging on his every word: he calmly informed those present that he was declaring a Public Health Emergency of International Concern, the highest alarm bell that can be rung at the global level. The clock started ticking. This was not a drill. This was the real deal.

The moment had come for countries to prepare, or not to prepare, for the coming tide. Some, like Senegal and Taiwan, had already been racing ahead even before WHO; others, like Britain, the US and Italy, weren't paying attention. To these countries, this was still a Chinese emergency and one playing out thousands of miles away, and of little relevance to their daily lives. This cavalier attitude to what was happening in China would come back to haunt them, as each fell into repeated lockdown cycles, and overloaded hospitals and morgues; and the early scenes that had taken place in Wuhan would recur in Lombardy in Italy and in New York City.

The Emergency Committee pushed countries to work towards containment. It told them that 'It is still possible to interrupt virus spread, provided that countries put in place strong measures to detect disease early, isolate and treat cases, trace contacts, and promote social distancing measures commensurate with the risk.' It warned that COVID-19 would be arriving soon, if it hadn't already, and to build the public health infrastructure rapidly and to share full data with WHO.

However, it did not recommend any travel or trade restrictions. Some countries like the US ignored this advice and introduced targeted border screening or closures to China. Taiwan had already closed its borders, and countries across the world rapidly moved to stop importing any cases at all from anywhere. Mali closed its borders to all countries before it had even had its first case. New Zealand also bubbled itself off from the world, as did China, once it had eliminated the virus. And Australia even shut down travel between states, with each state working to eliminate the virus one by one. By contrast, the UK left its airports completely open with no screening, testing or restrictions for passengers entering the country until late 2020. Some would point to the advice by WHO not to stop travel as a fundamental misstep that let the virus take off in country after country left open to importation.

The Emergency Committee also gave China tailored recommendations. It urged the government to work with WHO on a joint mission to investigate the evolution of the outbreak, measures to contain it and the documentation of its epidemiology. It also pushed China to identify the zoonotic source and share more information on the origin.

Crucially it encouraged China to conduct exit screening at international airports and ports, to detect early symptomatic travellers for further evaluation and treatment, while minimizing interference with international traffic. Had China tried to keep the pandemic within its borders, the virus could have been contained, or even delayed. Instead it took a more lax approach to export of the virus, given domestic economic considerations.

In addition, WHO was fixated both on symptoms and on the case finding of individuals reporting they were unwell. This has been traditionally used in infectious disease management, and with SARS and MERS outbreaks in particular. However, as would become well known in the months ahead, SARS-CoV-2 can also spread via asymptomatic (no symptoms at all) or presymptomatic (days before becoming ill) cases, which meant that just focusing on screening for symptoms would miss large numbers of cases and fail to break chains of transmission. Only mass testing using diagnostics would be able to detect these cases and highlight which individuals should isolate so as not to pass the virus on to others.

## *Early Scientific Data on COVID-19*

While governments were being informed directly by WHO about the Chinese experience, scientists were storming ahead with their own information-gathering and dissemination. The *Lancet*, for example, a major British medical journal, played a crucial role in rapidly publishing the first studies from China. Richard Horton, the Editor of the *Lancet*, would later point to the neglect of these studies by UK scientific advisers, calling it 'the greatest science policy failure for a generation'.

On the 24th of January 2020 the *Lancet* published a paper containing the first clinical description of the '2019-nCoV' disease from a 'novel betacoronavirus' in 41 hospitalized patients (as of the 2nd of January). Of those, 27 patients had been exposed to Huanan seafood market. Comorbidities were cited as a driver for worse health outcomes. The paper was full of warnings: 'The 2019-nCoV infection caused clusters of severe respiratory illness similar to SARS and was associated with ICU admission and high mortality'; 'Patients had serious, sometimes fatal, pneumonia'; 'We are concerned that 2019-nCoV could have acquired the ability for efficient human transmission' and '2019-nCoV has pandemic potential.'

This news was stark and the paper even made suggestions as to how to protect against spread: 'Airborne precautions, such as a fit-tested N95 respirator, and other PPE are strongly recommended.' It also detailed what doctors should be looking for in patients: 'most patients presented with fever, dry cough, difficulty breathing and abnormalities on chest CT scans'.

The *Lancet* also published the first report of possible human-to-human transmission and intercity spread, examining a family of six who travelled from Shenzhen to Wuhan between the 29th of December 2019 and the 4th of January 2020. Infections were found in 5 out of 6 family members, and one, who did not travel with them, became infected a few days after they returned. None of the patients had had any contact with animals or wet markets. The *Lancet* warned about asymptomatic transmission and urged public health officials 'to isolate patients and trace and quarantine contacts as early as possible because asymptomatic infection appears possible (as shown in one of our patients), educate the public on both food and personal hygiene, and alert health-care workers on compliance to infection control to prevent super-spreading events.'

WHO and most high-income countries were slow to identify asymptomatic transmission as a key driver of spread, and this would soon become the Achilles heel of response efforts, and part of the reason why infections arrived so quickly in care homes across the world: the virus was being spread by asymptomatic staff.

On the 30th of January 2020 the *Lancet* published a paper on the largest descriptive case series of clinical features to date, based on studies from an adult Wuhan hospital. Once again, comorbidities were highlighted as a risk, as well as age and gender. Older men were at the highest risk. Clinical features, including the rapid progression of ARDS (acute respiratory distress syndrome) and septic shock, followed by multiple organ failure, were detailed. ARDS is when fluid builds up in the lungs, preventing them from filling with enough oxygen. This leads to severe shortness of breath and organs not getting enough oxygen from the bloodstream. It is a life-threatening condition. Patients did better through 'early identification and timely treatment of critical cases' – a lesson that the UK did not take heed of during its first lockdown, leading to many dying at home of COVID-19 or arriving at emergency rooms already in a severe state.

The final important paper in January 2020 in the *Lancet* was a modelling study published on the 31st, which identified the risk of pandemic potential, emphasizing the role of both domestic and international travel hubs that could act as epicentres for outbreaks. The authors estimated $R_0$ to be 2.68 and warned that containment was still possible outside China, but 'To possibly succeed, substantial, even draconian measures that limit population mobility should be seriously and immediately considered in affected areas.' They also pushed for countries to begin reducing mixing within populations through 'cancellation of mass gatherings, school closures, and instituting work-from-home arrangements, for example.'

Countries were urged to get ready to deploy their preparedness plans, including ensuring adequate PPE supply chains, hospital supplies and sufficient numbers of trained health workers. They were, in effect, telling countries to prepare for battle.

Again, high-income countries were largely sleeping through these reports. The basic principle that the virus moves when people move seemed to be forgotten. Stopping the movement of those living in infected areas to those that are 'clean' areas was the most essential element of containment. These countries also delayed any kind of restrictions until the virus had already spread, instead of pre-emptively preparing their populations for reduced social contact. And, finally, they waited to deploy their preparedness plans, meaning that there was insufficient PPE when needed, and health care workers were sent on to COVID-19 wards wearing bin bags and unsuitable face masks, leaving them exposed to the virus while at work.

The *New England Journal of Medicine*, the flagship American medical journal, was also rapidly publishing key studies on COVID-19 that contained gold nuggets of information for scientific advisers to governments, who were peering through the fog and trying to assess how best to respond. On the 24th of January 2020 *NEJM* published a brief report on 3 adults (49, 61 and 32 years of age) hospitalized with severe pneumonia, looking at exactly how it affected their lungs. On the 29th of January it published a large study of 425 patients, the majority of whom (55 per cent) were linked to the Huanan market. Based on this sample, they estimated $R_0$ to be 2.2 and the doubling rate of case numbers in Wuhan to be 7.4 days.

Based on what they were seeing in this patient group, the authors estimated what an adequate quarantine and isolation period should be: 'Our preliminary estimate of the incubation period distribution provides important evidence to support a fourteen-day medical observation or quarantine for exposed persons.' Soon it would become standard across most countries for those who had tested positive for COVID-19 or been exposed to someone who had tested positive to go into isolation for fourteen days. But this was a large ask for many people, especially when isolation wasn't backed up with adequate financial support. And compliance was highly variable among countries, making it harder to break chains of transmission. The point of testing people and tracing contacts was to ensure they went into quarantine.

My research team and I did an international review on varying country policies for the *British Medical Journal* on this topic, published in March 2021. A series of online surveys conducted in the UK with over 30,000 participants found that only 18 per cent of those who were symptomatic complied with isolation and only 11 per cent of close contacts quarantined. When asked why they weren't complying, people listed issues such as childcare responsibilities, financial hardship, low awareness of what they should be doing and, finally, working in a key sector. And the fear of not having adequate support to isolate became a major barrier to testing uptake, especially in communities where household incomes are low. Compliance was higher in countries like Finland, which offered 100 per cent of lost income during isolation, and in Norway, which offered 80 per cent of salary up to a £50,000 salary cap. This support was accompanied by harsh penalties: in Norway a fine of £1,600 and up to 15 years prison for breaking isolation; and, in Finland, a fine depending on annual income or up to 3 months in prison.

The final important *NEJM* study was published on the 31st of January 2020 and was a detailed clinical case report on the first confirmed 2019-nCoV infection in the US, which was linked to a travel history to Wuhan. This case report provided important insights on symptom development and duration over fifteen days of illness. The patient had a fever, continuous cough, a runny nose, fatigue and nausea, as well as vomiting and diarrhoea for part of the time. The symptoms of COVID-19 overlapped with seasonal infections like flu, making it hard to identify just based on those reporting feeling unwell. Fever, body pain, a cough and feeling generally unwell are generic symptoms that apply to a number of infections. This meant that identifying who should be tested for COVID-19 was challenging. In the end the UK, for example, relied on either cough or fever, and, soon after, added on loss of taste or smell as distinctive features of COVID-19. These last symptoms were not identified in this early *NEJM* study.

## *Seeing through the Fog*

The studies mentioned above in the *Lancet* and the *NEJM* indicate just how much information was already in circulation about the novel coronavirus within the scientific community by the end of January 2020. We already knew within a few weeks about the clinical progression, who was most at risk, how it was spreading, what health systems should be prepared for, potential asymptomatic transmission, the incubation period, the best strategy for containing spread and estimates of $R_0$ for the virus. Adding this information to the emerging information from WHO on China's response provided a pretty robust picture on what other countries would be facing as flights arriving into airports seeded the virus across the globe.

Within the span of three months, China had eliminated the virus fully within its borders. In late March 2020, with the virus taking off exponentially in North America and Europe and with countries falling one after another into drastic lockdowns, the Chinese government shut its own borders to stop the virus returning.

When did scientists know this would be the next pandemic? By mid-January 2020 the warnings were clear, based on early evidence from Wuhan. China would not have locked down 60 million people over flu. I tweeted on the 16th of January 2020: 'Been asked by journalists how serious #WuhanPneumonia outbreak is. My answer: take it seriously bc of cross-border spread (planes mean bugs travel far & fast), likely human-to-human transmission & previous outbreaks have taught over-responding is better than delaying action.'

At this point it was up to countries and their scientific advisers to heed these warnings and pay attention to the clear data – as South Korea and Senegal did, as discussed in the next chapter – or to stick their heads in the sand and hope that somehow, possibly through divine intervention, the virus would not arrive at their shores, or be as dangerous as the scientific literature and China were claiming.

# 3. South Korea, West Africa and the *Diamond Princess*

South Korea was already on edge when news about a novel coronavirus in China was broadcast. As we saw in the Prologue, the South Korean government was able to mobilize quickly because of its experience with MERS, and, while it struggled with a high number of infections (almost 1,000 per day at the peak of the first wave), the country managed to bring the epidemic under control throughout 2020 using a mass testing, tracing and isolating programme.

The government seemed to do the impossible, which was to 'crunch the curve', rather than just flatten it, and to do it without a lockdown. This model would go on to influence other countries across the world that had to make rapid decisions on what to do, and could follow a tried and tested East Asian 2020 playbook.

In this chapter I take a closer look at South Korea's response to the virus, as well as the preparations that Western African countries, specifically Senegal, were making based on their experience of Ebola in 2014 – and what the world had learnt from Ebola in terms of pandemic preparedness. Africa's success in managing the first wave of COVID-19 was not down to luck: there were clear steps and actions taken by governments to do what they could, even with limited resources. We then turn to the extraordinary story of the *Diamond Princess*. This was a cruise ship that experienced rapid spread of the virus at an early stage of the pandemic. Passengers on board faced twenty-seven days of uncertainty as lab rats in a natural experiment that taught scientists about asymptomatic carriers, transmission of the virus and fatality rate.

## *The East Asian Playbook*

South Korea's first case was identified on the 20th of January 2020 in a 35-year-old Chinese woman, followed by another case in a

55-year-old man recently returned from Wuhan. The government alerted the public to these on the 24th of January 2020. Already on alert from early January, the government moved quickly to activate the Central Disaster and Safety Countermeasures Headquarters. Using strong pre-emptive action South Korea was able to keep cases in single digits before a large outbreak in Daegu on the 19th of February resulted in 909 newly confirmed cases by the 29th of February. The government managed to handle this first wave and bring case numbers down again rapidly using mass testing, robust contact tracing and strict isolation on a scale never seen before in the country.

Much of the response was driven by the experience of MERS in 2015, which resulted in clear policy directives and preparedness. More infection control staff were trained, isolation units were created, simulations were run regularly, and personal protective equipment was expanded to ensure adequate supply. The government also put in place an emergency process for approving new diagnostic tests, and the country's public health laws were amended to allow for contact tracers to use data from mobile phones, credit cards and public transport records.

Dr Ducksun Ahn, President of the Korean Institute of Medical Education, explained it like this: 'The way we dealt with MERS wasn't satisfactory. It was very embarrassing. It created a great fear in society. With that fresh in their memories, people hearing of this new epidemic were anxious and so followed what the government asked them to do.' Dr Kee Park, Lecturer at Harvard Medical School, agreed: 'There was a lot of criticism about the way the MERS epidemic was handled. The South Korean government and the Ministry of Health and Social Welfare, they learned, "Well we don't want to go through this again", and so had a plan in place.'

South Korea largely held off going into a national stay-at-home lockdown, and, unlike New Zealand – another country held up as a success story – did not aggressively pursue viral elimination. Life in South Korea wasn't normal, though. Social distancing measures were consistently in place, and mask wearing was required early on in all public spaces. Schools stayed open but in a blended model of mixed in-person and online learning, so that students could maintain space from each other in the classroom.

The South Korean Director for Epidemiological Investigations at the Korea Disease Control and Prevention Agency, Dr Park Young Joon, emphasized the need for balance: 'The basic principle we have is to keep a balance between preventive measures and economic growth . . . We call it long-term suppression – maintaining our society while suppressing the virus. The basic concept we are working with is living with COVID-19.'

While travelling to South Korea has been permitted in general throughout the pandemic (unlike Australia and New Zealand), those arriving into the country had to quarantine upon arrival from the 1st of April 2020 until the time of writing of this book, in August 2021. Before the 1st of April, the government advised against international travel and was testing all in-bound travellers from Europe; even earlier it was banning flights from Hubei Province. Procedures included designated entry lines, questionnaires, temperature checks, testing of all incoming passengers and mandatory fourteen-day quarantine. Those with Korean residency were provided with facilities for self-quarantine, but they had to pay the cost of their stay. In October 2020 travel restrictions were further tightened, and all foreign nationals needed visas to enter the country.

Because the core of the South Korean response has been the test/trace/isolate system, the country raced ahead of most others by months with its testing, bringing out its approved test-kits by the 4th of February 2020; and by March 2020 it had the highest per capita test rate in the world, with results back within twenty-four hours and sometimes even four hours. At this time Germany was returning results within a week, and the UK wasn't offering testing at all except in hospitals. In South Korea screening facilities were set up outside hospitals as well as in drive-through centres and photo-booth style testing stations.

Western countries seemed unaware of what South Korea was doing, and how it was working. To try to bring attention to its approach, on the 23rd of March 2020 I wrote an article for the journal *Foreign Policy* titled 'Without Mass Testing, the Coronavirus Pandemic Will Keep Spreading'. This was when the UK was still debating whether testing mattered, and I was hoping to convince policy-makers of the merits of expanding testing into the community.

In my article I argued that the 'Seoul' model of mass testing was one route to slowing spread without harsh lockdown measures. Testing was indeed important, because people could alter their behaviour (i.e., isolate) if they knew their status and this would break chains of transmission in the community; local hospitals could plan for patients based on how many cases were in the local community; and public health officials could know where hotspots were in the country, and how large a problem COVID-19 was in order to have more precise surveillance. But testing alone was just one step: the next was contact tracing, so the authorities could see who else might have been infected by a positive case.

Within about ten minutes of receiving confirmation that a patient was positive, Korean public health teams could trace exactly where that person had been over the past week and, based on those locations, identify who else could have been exposed. This was as a result of legislation passed post-MERS to acquire individuals' GPS data from their phones as well as quick access to credit card details and closed-circuit television. In June 2020 the KI-Pass, including a QR code – which is a check-in electronic log system to collect data on visitors – was introduced to help contact tracing efforts after outbreaks at several high-risk locations. Data was deleted after fourteen days.

Overall South Koreans accepted this infringement of their privacy as the necessary trade-off for freedom of movement. But this also revealed that they clearly trusted their government not to abuse these details for non-public health reasons. According to some lawyers the South Korean government stretched its authority too far, by, for example, using mobile phone location data to alert anybody within an area of potential exposure during a nightclub outbreak, as will be discussed later in this chapter. Kelly Kim, of Open Net Korea, said, 'Most Koreans are willing to compromise their privacy for their life.' A disaster alert app, which was used for typhoon warnings before the pandemic, would check a person's movements and flash if they had been anywhere near an infected patient; and subsequently it would tell them to get tested if they had been.

During the MERS outbreak there was criticism about authorities concealing too much information; in contrast to this, data about confirmed COVID-19 patients was made public on official websites. While names were concealed, the details presented – such as age, gender, occupation and location – were sufficient to allow identities to be discovered. This has resulted in revelations about affairs, plastic surgery and even patients' sexual behaviour.

Restaurants and shops visited by patients were made public, and, while they were closed for cleaning, owners were worried their association with COVID-19 would reduce future business. In one situation someone pretended to be infected with COVID-19 to blackmail the restaurants they had allegedly visited. Kim Na-hee, a restaurant owner in Seoul, said, 'I thought I only had to protect my health, but now I think there are other things more scary than the coronavirus.' Concerns were raised about the consequences of making this kind of private data public, and on the 15th of March 2020 the Korean government amended patient information disclosure guidelines so that any personal identifying details would not be included.

Isolation was also managed effectively. During the first wave, privately owned public facilities and retreat centres were transformed into temporary isolation wards, preventing transmission within households as well as relieving hospitals of bed shortages. Those Koreans who tested positive and preferred to stay home were checked on twice a day to report any symptoms; they also received food and toiletry deliveries as well as psychological counselling. A Self-Quarantine Safety Protection app was required for fourteen days, which included location tracking to identify when people broke quarantine. Fines for violation were high, at US$8,217.

Those who felt that they would rather not isolate at home or who needed medical support were moved into 'isolation facilities'. Here patients were closely monitored, and, if symptoms became worse, they were moved into hospitals. South Korea has attributed its extremely low fatality rate to early intervention for those with symptoms, instead of leaving hospital admission until an individual was in a severe state. Recognizing it had found a reasonably effective model of containment, the South Korean government developed protocols to share with other countries and tried to benefit others with its accumulated knowledge on containment.

South Korea has also faced challenges particularly around nightclub super-spreader events. In contrast to flu – which, on average, results in every infected person passing on the virus to the same number of people – COVID-19 spreads differently: it radiates in bursts from certain infectious individuals during the time when they have a high viral load. If these individuals are in a crowded bar or nightclub, the location ends up being the setting for a super-spreader event. These kinds of settings (indoor, crowded, poorly ventilated) mean that, even with only one infectious individual, almost everyone around them will also become infected. This is in contrast to, say, an outdoor exercise class in a park, where, if one person is infectious, it is highly unlikely they will infect others, given the good ventilation that exists outside and distancing.

On the 6th May of 2020 a young man was diagnosed with COVID-19 who, in the days before falling ill, had visited five nightclubs. Several media outlets reported that the venues at the centre of the outbreak were gay nightclubs. With gay men in South Korea experiencing stigma and discrimination, those at the clubs those nights were hesitant to come forward to get tested. Lee Youngwu, one gay man at the club, felt betrayed: 'My credit card company told me that they passed on my payment information in the district to the authorities. I feel so trapped and hunted down. If I get tested, my company will most likely find out that I'm gay. I'll lose my job and face public humiliation. I feel as if my whole life is about to collapse. I have never felt suicidal before and never thought I would, but I am feeling suicidal now.' Lee's anxiety is a clear indication of how government surveillance can dangerously infringe on personal privacy, even if it helps to contain a pandemic.

To encourage people to get tested, anonymous testing was introduced. Professor Sung-il Cho, of Seoul National University, noted the challenge: 'Many of the visitors [to] the club who have not been identified may be reluctant to come out, so tracing may not be as complete as before.' As the tracing system struggled, all nightclubs were shut on the 9th of May 2020, and mass gatherings were banned in several regions. Blanket restrictions like these happen when the tracing system breaks down and infectious individuals cannot be pulled out.

With the help of mobile phone data, credit card records and lists of nightclub visitors, 5,517 people were identified for screening and 1,257 were actively monitored. Additionally, 57,536 people who had spent more than thirty minutes near the nightclubs, as determined by their mobile phone location data, were sent a series of text messages to encourage them to get tested. This incredible workload for tracing has strained the system: 80 per cent of those working in tracing experienced high levels of burnout. Some logged more than a hundred hours of overtime each month. Jang Hanaram, a military doctor leading Incheon's contact-tracing operation, noted the challenge: 'I was the only person handling contact tracing work for all of Incheon.' Incheon has a population of 2.9 million.

In early 2021, as will be covered in Chapter 9, safe and effective vaccines were developed. This was the next stage of South Korea's strategy. While South Korea successfully managed the pandemic over the first year and a half, in early 2021 the next challenge was acquiring enough vaccine doses to start protecting the population, as a more sustainable and long-term solution. Without the market access of the US and European countries, the government struggled to get enough supply. While they managed to buy time in an efficient way, preserving lives and livelihoods, the clock was ticking to get jabs into arms before new, more transmissible variants could take off that would be harder to suppress.

Close to South Korea, a cruise ship named the *Diamond Princess* left Yokohama port in Japan on the 20th of January 2020, with nearly 3,000 passengers on board. One of these passengers, an eighty-year-old man from Hong Kong, had developed a cough the previous day but boarded the vessel anyway. Public health researchers joke that there are three things we avoid: mosquitoes, bats and cruise ships. While it might seem a random assortment, there's a clear logic behind each focus of anxiety.

## *Mosquitoes and Zika Virus*

Despite being small and ubiquitous in tropical countries, mosquitoes are terrifying. They carry parasites like malaria, and viruses like dengue, Zika and chikungunya. Dengue and chikungunya are unlikely to kill healthy adults, but they can lead to joint pain that lasts for years. While most people would name sharks or crocodiles as the deadliest animal, it's the tiny mosquito that wins that prize, killing 2.7 million people each year. That's like the entire city of Chicago being wiped off the face of the planet on a yearly basis, with the majority of those deaths due to malaria.

Zika virus was the last major outbreak pre-COVID, and showed the wily nature of viruses. Before the large numbers of Zika cases were seen in Brazil, it was considered a mild disease resulting in flu-like symptoms and not much to be concerned about. However, when case numbers exploded in the summer of 2016, it became clear that pregnant women infected with the virus could give birth to babies having a much smaller head than expected, because the brain had not developed properly during the pregnancy. This condition, called microcephaly, results in lifelong developmental difficulties. Because of its role in causing microcephaly in unborn babies, Zika virus was declared a Public Health Emergency of International Concern in 2016 by WHO.

But Zika was brought under control through a massive campaign against mosquitoes and their larvae. At least in rich countries, having a vector like a mosquito or a tick makes it easier to control a disease, because governments can just throw resources and technology at eliminating the insect or animal. Swine flu? Kill off the pigs. Avian flu? Mass cull of chickens. Zika? Spray insecticides from helicopters over cities. But what happens when the vector is other humans, and the most dangerous humans are the ones you work alongside, live with and love?

## *Bats and Ebola Virus*

Bats are also a significant concern to public health researchers because they are exceptional at hosting zoonotic infections – diseases that pass from animals to humans – and act as reservoirs for over sixty viruses that infect humans. As we have seen, coronaviruses like

MERS-CoV, SARS-CoV-1 and SARS-CoV-2 all came from bats originally. But bats probably acquired their notorious reputation from their role in Ebola virus, which is one of the most dangerous pathogens around. In December 2013 scientists suspect that a toddler in one of the poorest countries in the world, Guinea in West Africa, found and ate a dead bat, and became infected with Ebola. This was the spillover event that set off a chain of infections that would kill thousands of people and cause worldwide panic.

Without any medical intervention, Ebola kills over 70 per cent of people who get it, and, while it begins like the flu, within seven days the virus starts to destroy internal organs, leading to massive bleeding and eventually death. But it is an easier virus to contain than COVID-19, because it spreads only through bodily fluids, usually blood, saliva, sweat, urine, vomit or semen. At the time of the 2014 outbreak, there was no rapid diagnostic, no approved treatment and no vaccine.

As the virus spread quietly through communities, it wasn't until May 2014 (six months after the spillover event) that the outbreak was even identified as Ebola, and it took another three months for WHO to convene the Emergency Committee to decide whether the virus constituted a global health emergency. These delays – all down to limited ability to detect the outbreak and to identify the virus, to the slow response by WHO and the international community, and to breakdowns in coordination during the response – resulted in (at the time) one of the worst responses to an outbreak in modern history. Over 11,000 people lost their lives to Ebola, with thousands more affected by health care disruption and school closures.

Seventeen panels convened in 2014 and 2015 sought to identify what exactly had gone wrong; how the mistakes could be avoided in the future; and how WHO, including its Health Emergencies Programme, would need to reform and highlight the importance of investing in pandemic preparedness. I co-chaired one of these panels with Dr Peter Piot. Piot is best known as the co-discoverer of the Ebola virus, the politically savvy former head of the Joint UN Programme on HIV/AIDS (UNAIDS) and former Director of the London School for Hygiene & Tropical Medicine. In 2015 Peter and

I co-chaired the Harvard/London School of Hygiene & Tropical Medicine Independent Panel on the global response to the 2014 Ebola outbreak and have collaborated closely over the years on global health research and analysis. From my work on Ebola, I learnt the importance of early response, mass testing, robust tracing and isolation policies, as well as clear messaging from government. My work on Ebola helped to build my expertise in offering public health advice in response to COVID-19. Sub-Saharan African countries were also burned by their experience with Ebola, and used their Ebola structures to move quickly in their response to COVID-19.

For those of us working on pandemic preparedness, analysis has consistently focused on low-income and fragile settings, as these fit with the pattern of outbreaks over the past few decades. Rankings of 'global health security' capacity have repeatedly put high-income countries like the US, UK, Netherlands and Canada near the top of the list, while poor countries like Yemen, Haiti and Guinea are near the bottom.

## *2014 West Africa Ebola Outbreak*

Ebola is a classic example of how outbreaks usually emerge. The virus was first identified in the 1980s by a team of international researchers sent into the Congo to investigate a new deadly illness, and to figure out how people were becoming infected and what the key symptoms were. As noted above, while Ebola is a frightening disease with a very high mortality rate, the way it transmits means that outbreaks have been limited to rural areas and the number of people infected has been relatively low.

As mentioned above, in early 2014 a large outbreak occurred in West Africa, a region that had never before seen this disease. By the time it had been recognized as Ebola, it had spread considerably through Liberia, Sierra Leone and Guinea. The crisis was exacerbated by the lack of functioning health systems, little-to-no laboratory capacity to identify pathogens, inadequate financial resources, and not enough skilled people to run labs and response efforts. Similar stories can be told about cholera in Haiti, plague in Madagascar and Zika virus in South America.

Over the course of the Ebola outbreak in 2014, over 10,000 people would die, cases would be identified in many parts the world, and global panic set in to the point that 'Ebola' was one of the top googled terms in almost every country. After an initial sluggish response by WHO and the World Bank, partnerships among key donor countries like the US, UK and France were rapidly established to aid the region and contain spread as rapidly as possible. This involved a NATO military response, with the US taking responsibility for Liberia, the UK for Sierra Leone and France for Guinea. Research funders like the Wellcome Trust started the first human trials of an Ebola vaccine during the outbreak.

In addition, an experimental treatment called ZMapp, which was approved for emergency use in extremely ill patients, was seemingly effective at improving survival. This triple monoclonal antibody drug had been tested only in eighteen rhesus monkeys, and limited doses were available. Given the drug was produced through plant-based technology, which is a time-intensive process, a steep rise in doses would have taken at least six months – too late to have an impact on the outbreak at that time. How were decisions made on how this drug would be distributed? It came down to citizenship. Citizens of rich countries like the US, UK and Spain were offered ZMapp, and those from Liberia, Guinea, and Sierra Leone were not. Two doctors working side by side during the outbreak both became infected with Ebola. One was flown back to the US and survived; the other was left to die in Liberia – a gruesome example of how global politics shape our health.

## *COVID-19 in West Africa*

Based on this recent experience, West African countries were able to move rapidly to reconfigure their post-Ebola and other infectious disease recovery structures once the threat from COVID-19 became known. These governments knew the pain and devastation that infectious disease outbreaks could bring, and that their only defences were to introduce border screening to limit importation of the virus; to implement early physical distancing; and to try to build diagnostic capacity as rapidly as possible. As Dr Rosemary Onyibe, who had worked on polio eradication in Nigeria, said, 'Once I heard the news, we immediately mobilized the existing polio personnel, tracking contacts and conducting follow-up visits.' They also introduced environmental hygiene measures like public handwashing facilities, use of face coverings and masks (even cloth ones), and temperature screenings in train stations and airports.

Early into the pandemic the African Centers for Disease Control and Prevention (Africa CDC) also started weekly briefings for health ministers from the region on how to prepare their countries for the virus. But, even as they raced to make the arrangements, to some of these leaders it was clear that a tidal wave was coming, and there was little they could do to stop it.

The idea of creating a continent-wide public health organization was first considered in 2013, and the Africa CDC was officially launched in January 2017. The 2014 Ebola outbreak strengthened the need for an initiative to support Africa's public health systems, signalling the need for early warning, response surveillance and improved capacity-building to increase their overall response to health emergencies. In July 2015 the African Union ministers urged the fast tracking of the agency at a health meeting in Malabo, Equatorial Guinea.

The majority of Africa CDC's funding comes from grants and donations. Following the outbreak of COVID-19, grants have been awarded to support the implementation of public health strategies from the UK Department of International Development and the Wellcome Epidemic Preparedness for Coronavirus.

The Africa CDC is led by Dr John Nkengasong, an experienced and highly articulate Cameroonian virologist, who sees a self-sufficient and resilient future for African countries. While doing his undergraduate degree at the University of Yaounde, Nkengasong met Piot, who encouraged him to go to Belgium for further studies and to work with his research group on HIV in Africa. Piot is known for his mentorship and support of younger African scientists and his investment in building future leaders.

In 1993 Nkengasong joined WHO as Chief of Virology, with a focus on HIV. He soon moved to the US CDC, and, in 2011, returned to Africa to establish the African Society for Laboratory Medicine. He was an easy choice to become the inaugural Director, in 2016, of Africa CDC and has played a pivotal role in leading African preparedness and response to COVID-19. African countries would have fared much worse without his competence and leadership.

Nkengasong outlined five pillars for African countries trying to cope with the virus: surveillance, infection prevention, management of those infected, lab diagnosis and community engagement. As he said, 'In close collaboration with WHO and other partners, we are preparing the continent as quickly as possible to respond to this threat.'

In response to the WHO Emergency Declaration on the 30th of January 2020, Africa CDC moved quickly to establish the Africa Task Force for Novel Coronaviruses on the 3rd of February 2020, which worked to support member states in their rapid detection and containment of the virus. Africa CDC supported WHO advice for all countries to prepare for a high number of cases and ensure they had adequate testing and tracing for the approaching storm. General public health measures of handwashing, wearing face coverings, quarantining those infected and limiting incoming travel were also stressed.

Africa CDC launched the Partnership to Accelerate COVID-19 Testing, which included setting up distribution centres for medical supplies, pooling procurement of diagnostics, support for the testing of one million Africans in ten weeks, support for health care workers and clear sharing of testing data on public platforms to ensure trust in government.

A press briefing by Africa CDC noting the progress made in testing was delivered on the 14th of February 2020; more than 16 countries were ready to test for the virus and 20 more would be ready by the 20th of February. Reinforcing Tedros's message to 'test, test, test', Nkengasong also stated that, for the risk to be minimized, there would need to be the same standard of public health systems across all countries, regardless of wealth. While unrealistic in the short term, it provided a good goal to aim towards. Governments therefore were encouraged to invest in public health immediately in collaboration and partnership with their neighbouring countries.

On the 22nd of February 2020 the African Union Commission and Africa CDC had an emergency ministry meeting to provide disease updates to the states, concentrating on strategies to better respond and prepare. Top of the list was how to coordinate similar travel policies regarding the movement of people across the continent to prevent cross-infection. The outcome of this meeting was the formal *Guidance for Assessment, Monitoring and Movement Restrictions of People at Risk*. Technical assistance on infection prevention and control protocol was provided, as was information on surveillance, contact tracing and quarantine. Africa CDC and WHO also worked closely to accelerate stockpiling of PPE and quality-assured diagnostics. Africa CDC started weekly briefings on the spread of the virus, and on changes in scientific knowledge, public health policy and guidance from WHO. The region was doing what it could to prepare for the onslaught of infections. By contrast, Europe, Britain and the US were still largely asleep.

## *Testing in Africa*

Lab testing is vital to identifying infectious individuals. In January 2020 only two African countries had diagnostic capacity. Even in February 2020 many labs lacked the resources needed to effectively test the population, with many having to send samples to other countries, causing a backlog of results. By April 2020 forty-three African countries had lab capacity, but testing per capita still remained incredibly low across Sub-Saharan Africa.

By the 24th of June 2020 Tedros stated that all African countries had developed lab capacity to test for COVID-19. WHO largely contributed to this advancement by funding training and by providing technical aid, including the salaries of thirty lab scientists for six months. Professor Sahr Gevao, Director of Lab Services in Sierra Leone, said, 'The laboratory sector has been very challenged over the years. From the Ebola epidemic to the COVID-19 pandemic, laboratory capacity has consistently proven to be low. We are therefore pleased and grateful for the support that the WHO is providing to ensure improved laboratory capacity that goes beyond the COVID-19 pandemic.'

Rapid test-kits were also a major problem for countries, with the added difficulty that African countries had to compete with richer nations for test-kits and other supplies. In April 2020 WHO, in collaboration with Africa CDC, delivered essential supplies to 52 African countries. A system was created to allow more than 1,140 orders to be processed and delivered to 47 countries. WHO also formed a global procurement association to leverage their networks, supporting countries that have limited access to markets.

On the 3rd of June 2020 Nkengasong stated the need for 20 million new test-kits within a hundred days. To meet that goal, countries pooled their resources to place large joint orders. Africa CDC also agreed the purchase of 90 million test-kits over the following months with manufacturers, to help combat the problem of competition with richer countries placing large orders.

As is clear from the decisive action by Africa CDC, there was no complacency around this new threat. Nkengasong understood the challenge well from the start: 'This disease is a serious threat to the social dynamics, economic growth and security of Africa . . . If we do not detect and contain disease outbreaks early, we cannot achieve our developmental goals.'

While African countries did reasonably well against the first wave, the second wave in spring 2021 took a far greater toll, especially with the seeding of the more transmissible Alpha and Delta variants, which I discuss in Chapter 8. This coincided with the development of effective vaccines. While rich countries purchased enough doses to vaccinate their population multiple times over, African countries were left behind, with not enough vaccines even for health workers and elderly and vulnerable groups. Nkengasong pushed heavily for intellectual property rights to be lifted so African countries could manufacture and distribute their own vaccines, and noted that Africa should not have to depend on the benevolence of the West in donating extra doses. I return to this 'vaccine nationalism' in Chapter 10.

## *West Africa's Shining Star: Senegal*

If one African country stands out for its response, it is Senegal. Senegal reported its first imported case on the 2nd of March 2020, making it the third African country to report a confirmed COVID-19 infection. The Senegalese government acted fast, releasing information and enforcing measures to prevent the spread. Dr Abdoulaye Bousso, the Director of Public Health Emergency Operation Centre in the Ministry of Health, said, 'The first response to this pandemic was to set up an emergency operations centre to lead and coordinate all operations.'

President Macky Sall understood the necessary public health measures and recognized the need to help health workers to identify, isolate and treat infected people. He said, 'It is in response to an imminent peril. The emergency gives us the means to strengthen our ranks and intensify our efforts to defeat the common enemy.'

President Sall knew to go early, go hard and keep it simple. He announced school closures on the 16th of March 2020, making it the first Sub-Saharan African country to take this measure. On the 20th of March international air travel was suspended, and on the 23rd of March a national state of emergency was declared for three months, subject to review. At this time there were only eighty-six cases. Measures taken included mandatory testing for arriving travellers and isolation for individuals who tested positive, with a mandatory fourteen-day quarantine. Additional restrictions were placed on public gatherings, including in places of worship, and face coverings were mandatory in all public spaces and on public transportation.

Senegal is 95 per cent Muslim, and religion plays a vital role in daily life. With a curfew lasting throughout the whole month of Ramadan, people could no longer go to the mosque for prayer, congregate and breakfast together, and had to rely on home and online prayers.

Despite increased pressure from the Muslim majority to reopen mosques, religious leaders remained vigilant, urging their followers to comply with the government orders to stay home. Imam Amadou Kanté of Grand Mosque Point E says, 'We prefer to be careful, so that by Ramadan we would see how it evolves and what other receivers tell us. If we are told there's a decline, that the dynamic is not on the rise, we could satisfy this religious demand of a few believers to resume prayers.' A survey found that 90 per cent of the people interviewed complied with restrictions and stopped going to mosque. And even when mosques were open for worship in the last week of Ramadan, many still chose to pray from their house and celebrate Eid al-Fitr with their own families at home. This is similar to compliance in the UK over the first lockdown in the spring of 2020.

Senegal's years of effective infectious disease management for measles, rubella, meningitis, TB and rabies placed it in a better position from the outset, compared with its regional neighbours. Dr Amadou Alpha Sall, Director-General of the Institut Pasteur in Dakar, said, 'We were able to enhance surveillance systems for all kinds of diseases across Senegal. The equipment provided will allow us to rapidly detect and improve our capacity to mobilize resources and respond to disease outbreaks like coronavirus.' WHO Senegal Coordinator Dr Lucile Imboua agreed: 'Senegal, like all other countries in Sub-Saharan Africa, is used to managing outbreaks and has the experience and capacity to respond. The experience gained from the Ebola outbreak has been useful in triggering preparedness and response interventions.'

Overall, Senegal is considered a successful model of response. This involved its public health messaging, its testing strategy and its economic packages to support restrictions – the last provided an income and food support to those most vulnerable. Judd Devermond, Director of the Africa Program at the Center for Strategic and International Studies, in Washington DC, pointed this out: 'You see Senegal moving on all fronts: following science, acting quickly, working the communication side of the equation and then thinking about innovation . . . [it] deserves to be in the pantheon of countries that have . . . responded well to this crisis, even given its low resource base.' The magazine *Foreign Policy* developed a COVID-19 global response index assessing the strength of each country's strategic response in 2020. Senegal (97.2 per cent) ranked second after New Zealand (100 per cent).

In its public health messaging, Senegal, like other West African nations, adopted a comprehensive communications strategy to raise awareness about the disease and how to curtail its spread. To improve people's understanding of the pandemic, the government resorted to community mobilization and dissemination of vital information. Social influencers such as musicians and religious leaders were used to drive its messages to their followers. The strategy was highly innovative: musicians teamed up to release a single called 'Daan Corona', meaning 'defeat coronavirus' in the local Wolof language. They used social media channels, radio, television and flyers to communicate with people in the language they understood. In *Foreign Policy*'s assessment, Senegal was given 100 per cent for public health directives and 100 per cent for fact-based communications.

Testing in Senegal has also been a major success. In February 2020 the Institut Pasteur in Dakar was one of only two laboratories in Africa that was able to test for SARS-CoV-2. It then shared expertise and trained staff from dozens of other countries on how to test. By April 2020 forty-two African countries had the capability to diagnose COVID-19 effectively. The Dakar laboratory created a 24-hour-testing facility, which was expanded across the country. Symptomatic testing was provided free of charge with results usually released in 8 hours.

The Institut Pasteur's Dr Amadou Alpha Sall said it partnered with the UK-based firm Mologic, as mentioned above, in late January 2020 to develop two home test-kits: one to test if someone has the virus, like a PCR test, and the other to check if antibodies from previous infection are present in an individual. The antibody test was priced at US$1 with results within ten minutes.

As noted above, restrictions were put in place to limit mixing. Almost half the population in Senegal live under the poverty line, which means prolonged lockdown can lead to particularly acute financial distress and social disorder. If the choice is between going hungry and staying home or going out to work and risking COVID-19, the latter will win out even with a government ban.

To extend support to marginalized groups, the government approved 50 billion CFA francs (US$85 million) to ensure economic support – for food, water and electricity – was available to the poorest living under restrictions. Senegal also received around US$880 million in direct support from donors such as the World Bank, the IMF and the Islamic Development Bank. The role of the World Bank will be discussed in Chapter 10.

The government also allocated 64.4 billion CFA francs (US$115 million) to support the health sector. This included immediate deployment of health experts, medications and other necessary medical supplies. Senegal has a fragile health system, like most African countries. With only 7 doctors per 100,000 people (compared to 277 per 100,000 people in the US) and limited resources, the continent's situation was precarious even at the best of times.

What Senegal's story shows is that, even in the context of limited resources and scientific uncertainty, certain countries reacted quickly and effectively to prevent a crisis. In a population of 16.3 million, Senegal has only had 45,000 confirmed cases and 1,190 deaths as of July 2021, even after several small waves of infection. The government saw the spread of SARS-CoV-2 as preventable and took the measures necessary to contain it with as little economic damage as possible. Unfortunately, most of the Western world was ignoring the preparations and leadership in East Asia and West Africa.

## *The Natural Experiment on the* Diamond Princess

Which brings us back to cruise ships, and one of the most harrowing human stories of the COVID-19 pandemic: the imprisonment of both healthy and ill passengers for twenty-seven days on the *Diamond Princess*. As already mentioned, it's rare to see public health researchers on cruise-ship holidays, because they are perceived from an infectious disease perspective as floating germ factories. They're notorious for many reasons: norovirus (vomiting bug) and flu outbreaks flourish, which is understandable, given the crowded and largely enclosed space in which people from across the world live, entertain and eat together; the ships stop at various ports where passengers disembark, easily pick up a local bug and hop back aboard and share it with the rest of the ship; and crew members work through multiple trips. All of this points to the constant circulation of viruses among passengers. Cruises also tend to attract retired and elderly individuals, with weaker immune systems. For COVID-19, it was a recipe for disaster.

On the 7th of February 2020 I tweeted, 'Things that #public-health researchers stay away from: 1. Cruises – floating germ factories. 2. Bats – exceptional at hosting zoonotic infections & reservoirs for >60 viruses that infect humans 3. Mosquitos – world's deadliest animal, kill 2.7M people/year.' This tweet prompted a representative from the cruise industry to reach out and push back against this characterization of the industry. Just another example of my bluntness on Twitter inadvertently getting me into hot water.

In February 2020 a passenger who had recently departed the *Diamond Princess* ship in Hong Kong tested positive for coronavirus. The eighty-year-old man was a Hong Kong resident and had visited Shenzhen in China's Guangdong Province for a few hours on the 10th of January. He took a flight from Hong Kong to Tokyo seven days later, and started coughing. Despite feeling unwell, he boarded the *Diamond Princess* at Yokohama Port on the 20th of January and spent five days on the ship before disembarking in Hong Kong with a fever. He tested positive for COVID-19 on the 1st of February 2020.

This immediately set off alarm bells for the crew and the Captain of the ship docked in Tokyo Bay, as they scrambled to figure out how many people had already been exposed. The ship carried 2,666 passengers (1,281 of whom were Japanese) and 1,045 crew from a combined total of 56 countries. Hundreds became unwell and were confined to their cabins. The number of confirmed cases on board quickly jumped to 454 by the 18th of February

The *Diamond Princess* became the responsibility of the Japanese government, which wanted to tightly contain the situation. Unlike the US, Japan does not have a central disease-control agency. Professor Christos Hadjichristodoulou, leader of a WHO team of advisers, offered to organize a group of experts to board the ship and advise, but his offer was declined by government officials, who wanted full control of the situation.

The incubation period – that is, the time in which people might test negative but soon become infectious as the virus replicates in their body – of SARS-CoV-2 is roughly 10–14 days, so the Japanese government could not just let the people off the ship or have them fly back to their home countries, because of the risk of their carrying the virus and spreading it further. On the other hand, keeping the passengers on the ship was equally risky: they were unsure how many people might have it and how exactly it could spread. Passengers were allowed to mix on the ship, including participating in exercise classes, quiz games and dances, until the 5th of February 2020. Dr Esther Chernak, an infectious disease physician, commented, 'Quarantine on the ship may be increasing the risk of infection and creating a barrier to accessing medical care.' Those who were infectious were passing it on to the others who were kept in close conditions with them, and no attempt was made to ensure people isolated from one another.

The decision was made to keep all the passengers on the ship for fourteen days (referred to as a cordon sanitaire) with an end date of the 19th of February, and to ask anyone who was symptomatic to come forward for testing. Out of 31 swabs taken, 10 were positive. But, other than asking passengers to wash their hands, disinfecting surfaces and handing out hand sanitizer, not much changed for the passengers' daily routine – except that the numbers of coronavirus cases kept increasing. First 10, then another 10, then 41, 66, 218, 531 . . . By contrast, the total number of cases in Singapore at that time (the most affected country after China) was only 77. On the 20th of February 2020 WHO announced that the largest number of COVID-19 cases outside China was on this single ship.

As the numbers increased, the Japanese government brainstormed how to introduce a more nuanced quarantine on the ship: ideally a 'red zone', where the virus was known to be circulating; a 'yellow zone', where medical personnel or staff changed and disinfected themselves; and a 'green zone', which would be virus-free.

While it sounds great in theory, this kind of quarantine arrangement had never been tried before on a cruise ship. The Japanese infectious disease doctor Dr Kentaro Iwata went on board as part of a disaster management medical team to investigate how the outbreak was being controlled. Iwata had studied SARS, cholera and Ebola and noted, 'I wanted to enter the cruise ship and wanted to be useful in helping to contain infection there.' He reported in a YouTube video how shocked he was that there was no clear distinction between the zones, and that, while he had worked on the frontline of the Ebola response, he felt more afraid for himself on the ship than he had during the Ebola outbreak. The lack of standard infection control practices meant there was no clear way of knowing where the virus was. He noted, 'There was no distinction between the green zone, which is free of infection, and the red zone, which is potentially contaminated with the virus.'

Cruise ship staff became the managers of these zones, but, because they'd had no training in infection control, they carried on moving between rooms of the ship delivering food, packages and towels, as well as distributing thermometers. Iwata noted: 'There was no single professional infection control person inside the ship, and there was nobody in charge of infection prevention as a professional, the bureaucrats [Japanese officials] were in charge of everything.' If someone had a fever, they were offered a COVID-19 test. If not, they were considered healthy and not infectious. No routine testing was offered for the first week.

Whether the passengers knew it or not, they were taking part in a giant experiment, with the Japanese, German and American governments watching closely to find out more about how the virus transmitted, who was affected, and how to bring their nationals off the ship and back to their home country.

Each day confirmed cases among passengers and crew increased, and questions were raised about the ethics of imprisoning healthy people with those infected rather than separating them. In fact, a later study suggested that if everyone had been taken off the ship on the 3rd of February (when the Captain was first alerted), fewer than 80 people would have been infected instead of the total number of 712. Of those, 14 passengers died.

Suddenly it dawned on scientists, who had solely focused on fomites as a means of transmission (bits of virus on surfaces and hands) – it's the usual route for many infectious diseases – that the virus could also be spreading through droplets or possibly even through the air. The emphasis on symptomatic individuals (those who had a fever or cough) also missed out asymptomatic carriers of the virus, who could be passing it on to others unknowingly. Finally, on the 11th of February, it was decided that they would begin to test everyone, and to start letting those who tested negative disembark so they could return to their home countries.

The natural laboratory of the *Diamond Princess* revealed several key scientific aspects of the virus. Because Japanese officials eventually tested almost all passengers and crew, with over 3,000 tests taken, this helped scientists to understand 'a key blind spot in many infectious-disease outbreaks – how many people are actually infected, including those who have mild symptoms or none at all. These cases often go undetected in the general population', as journalist Smriti Mallapaty reported in *Nature News*.

First, scientists discovered that those carrying the virus could feel completely fine and yet still transmit the virus to others, who may then develop severe symptoms. During one round of testing 21 per cent of those on board tested positive, with 51.7 per cent of these representing asymptomatic infections. A paper in the journal *PLoS ONE* suggested 50.5 per cent of passengers and crew members were pre/asymptomatic and around 17.9 per cent of the infected individuals never developed symptoms. Scientists at the London School of Hygiene & Tropical Medicine estimated that, while 74 per cent of people infected on board remained asymptomatic, they were responsible for 69 per cent of all infections during the outbreak; although it is worth noting these are estimates with considerable uncertainty over exact figures.

Second, they learnt that the virus was able to move through the air, whether at short distances (droplets) or longer ones through ventilation (airborne), and that just stressing handwashing and disinfecting surfaces was not enough. One study of transmission on the ship noted, 'Aerosol inhalation was likely the dominant contributor to transmission among passengers, even considering a conservative assumption of high ventilation rates and no air recirculation conditions for the cruise ship.' The paper also noted that proper ventilation alone was not enough to prevent transmission, as the ship was considered very well ventilated. It instead reinforced the importance of mask wearing, either proper medical-grade masks or double-masking (both a surgical and a cloth mask).

Aerosol transmission would become one of the most heated topics within the scientific community. Experts were divided on the role and relative importance of it. Professor George Rutherford said that, outside hospital settings, 'large droplets in my mind account for the vast majority of cases. Aerosol transmission – if you really run with that, it creates lots of dissonance. Are there situations where it could occur? Yeah, maybe, but it's a tiny amount.' On the other hand Professor Julian Tang disagreed: 'If I'm talking to an infectious person for fifteen to twenty minutes and inhaling some of their air, isn't it a much simpler way to explain transmission than touching an infected surface and touching your eyes?'

Several scientists, such as Professor Trish Greenhalgh, criticized the CDC and WHO for acknowledging airborne spread only in May 2021, over a year after the first findings from the *Diamond Princess* had been identified. The elevation of fomites over aerosols would mean heavy emphasis by certain governments on handwashing instead of ventilation and face coverings. This would result in mixed messages to the public on how to avoid the virus, as well as the late introduction of face masks and ventilation into public policy in countries such as Britain, the US and Sweden.

The ship also provided a rough estimate of the case fatality rate, around 2 per cent, but with the understanding that this was on the higher end because of the elderly age profile of the passengers. And it also became clear that cruise ships are not hospitals or appropriate quarantine facilities. The next cruise ships that reported confirmed coronavirus cases were immediately disembarked and passengers put into proper facilities on land. In the meantime WHO developed a protocol for handling coronavirus outbreaks on cruise ships.

## *Lessons Ignored*

From a public health perspective South Korea, Senegal and the *Diamond Princess* each imparted important information to the world. South Korea provided a blueprint for suppression without harsh lockdowns or restrictions. Senegal showed that even poorer countries with limited resources could suppress with proactive messaging, investment in testing and economic support packages. And the *Diamond Princess* revealed important information about asymptomatic transmission, aerosol spread and rough fatality rates. Each of these examples was a key piece in the COVID-19 puzzle, which the scientific advisers to governments needed in order to give advice and direction.

Adding this new knowledge to all the existing information that had come from China in January 2020 meant that countries had an increasing understanding of what COVID-19 entailed: how it spread and some of the tricky challenges in stopping it. Of particular use was the knowledge that people who felt healthy could infect others, and that it could spread through the air.

And yet rich countries were still not paying attention. Europe, Britain and the US were consumed by their own internal issues. The policy response of Senegal, and its success in containing COVID-19, point to the fact that it was competence, not wealth, that determined the impact of the first wave of COVID-19 on lives and livelihoods. It wasn't until Europe was badly hit in Italy, with a tsunami wave of infections and hospitals on the brink of collapse, that the Western world started to wake up to the threat COVID-19 posed. We now turn to Europe's response.

# 4. Europe's Tsunami

Italy faced a dramatic rise in cases in the Lombardy region, and initiated the first regional lockdown in Europe in late February 2020. As Italian hospitals started to become overwhelmed with COVID-19 patients in early March, one leading doctor compared the situation to an 'apocalypse'. One of the wealthiest health systems in the world struggled to have enough beds, oxygen and ventilators, as well as to protect staff with adequate PPE.

The most surprising aspect of the crisis to Italians was that what had happened in China could indeed happen in Europe – both in terms of the way COVID-19 spread and the restrictions imposed. Italy went into full lockdown in March 2020 as the country attempted to manage the virus, with some regions not even letting people out of their homes for walks or exercise. Mayors went through the streets with police, arresting individuals who broke these rules and fining (from €400 to €3,000) and publicly naming those who disobeyed. The only permitted excursions were for food or medical necessity, and slowly but surely the number of cases in Italy dropped. But the death toll in some parts of the country stood as the highest in Europe in spring 2020. This was a stark warning to other countries of the danger of complacency and reacting late.

Modelling on spread usually relies on air traffic passenger routes, so Lombardy wasn't pulled out initially as a potential hotspot by scientists. Neither was the Austrian ski resort Ischgl, which became another hotspot and caused a surge of infections across Europe. These examples, again, show the randomness of where and how hotspots emerge.

Surprisingly, poorer countries within Europe, which knew they could not treat their way through the pandemic, contained the virus more aggressively during the first wave than their wealthier neighbours and had smaller case numbers and deaths. Two underdogs performed better than expected initially: the Czech Republic and Greece. In this chapter we will look at how they managed to contain the virus at the start of 2020. Their experience contrasts with that of France and Spain, countries that reacted more slowly, and paid a heavy price.

In short there was huge variety in the European response – despite its being a wealthy bloc with established 'plans' and strong institutions. Some governments imposed policies like mandatory face masks early using the 'precautionary principle' (moving ahead with policies that carry little harm but potentially large benefit), while others delayed making decisions until firmer evidence was available. To most of us who witnessed this pandemic, the better approach was obvious. Sadly, it wasn't obvious to all governments – and still isn't.

## *The Fall of Lombardy*

Lombardy is one of Italy's most wealthy and populous regions. Well known across the world for Lake Como, where glamorous celebrities like George and Amal Clooney are spotted, the area is a tourist destination for its stunning architecture, mountains and lakes. Even with the best of models on air traffic and global virus spread, no one could have predicted this region would become ground zero for Europe's COVID-19 epidemic. Until May 2020 Lombardy was the hardest hit region in Europe. As of the 13th of October 2020 Lombardy had one third of all cases in Italy and half of all the country's deaths. The images of hospitals being overwhelmed, and newspapers filled with obituaries of those who had died, would be covered by the media across the world.

The first discovered case in the region was found in Castiglione d'Adda, on the 20th of February 2020, when a 38-year-old man tested positive. He is believed to have spread the disease widely before developing severe symptoms. An additional 133 cases were detected within three days. A day later, on the 21st of February, an emergency task force was formed by the government of Lombardy and local health authorities to lead the response to the outbreak.

Two different control strategies were implemented in northern Italy. Veneto (a neighbouring region) opted for strict containment of the outbreak and piloted targeted mass testing, whereas Lombardy strengthened hospital services to meet an increased demand for hospitalization and ICU beds. In effect, Lombardy's plan was to treat its way through the epidemic. While other factors might have been at play, as of the 15th of April 2020 the case fatality rate in Lombardy (18.3 per cent) was almost three times greater than in Veneto (6.4 per cent) and nearly twice as high as in the rest of Italy (10.6 per cent). Already this pointed to the advantages of a containment strategy (keep finding cases and breaking chains of transmission) versus a mitigation strategy (allow the virus to spread and prepare for a huge influx of patients needing care). Throughout the pandemic and across the world mitigation strategies (pre-vaccine) have always resulted in lockdown measures to stop an exponential growth in hospitalizations and to prevent health services collapsing. Yes, even in Sweden, which I discuss in Chapter 7.

That was also the case for Lombardy on the 23rd of February 2020, when harsh restrictions from the local government and health authorities were introduced. The plan was to completely lock down and close all social and economic activities in an area of 169 $km^2$. This was called the 'Lodi Red Zone', after the name of a Lombardy province, and covered ten municipalities and 51,500 inhabitants. It was similar to the Wuhan lockdown (in severity, if not scale) and should have been a sharp reminder to all countries that what had happened in China would play out elsewhere as well, even in a cultural context as different as that of northern Italy. Italians are known for being 'freedom-loving' people who are used to living in a more 'chaotic' political system and without strong top-down directives.

The Lodi Red Zone lockdown was strict: no entry or exit from the area without special permission; schools and businesses closed and all public events stopped; anyone in direct contact with a case needed to quarantine for fourteen days; and all carnival celebrations were cancelled. Italy also suspended direct air traffic with China, which was largely pointless at that stage, given that the virus was already seeded in Italy and in popular tourist destinations across Europe. This recurrent policy measure of stopping flights directly to and from China failed to recognize that the virus could come in from any number of countries, and that people would still travel from China but just change planes in Singapore or Dubai. I talk more about the role of travel restrictions in pandemics in Chapter 10.

A draft decree banning people from entering or leaving the region was revealed by the newspaper *Corriere della Sera*. Someone in government leaked to the paper that, under the decree, police and armed forces would have the power to patrol Lombardy's access points (train stations, motorway entrances and exits, border areas). This kind of lockdown on movement was unprecedented in recent history, but was seen as unavoidable by the government and their advisers. With hospitals already strained and case numbers rising exponentially, Dr Walter Ricciardi, an adviser to the Italian Ministry of Health, defended it: 'The fact that the epidemic is still increasing substantially obliges us to take these measures to limit the freedom of people, which of course are very extreme measures that I don't think have ever been taken in any other democratic country.'

Unfortunately, when the imminent imposition of a cordon sanitaire was leaked, thousands attempted to flee to the south of Italy. Unlike the Chinese context, within Europe it was impossible to stop people reacting and moving as they wanted to. And, in fact, these restrictions on movement caused a rush of people to trains and motorways to escape the lockdown. As Roberto Burioni, Professor of Microbiology and Virology at the Vita-Salute San Raffaele University, Milan, said, 'What happened with the news leak has caused many people to try to escape, causing the opposite effect of what the decree is trying to achieve. Unfortunately some of those who fled will be infected with the disease.'

For example, when news was leaked, Stefano Poggi, a web designer, and his girlfriend headed to Milan's central rail terminal to catch the next train out of the city, fearing they would be stuck if they did not. He said, 'When we heard about the lockdown, we rushed to the station with just the essentials. We didn't want to risk being stuck here forever. We decided against seeing our parents to avoid spreading the virus, in case we are sick and we do not know.'

Similarly, Lorenzo Scalchi, a social worker living in Milan, also decided to escape to his home city of Vicenza as quickly as he could: 'I thought I'd be stuck in Lombardy so I went back home, but now I'm living in a state of uncertainty. I left all my belongings in my Milan flat and I don't know when I'll be able to go back and get them.'

Michele Emiliano, the President of Puglia, urged people to turn back. 'I speak to you as if you were my children . . . stop and go back. Get off at the first train station, do not catch planes, turn your cars around . . . Do not bring the Lombardy, Veneto and Emilia-Romagna epidemic to Puglia.'

Police officers and medics, protected with masks and hazmat suits, waited at train stations for passengers arriving from Lombardy, trying to get them to turn back. Emiliano signed an order obliging all those fleeing northern regions to quarantine. 'You are carrying the virus into the lungs of your brothers and sisters, your grandparents, uncles, cousins and parents.' It raises the question why only advice was given, and why trains were not actually stopped from leaving. This pattern of advice, rather than enforcement, in Europe stands in contrast to the more strictly enforced East Asian approaches.

The situation in Lombardy would play out across Italy, and then in France, Spain, Germany, Austria and the rest of Europe. No country would escape lockdowns or harsh restrictions as they struggled to keep a handle on its spread while also having to take actions that infringed on freedoms of movement and individual rights. The question of whether the state could decide about the closure of religious spaces, such as churches, the shutting down of entire sectors, such as hospitality, and the limits on people's movements would be an ongoing issue and one that people across Europe would begin to resist.

And, while European countries were screening passengers from China in February, they should have been paying attention to skiing holidays in Austria – as Ischgl showed.

## *Ischgl Ski Resort in Austria*

Ischgl is a popular ski town of 1,600 inhabitants in the Austrian Alps. Located in the Tyrolean Paznaun Valley, it is often called 'Ibiza on

Ice' or 'Ibiza of the Alps'. In the middle of March 2020 thousands of tourists (at least 6,000 based on self-reporting), including roughly 180 British travellers, likely became infected and travelled back to their respective home countries. This hub was responsible for the unusually large numbers of imported cases from Austria across Europe; when cases were simultaneously reported in country after country, all eyes turned towards Ischgl. Andreas Steibl, Ischgl's head of tourism, noted that the March 2020 event was a 'mega-hit comparable to a tsunami'.

Most of the imported cases were male, probably because Ischgl has a reputation as a boys' holiday resort. As one tourist described the many après-ski bars: 'They are a bit like discos for teenagers, but full of men in their fifties.'

From the 9th to the 16th of March 2020 the number of positive cases rose dramatically at the University Hospital Muenster, north-western Germany. A study at the hospital found that almost 40 per cent of the patients had visited Ischgl recently. Several patients without direct travel history had had contact with those who had been to the ski resort.

As there was a concerning number of imported cases, Iceland declared Ischgl a risk region, in the same grouping as Wuhan and Iran. Norway reported that 57.1 per cent of imported cases had been traced to the resort, while Denmark had a 50 per cent rate of imported cases from there. As a result of the continual rise in cases, on the 13th of March 2020 a shutdown was declared in Ischgl by the Austrian government, forcing the termination of the ski season, which then led to an independent investigation to establish the route of the virus.

It soon became clear from this investigation that transmission had originated in a cluster in this town. The first COVID-19 diagnosed patient was a bartender working in an après-ski bar, and subsequently many cases in Iceland, Norway and Denmark were traced back to this case. In fact, a study in the leading journal *Science* confirmed that the ski resort was a super-spreading location that had brought the virus to much of Europe: 'Our results provide fully integrated genetic and epidemiological evidence for continental spread of SARS-CoV-2 from Austria and establish fundamental transmission properties in the human population.' Genetic sequencing of positive COVID-19 patients across Europe identified how different strains of the virus spread; and, in this case, it showed how much of Europe's spread came from Ischgl.

A study between the 21st and 27th of April 2020, across 79 per cent of the full Ischgl population, found considerable transmission. For example, 45 per cent of adults tested were positive for antibodies, as were 27 per cent of children under eighteen. Of those who were seropositive (meaning tested positive for antibodies), 83.7 per cent had not been previously diagnosed with infection. These people had had COVID-19 and not known about it, unwittingly passing it on to others. This study validated that Ischgl was hit hard and early with a large wave of infection imported by international, mobile tourism.

What Ischgl taught us was the importance of super-spreading events. Usually with epidemics $R_0$ is used to characterize transmissibility, and $R_0$ tells us how many people one person is likely to infect. But it is a misleading metric with COVID-19, because it hides the fact that transmission is stochastic, which means often dominated by a small number of individuals and heavily influenced by super-spreader events. This kind of randomness makes mathematical forecasting of growth difficult.

Dr Monika Redlberger-Fritz, head of the Influenza Department of the Medical University of Vienna, told CNN that it can be assumed there was at least one patient with a very high viral load (how much virus they have in their body) who transmitted the virus to a range of 40 to 80 people. Understanding that COVID-19 spreads in bursts, rather than slowly and steadily – it can simmer for a while before taking off quickly – was an important element in controlling its transmission.

Super-spreading events seem to be the confluence of someone carrying a high viral load (at that particular moment in their illness) with an environment in which a great many people are forced into close proximity to one another. They are characterized by several key factors.

The first factor is biological. For SARS-CoV-2 individual-level viral loads are dependent on the amount of time since onset; viral load peaks at or just before the onset of symptoms and decreases quickly to near the PCR detection threshold within a week. The second factor is behavioural: individuals who tend to cause super-spreading events usually visit venues where they mix closely with large groups – think of nightclubs, cruise ships, crowded public transportation, parties, choirs and conferences.

The third factor is occupational: certain high-risk facilities – such as meat-packing plants (the virus thrives when there is a lack of ventilation and in cold temperatures), workers' dormitories, prisons, long-term care facilities and health care settings (where there is prolonged contact with other people at close quarters, sometimes in unsanitary conditions) – are conducive to large outbreaks. The environment plus the nature of interactions in these places seem to repeatedly place individuals at higher risk of acquiring and transmitting infections. Importantly, these institutionally based outbreaks often leak into the wider community.

This transmission dynamic is considerably different to how flu spreads – which meant that strategies to aggressively identify and contain super-spreading events were slow to evolve. As mentioned above, roughly each individual who is infectious with flu spreads it to a similar number of people; clusters and super-spreading events seem to be less important with flu transmission. Japan managed to avoid major lockdowns during its first wave by focusing on 'cluster-busting'.

Ischgl was clearly a super-spreading event that had ramifications across the world. It likely contributed to COVID-19 spreading in forty-five countries on all continents, with more than 3,500 Germans stating they believe they contracted it there, based on having recently visited. There is also an ongoing class action lawsuit by over 5,000 ski tourists against the Austrian government over a suspicion of negligent endangerment of people by communicable diseases.

An independent commission found that the Austrian and local governments were responsible for 'momentous miscalculations': first, by hesitating to shut down the resort, and then by rushing to evacuate the alpine ski resort without testing or quarantine procedures for tourists. Rolan Rohrer, the commission's Chair, noted, '[the Austrian government's response was the] wrong decision from an epidemiological perspective' and the government had acted 'untruthfully and therefore badly'. Andreas Steibl, the Head of Tourism, defended their response: 'Many things were unclear at the beginning . . . Neither the authorities nor the virologists knew at the time what the insidious virus was doing.'

That a ski resort would become the main hub of infection in Europe was, again, not predicted by modellers or experts. This is because, in previous epidemics, cities that were airport hubs, like London and Frankfurt, were considered to be at the highest risk and the sources of further spread. This just highlights how difficult it is to prevent and predict super-spreading events in an ongoing pandemic.

## *Czech Republic*

One country that was paying attention to China, Italy and Austria was the Czech Republic in Eastern Europe. Prime Minister Andrej Babis moved early, driven by what he had seen happen in Italy: 'I understand it is a great burden for everyone, we are not happy about it. But if we don't do [anything], the whole world will see Bergamo.' The Czech national and local leaderships proactively moved and anticipated challenges. For example, as early as the 28th of January 2020 they were testing all suspected cases. Even in late March 2020 the UK was debating whether testing mattered, and the US was struggling to get test-kits to hospitals and clinics.

Two unique aspects of the Czech response were early border measures and the early adoption of face coverings. On the 6th of March 2020 an obligatory fourteen-day quarantine for returning travellers from assumed high-risk countries like Italy and China was adopted. When this was seen as insufficient in stopping the importation of cases – an increased number were detected as coming from other places – the government, on the 12th of March, introduced a mandatory fourteen-day quarantine for all individuals entering the country. From the 14th of April 2020 citizens could travel abroad only for specific reasons, such as seeing family, essential work, funerals or serious health concerns.

Face masks were also identified quickly as an important intervention by Prime Minister Andrej Babis, who set a public example by wearing masks during national television appearances. By the 18th of March 2020 face coverings were made mandatory, making the Czech Republic and Slovakia the only two European countries to impose this measure. It would be five months later, in July, that Britain adopted a similar policy. Czech citizens were prohibited from being without protective respiratory equipment (surgical masks, respirators, face masks, scarves) outside their homes; local law enforcement were empowered to fine anyone 10,000 Czech korunas (US$470) if they didn't comply.

Unlike Taiwan, which provided free masks to every household, the Czech government did not supply any. Most shops were closed and pharmacies ran out of stock – and all this was within a context of PPE shortages for health care workers. Public efforts were described as 'universally embraced', with people designing and stitching their own masks. Politicians and celebrity figures made a point of wearing them reliably. The Prison Service distributed sewing machines and equipment to prisoners to make their own protective masks; inmates produced an estimated 50,000 masks.

After pleas were posted on social media, fashion students at Umprum Academy of Art, Architecture and Design in Prague stitched masks for maternity wards. Alice Klouzkova, an assistant teacher at the academy's fashion design studio, remarked that 'The students are making hundreds and hundreds of masks. Most of them have sewing machines at home and are happy to work with their hands. It's important that the material is made with 100 per cent cotton so it can be sterilized.' Face coverings were seen as one way each citizen could contribute to the cause, fostering a sense of solidarity.

In a widely shared face mask video that reached 5.8 million views (in a population of 10.6 million), the Minister of Health urged the wearing of face masks, with the narrator noting, 'The Czech Republic is one of the few in Europe that has significantly slowed down the spread of the virus. The main difference is that everyone who has to leave their house has to wear a mask.'

What the Czech Republic leadership showed was the importance of moving early and quickly to contain spread, even before having all the evidence on whether interventions were effective or not. While scientists in the UK were embroiled in tense debates about studies on face masks, and the US was politicizing face masks based on party affiliation, the Czech Republic just made a decision that face coverings carry little harm, and possibly huge benefits, and therefore it made sense to mandate them. This decision would pay off in letting the country stay open for longer, while containing spread during its first wave. By April 2020 Spain had recorded 517 deaths per million people, Italy 453, the UK 325 and the Czech Republic only 21.

## *Greece*

In contrast to other European countries and like the Czech Republic, Greece successfully managed its first wave. The leadership were praised for acting proactively: Prime Minister Kyriakos Mitsotakis stated that 'swiftness' was needed, especially given the country's constrained health care systems and fragile economic stability. He described the government's crisis management as 'state sensitivity, coordination, resolve and swiftness'. Another crucial difference from other countries was Greece's efficient and systematic border control, as described below.

The Greek response relied on experts who understood that the way to avoid prolonged problems was to react early and sharply. Alex Patelis, the Chief Economics Adviser to the Prime Minister, said, 'There are problems you can solve through spin and others that require truth and transparency. It was very clear we needed experts and we needed to listen to them. That said, Greeks have been through crisis; they know what it is. I think that also enabled them to adapt and be stoic.'

This reliance on science was reinforced by George Pagoulatos, political economist and Director-General of the Hellenic Foundation for European and Foreign Policy: 'The lockdown was imposed much earlier than in most countries in the Western world. The government reacted in a very competent manner, listening to the right people and making the right judgement, especially given the lack of a strong precedent but also in terms of communicating the decision and keeping people in their homes.'

Greece was struggling with health and economic challenges even pre-COVID. Public health expenditure in Greece is 5 per cent of GDP, compared with the European average of 7.2 per cent. Out-of-pocket payments compromise 35 per cent of total health spending, over twice as high as the EU average of 15 per cent. Since 2010 the public health care system has been severely affected by the austerity measures driven by the Troika (a decision group composed of the European Commission, the European Central Bank and the International Monetary Fund), designed to reduce public spending in the wake of bailout loans. The country has also faced major economic contraction (GDP decreased by 7.1 per cent in 2011) and rising unemployment (around 25 per cent in 2014). The government knew it could not afford the collapse of health care services that were already underfunded and overstretched, and so had to focus on containment and protecting hospitals.

In early March 2020 the Ministry of Health and Welfare established a coronavirus task force. Led by the Prime Minister, its official name was the 'Commission for the Management of Emergency Events Due to Infectious Diseases'. This was an ad-hoc scientific committee with top epidemiologists, virologists and infectious disease experts, as well as representatives from the Ministry of Citizens Protection. This scientific committee monitored local progression of the virus, advised about public health risk and mandated necessary actions. The Ministry of Citizens Protection was given the power to implement lockdown measures, without having to face objections from other authorities. The public were updated daily through national television networks.

Unlike the UK's SAGE or the US CDC COVID-19 Response team (which I discuss in the next two chapters), the Greek taskforce was both an expert advisory and a decision-making panel. This was a controversial decision with some accusing the government of not just entrusting the advice on the pandemic to scientists, but of handing over responsibility too. This backlash against scientists and their role in public policy occurred in every country of the world, as we will see in Chapter 5.

By the 27th of February 2020 Greece had cancelled Carnival, one of the most important holidays and celebrations of the year, even though there had been only one confirmed case. Soon afterwards, schools and universities shut down, followed by cinemas, gyms and courtrooms. On the 18th of March much of the economy was locked down (with 418 confirmed cases), and on the 23rd of March a nationwide restriction of movement was imposed. As Patelis, the Economics Adviser, noted, 'The faster you deal with a health crisis, the greater the short-term economic costs, but then the greater the long-term benefits too.'

Greece's success in managing its first wave without hospitals being overrun or major loss of life is also attributed to other factors outside of the government's response. For example, there is no major manufacturing industry requiring trade and major travel of workers, no direct links to China, no major cities aside from Athens, and a population spread across islands, which all afforded degrees of insulation. On the flip side, the country was vulnerable in that it has a significant elderly population who live closely with younger generations.

But the debates about sacrificing the old for the young, which happened in the US, Britain, France and Sweden, just didn't take off in Greece's cultural context. As Dr Alexia Liakounakou, an anthropologist, explained, 'The observance of rules is aimed primarily at helping the elderly stay alive, since most Greek families have active elderly members living close by. The majority of the Greek population truly grasps the extent to which the country is ill-equipped to handle such a crisis if it were to get out of hand.'

Test/trace/isolate was also quickly established at quite a robust level in order to find cases quickly and ensure isolation. While this test and trace process was similar to that of other European countries, the stringency of isolation was closer to the type of quarantine found in East Asia. On the 16th of March 2020 two villages in Western Macedonia, Damaskinia and Dragasia, were quarantined after several cases among their residents were confirmed. No one was allowed to go in or out of the villages except for medical and municipal staff delivering medication and food supplies. On the 30th of March 2020 five more municipalities in northern Greece were placed on lockdown for a period of fourteen days. Citizens who did not abide by the quarantine restrictions were fined €5,000 each and also faced criminal charges.

Another crucial difference in the Greek response were border checks. From the 20th of March 2020 every person entering Greece has been tested at the airport (at the time of writing). The process of testing was considered quick and efficient, as detailed by a reporter:

> First, people were escorted off planes and asked to fill in personal details on a contact tracing form. Next, they were taken into an arrivals lounge booth, where officials in PPE would take a throat swab. Then, they would collect luggage and board a bus to a designated hotel, where travellers had to stay until test results came through the next day. If negative, they were allowed home, but were still required to quarantine for fourteen days. Positive cases were told to stay in the hotel under medical supervision by a doctor. Monitoring adherence was said to be stringent; strict impromptu state surveillance visits were in operation.

Only Greeks were allowed into the country during this time, reducing the number of people entering the country to around 400 passengers a day (compared to about 93,000 a day in 2019), allowing the process to be more manageable. The UK, by contrast, still had around 15,000 passengers a day entering (compared with about 290,000 a day in 2019), only with no testing, monitoring or checks. On the 23rd of March 2020 the Greek government enforced a ban on non-essential travel until the 4th of May, with identification documents and permit of movement required.

The other area where Greece moved quickly was the preparation of hospitals, including efforts to prevent nosocomial transmission – this is when infected patients and/or staff infect others in the hospital. Hospitals themselves can become super-spreading locations: vulnerable patients are easily infected; staff are in close contact with each other, often in stressful situations when decisions must be made quickly over care and procedures; and discerning whether patients arriving are COVID-19 positive or negative takes time and isn't clear cut.

In Greece hospitals were prepared for COVID-19 cases: they massively reorganized the existing infrastructure to separate COVID-free areas (green zones) and COVID-involved areas (red zones); they strictly defined and separated COVID-free and COVID-exposed staff; they quickly increased the pool of medical, nursing and paramedical staff by putting them through a fast-track training process, with awareness of how health worker shortages would affect care. And, finally, they minimized crowding in hospitals by moving administrative staff to 'work at home' settings and shutting down outpatient clinics and services except for emergency care.

On the 4th of April 2020 primary care services were restructured to enable health care to remain running even with the COVID-19 crisis. 'Specific Health Centres' in five major urban areas (Athens, Thessaloniki, Patras, Larissa and Heraklion) were exclusively designated for the screening of patients with respiratory infection. These 'COVID-19 Health Centres' were involved in early detection, monitoring and management of possible and confirmed cases with mild symptoms that did not require hospitalization, and even telecounselling services were offered for those ill at home.

Kyriakos Pierrakakis, Minister of Digital Governance, noted, 'When the pandemic broke, the need to simplify government processes became paramount. One of the first things we did to limit the incentives for people to exit their homes was to enable them to receive prescriptions on their phones. That, alone, has saved 250,000 citizens from making visits to the doctor in the space of twenty days. It has dramatically helped reduce the number of people exiting their home, which can only be a good thing.'

Greece's early aggressive response paid off. As of the 10th of June 2020 it managed to keep deaths per 100,000 at 1.71, and cases at 29 per 100,000. This made it one of the better achievers, on a par with New Zealand, which had 0.44 deaths per 100,000 and 23 cases per 100,000, and Singapore, which had 0.44 deaths per 100,000 and 683 cases per 100,000. Compare this with the UK, which had more than 40,000 COVID-19 deaths at this point in time, and the US, which had 108,000 deaths.

What Greece revealed was the importance of humility, acknowledging uncertainty and moving early in the face of an infectious disease threat. The country had been in crisis mode for years, so there was no complacency among politicians. No one underestimated the devastation yet another hit to the health system and economy could bring.

## *Be Fast, Have No Regret*

What became clear around the world is that, in the face of incomplete information and when trying to assess a fast-moving situation, countries that reacted pre-emptively and in a cautious risk-averse way did better in their first waves in terms of fewer infections, fewer deaths and less economic contraction. By contrast, those that waited for all the information and had overly complicated and layered decision-making processes were late to react and thus suffered heavier losses. In a pandemic, once the data is clear that growth in the number of cases is exponential, it's too late to intervene without harsh restrictions such as stay-at-home lockdowns and closing all non-essential businesses and without considerable infection having already occurred.

Dr Mike Ryan, Executive Director of the Health Emergencies Programme at WHO, gave this blunt advice to countries. He warned on the 13th of March 2020: 'Be fast, have no regrets. You must be the first mover. The virus will always get you if you don't move quickly. If you need to be right before you move, you will never win. Speed trumps perfection. And the problem we have with society at the moment is everyone is afraid of making a mistake. Everyone is afraid of the consequences of error. But the greatest error is not to move. The greatest error is to be paralysed by the fear of error.'

Dr Ryan is a straight-talking Irish former trauma surgeon, who has spent decades in the field working to manage outbreaks as varied as Ebola, measles, cholera and SARS. In July 1990 he moved from Galway to Baghdad to train Iraqi doctors with his then girlfriend, later wife. Three days later, on the 2nd of August 1990, Iraq invaded

Kuwait, and he and his wife were held captive by Iraq as human shields and forced to do their work as doctors in harsh conditions, often with a gun pointed at their faces. After being in a bad car accident, Ryan and his wife were allowed to leave Iraq because of his severe injuries (including his spine being crushed) and he moved to Geneva to join WHO. He reflected in an interview, 'Everything in life is like sliding doors. I mean, my life has been one of those lives where nothing I've planned has come true and nothing I've intended has happened.'

While at WHO, Ryan developed measles outbreak response guidelines; from 2000 to 2003 he was Coordinator of the Epidemic Response; and he led the field response to contain the SARS outbreak of 2003. He also worked on the Global Polio Eradication Initiative in Pakistan and Afghanistan, and during the 2014 West Africa Ebola outbreak was the Field Coordinator in Guinea, Liberia and Sierra Leone. He's a boots-on-the-ground kind of person. He says himself, 'Getting to the field regularly keeps you honest and it keeps you focused on why you're doing what you're doing. I like that feeling of being out with the teams, you know?'

You'd be hard pressed to find someone more experienced in outbreak response and management, and yet his advice fell on largely deaf ears in richer countries. While Senegal was listening, and the Africa CDC was liaising regularly to keep African governments updated, Europe was asleep. Lombardy was the first casualty.

## *The Precautionary Principle*

One of the reasons why richer countries were so slow to move was the constant scientific uncertainty. In these situations some scientists pushed the 'precautionary principle'. This allows measures to be implemented in situations where scientific evidence is deficient, but inaction threatens permanent harm. As Professor Trish Greenhalgh, of Oxford University, said in the *British Medical Journal*, 'Absence of evidence is not evidence of absence.'

The principle first emerged in the 1970s, and since then has been used in a range of declarations and treaties. For example, the 1992 UN Rio Declaration on Environment and Development was driven by this principle: 'Where there are threats of serious or irreversible damage, lack of scientific certainty shall not be used as a reason for postponing cost-effective measures.' Since then it has been identified in the 1992 United Nations Framework Convention on Climate Change, the 1993 Maastricht Treaty and the 2000 Cartagena Protocol on Biosafety.

The scientific response to this principle is divided: some see it as unscientific and unsafe to move before firm information, while others view it as necessary early intervention needed to protect people. Loosely based on the principle of *Vorsage* (German for 'foresight'), it means taking the future into account. After it was formally introduced within the 1984 German Federal Government report as a planning tool, it became increasingly popular. The European Commission has stated that the principle should be 'based on an examination of the potential benefits and costs of action or lack of action (including, where appropriate and feasible, an economic cost/benefit analysis) and subject to review, in the light of new scientific data'.

The precautionary principle came to the forefront during the response to COVID-19, particularly around the aggressive and early measures adopted by Greece as well as the use of face coverings as mandated by the Czech Republic. One can contrast the precautionary approach to the pandemic taken by countries such as Taiwan, Singapore and South Korea – which implemented rapid testing, social distancing and isolation of the infected early on (treating SARS-CoV-2 like a SARS event) – with the British approach, which underestimated the virulence of the disease and assumed it would be similar to flu. As British journalist Harry Eyres said, 'The UK approach was not precautionary.'

At later points in Britain and Europe the precautionary principle was applied in various ways. It was not adopted for face coverings, with the UK government claiming, until summer 2020, that there wasn't enough clear data that they stopped transmission, but it was used by European countries when they suspended the use of the

Oxford/AstraZeneca vaccine. Norway, for example, paused the vaccine's rollout to assess whether it could potentially cause blood clots in young, healthy people. Yet this was criticized by David Spiegelhalter, Winton Professor of the Public Understanding of Risk at Cambridge University, who concluded that, in this instance, the action may have been misapplied, as the blood clotting events were fewer than average, even without the vaccine. In the end the vaccine was indeed linked to blood clots, with different countries making different decisions on its rollout among different age groups. The chance of a 55-year-old having a vaccine-linked blood clot was estimated to be 4 in a million, while being hit by lightning in 2021 was 1 in a million, and dying of coronavirus (which also causes blood clots) was 800 in a million.

## *Do Face Masks Work?*

Few issues during the early stages of the COVID-19 pandemic caused as much strife as that of face coverings. While already regularly used in East Asian countries for protection against other respiratory infections, richer countries were hesitant to suggest their use. In fact, the UK government, the US Centers for Disease Control and WHO all advised against wearing face coverings for the general public, because of concern this would adversely affect supplies for health care workers. Even in July 2020 there was discussion in Britain about face masks. I tweeted my frustration on the 5th of January 2020: 'I hope I never have to sit through another meeting (hours of my life) discussing whether face masks are effective or not. One issue that has caused paralysis for months.'

Scientists had vitriolic debates in meetings, on TV and on Twitter about whether face masks actually work in stopping transmission, which also linked to whether the virus was truly airborne or spreading only through droplets and fomites. Those against face masks pointed to a 2007 systematic review (which examines all relevant studies and summarizes the evidence) that said, 'With the exception of some evidence from SARS, we did not find any published data that directly support the use of masks . . . by the public.' Drawing on this study and several others, a 2020 study concluded, 'The evidence is not sufficiently strong enough to support the widespread use of face masks as a protective measure against COVID-19.'

Concerns over face coverings from academics included that they were rarely worn correctly: many people touch their masks; and many wear them for long periods of time, which creates excessive moisture and results in the loss of the mask's protective function. Some behavioural scientists warned that wearing a mask could make people feel fully protected and take part in riskier behaviours, causing the disregard of other public health measures. It was also said that masks could cause skin problems and acne, particularly in teenagers (which they do), and could harm child development for babies and young children who need to see faces as part of their psychological and speech development.

On the other hand scientists working as part of the Royal Society COVID-19 research initiative called DELVE (which I was part of) wrote a report in May 2020 that noted the important role face masks play in stopping transmission. They relied on mechanistic (physical) evidence: quite simply, that a face mask stops droplets spreading from an infectious person speaking, coughing or breathing, and therefore should be introduced widely, especially in indoor or crowded settings. It had been found that the virus particles remained in the air for hours, and laboratory experiments had discovered that surgical grade masks blocked these aerosols. As asymptomatic transmission was a problem, meaning people could infect others without having symptoms, wearing masks would therefore be more useful in affording protection to others rather than to oneself.

Professor Trish Greenhalgh, a spirited Oxford don who goes wild swimming each morning in the Thames, was one of the main advocates for early adoption of face coverings. For her efforts she received a torrent of abuse as masks became increasingly identified with the loss of liberties. Despite this, Greenhalgh has continued to stand up for evidence and public health and has a vast following on social media. 'The anti-mask' community became as vocal as the 'anti-vax' community, which I will return to later. In fact, I received two threats from members of the 'anti-mask' movement on Facebook, which police picked up because they contained threatening and personal information.

Greenhalgh published a paper saying that double-blind experimental evidence (i.e., randomized controlled trials, or RCTs: where there are two groups, one with an intervention and one without, and then scientists assess any differences in outcome) are not the sum total of scientific judgement. She argued that there was enough circumstantial and natural experimental evidence that face coverings protect, including the case of a passenger who flew from China to Canada wearing a face covering and tested positive for COVID-19 the next day, but without infecting a single passenger or member of the crew.

Greenhalgh has also been critical of WHO, noting that it was slow to acknowledge airborne transmission, instead focusing on large droplets or fomites. The consequence has been that public policy around the world from January 2020 until midsummer 2020 emphasized handwashing and distancing rather than ventilation and face coverings. For example, restaurants and workplaces in Britain became fixated with one-metre versus two-metre distancing as they reopened after the first lockdown in the summer of 2020. What became clear from a later 2021 US study on transmission in schools is that ventilation (adequate air flow) and face coverings were more important at stopping transmission than distancing alone.

In an indoor setting SARS-CoV-2 can travel further than one or two metres, which is why in a crowded bar with only one infectious person many people also become infected. But if the bar is well ventilated this reduces the chances of infection. The message should have been: wear face coverings, avoid crowded indoor settings, and open windows and allow air through. Instead people in richer countries were told: wash your hands and stand two metres away from anyone else, indoors or outdoors.

The importance of ventilation to staying healthy can be found even in guidance from Henry Mac Cormac, a Belfast doctor who, in 1865, wrote: 'Indoor life, it is, coupled with cramped postures in an unchanged stagnant atmosphere, that proves so hurtful to man . . . Do not breathe the same air again and you cannot incur tubercle, breathe the same air again and you cannot in the long run avoid tubercle. In fine, by avoiding rebreathed air, tubercle and tubercle induced maladies may be superseded now, and suppressed for ever.' As discussed later in this chapter, similar guidance about getting outside for fresh air, keeping windows open, wearing face coverings and avoiding crowded places was given during the 1918 flu pandemic. This public health advice was based on their observations on what interventions seemed to reduce transmission.

In the first year of the pandemic WHO recommended masks only for those with symptoms of COVID-19, because, as already mentioned, it was concerned about supply for health care workers. In its December 2020 report it shifted to advising those suspected of having COVID-19 to wear a medical mask, and acknowledged that when the public wear masks, the transmission of COVID-19 is hindered.

While the UK government waited for clear evidence before advising face coverings, other countries moved much faster. After its first cases, Germany made face masks mandatory in April 2020, and this was readily accepted by the population. Public opinion shifted towards face masks being an important tool in prevention. The city of Jena had an important part to play in these perceptions, as it was the first German city to make wearing a face mask mandatory, in three steps between the 1st and 10th of April 2020, causing the number of new infections to fall drastically, nearly reaching zero. A study that analysed the effect of face masks on the spread of COVID-19 found that the daily growth rate of reported infections had been reduced by 70 per cent just twenty days after the introduction of masks.

The UK made face masks mandatory only on the 24th of July 2020, by which time public opinion had already shifted to favouring mandatory policies on masks. By January 2021 Germany decided that cloth masks were not protective enough; instead, surgical masks became mandatory in workplaces, shops and on public transport.

As I write this book in August 2021, the debate on face covering continues. England has stopped all legal restrictions mandating masks in indoor public spaces, while Scotland has kept face coverings in shops, public transport and indoor public buildings. Scientists who formed their positions early on masks continue to advocate the same positions, with schools becoming the latest front of the battle on whether masks should be mandatory in classrooms. I'll discuss more on schools in Chapter 6. And people continue to use masks in their own way, regardless of what government says: some wearing it under their nose, some wearing it on the chin and some carrying it in their hand but not putting it on. When I mentioned to a colleague in Hong Kong in the summer of 2020 that there was 'an anti-mask' protest in Edinburgh that day, he asked me, 'Do people also protest against sunscreen?' Others have likened the debates on masks to those around seatbelts in cars when they were first introduced and banning smoking in indoor places.

## *1918 Flu Pandemic*

The debates on face coverings bear a close resemblance to the debates that ensued during the 1918 influenza pandemic. Seen as the last major pandemic, it lasted from February 1918 to April 1920, infecting approximately 500 million people. At the time there was no vaccine, no available medicine to treat the infection and limited communications among affected countries. Control efforts were restricted to non-pharmaceutical interventions such as: limited public gatherings, isolation (for those feeling unwell), quarantine (contacts of those who were unwell, i.e., potential cases) and good hygiene.

Masks were widely used during the 1918 flu pandemic; however, as many were made of gauze (which could let a lot of air through), their effectiveness in stopping the spread of the virus was questionable. As pandemic fatigue set in during the second wave, the resistance to mask wearing was high. It did not help that scientists at the time could not identify whether the disease was caused by a bacterium or a virus. Exacerbated by the First World War, this influenza pandemic had four waves, with the second being the worst. The waves were linked to peaks within the winter months, attributed to people spending more time indoors with others and by dry skin allowing more virus entry points. This contrasts with COVID-19, which spreads easily in summer and winter, and in cold and warm climates.

How did the 1918 pandemic end? Most likely, after enough people died. There were tens of millions of dead bodies. Alexander Navarro, the Assistant Director of the Center for the History of Medicine, gave his theory in October 2020 on how it ended: 'The end of the influenza pandemic occurred because the virus circulated around the globe, infecting enough people that the world population no longer had enough susceptible individuals in order for the strain to become a pandemic again. When you get enough people who get immunity, the infection will slowly die out because it's harder for the virus to find new susceptible hosts.'

## *Herd Immunity*

Because of this experience in 1918, some researchers assumed at an early stage that the COVID-19 pandemic would not end until there was a vaccine or enough population exposure. In April 2020 I wrote a column for the *Guardian* titled 'The coronavirus crisis could end in one of these four ways': one involved elimination country by country (implausible once the virus was seeded across the world); the second intermittent lockdowns until a vaccine; the third robust test/trace/isolate until a vaccine; and finally the discovery of a treatment that allowed COVID-19 to become a manageable health issue. Three of the above scenarios were dependent upon most people being exposed to the disease, albeit after scientists had developed tools such as a vaccine or treatment, to prevent mass death and disability. But the idea that a fast-moving respiratory pathogen like COVID-19 is uncontrollable or unstoppable is linked to the idea of 'herd immunity'.

Given that the two possible long-term outcomes for a highly infectious respiratory pathogen are eradication (no one gets it) and mass exposure (almost everyone gets it), several prominent scientists didn't consider there might be advantages to buying time for science and having better tools in hand before widespread exposure; they suggested the only option was to go for natural 'herd immunity', or letting enough of the population become infected and trying to ensure health services didn't collapse in the process. But the lesson from Lombardy was that this kind of approach would collapse any health service, however rich or well funded. And just to stay within health services capacity (so people would not die of lack of oxygen or a heart attack because there was no hospital bed or doctor for them) would require harsh lockdowns, which have major negative effects on jobs, the economy, and people's mental and social wellbeing.

Dr Alberto Mantovani, a Milanese hospital director, noted, 'In March, we were the first in Europe and the other countries did not take us seriously. I think that some European countries have a huge responsibility in not having learnt the lesson from the March outbreak in Lombardy. Now we are repeating the same mistakes.'

Where was Britain in February 2020? Still fast asleep, with the Prime Minister focused on delivering Brexit – the UK's divorce from Europe – and on his own divorce – from his wife. In the next chapter we'll see how this disaster unfolded.

# 5. Britain, 'Herd Immunity' and 'Following the Science'

In this chapter I delve into the British response to COVID-19, which I know especially well because of my direct advisory role to both the UK and the Scottish governments. Before the pandemic Britain was ranked near the top of the league for pandemic preparedness in terms of capacity and planning, and considered one of the most technically savvy and resource-rich countries of the world. No one could have anticipated the large death toll the country would have, nor the extended period of restrictions that would be needed to hold the wave of hospitalizations within National Health Service (NHS) capacity. To explain what went 'wrong' is far from simple. As one of the UK government's closest advisers, Dominic Cummings, stated in a hearing in 2021, 'It was a total systems failure.'

In the first part of this chapter I explore how the UK started off with a similar approach to other countries in rapidly developing a test-kit and offering testing to those arriving on flights from China from January to early March 2020. This 'contain' policy, described as the first phase in a four-pronged strategy, seemed to work, as the UK held off a rise in infections longer than neighbouring countries. However, even as cases increased and the first deaths occurred in February 2020, the UK government's message continued to be in the spirit of 'Keep calm and carry on.' Prime Minister Boris Johnson openly admitted that he went to a hospital and shook hands with everybody, including suspected COVID-19 patients, on the 3rd of March 2020. This was because, Cummings believes, Johnson feared overreacting to the virus was a bigger threat than the virus itself. Choosing to adopt complacency meant that the UK didn't move early or prepare as other parts of the world – even European neighbours like Greece or the Czech Republic – were doing

As we will see, part of this complacency came from their experience with swine flu, which many European governments later complained was overblown by WHO, and didn't result in a devastating pandemic that killed tens of millions. Sadly, successful public health interventions seem to fade quickly from memory.

## *Near-miss of Swine Flu*

Britain's last major experience with a pandemic was in 2009/10 with swine flu, a disease caused by a subtype of Influenza A, also referred to as H1N1. The virus was first identified in the US and Mexico in April 2009 and spread rapidly across the world even in the summer months of the Northern Hemisphere. WHO declared it a pandemic on the 11th of June 2009.

Swine flu was given its name because the H1N1 virus had segments that were similar to influenza viruses that had been recently identified in, and known to circulate among, pigs. However, it was later found that the 2009 H1N1 virus was substantially different from viruses that normally circulate among pigs. So, although 'swine flu' was not the most accurate description for H1N1, it took off colloquially and became the main way it was described. Like COVID-19, it was an acute respiratory infection spread by exposure to droplets expelled by coughing or sneezing, or from contaminated hands or surfaces. Patients presented with the typical flu symptoms of fever and chills, cough, sore throat, shortness of breath, headache and decreased appetite. It also hit those harder who were in elderly age groups, pregnant or with underlying health issues.

Initial signs from the first outbreak in Mexico were alarming. A study of 899 patients showed that 58 (6.5 per cent) became critically ill and 41 per cent of those patients died. Mortality among young children, adults and pregnant women was much higher than a typical influenza season, while older adults fared relatively well. In Mexico all educational facilities closed, and on the 29th of April 2009 the government declared a suspension of all non-essential public affairs and economic activities, including working from home, and closure of non-essential shops.

The virus spread rapidly throughout the world, with countries aware that flu symptoms could be linked to H1N1. In China a group of 21 American students and 3 teachers were confined to their hotel rooms after a passenger on their flight was suspected of having swine flu symptoms. In Hong Kong an entire hotel, including 240 guests and more than 100 staff members, were quarantined after one of the guests was diagnosed with H1N1. In Australia a cruise ship was not allowed to dock at Port Douglas, north of Cairns, as passengers began to show flu symptoms. In India all airline passengers were screened and those with symptoms had to quarantine for at least three days. Countries did their best to contain, and when this failed to stop community transmission, they moved to a treatment phase with the help of antivirals.

In response to outbreaks in the US and Mexico, WHO held an emergency meeting on the 25th of April 2009 to determine the severity of the pandemic and to announce a PHEIC. On the evening of the 27th of April the influenza pandemic alert level was raised from Phase 3 to Phase 4, and then to Phase 5 on the 29th of April. These phases are linked to how severe a pandemic is estimated to be, similar to the levels system for hurricane warnings. On the 11th of June 2009 the highest alert level, Phase 6, was declared for the first time in forty-one years. At that point in time community outbreaks were ongoing in various parts of the world, and more than seventy countries reported cases.

WHO assessed the risk of the H1N1 virus in the early days and compared the novel virus to the flu pandemic in 1918 that killed millions of people. It warned countries that this could cause mass morbidity and mortality, and that this was more contagious than seasonal flu, that young people were more susceptible, and that it would be better to overreact and prepare for a huge increase in cases than wait and watch.

The UK reacted quickly to the news of swine flu and activated its pandemic preparedness plans. This included setting up a senior Cabinet committee to ensure communication between departments and coordination over how to prepare and respond to this new threat.

On the 27th of April 2009 the first swine flu cases, passengers returning from Mexico, were detected in the UK, with the first case of community transmission soon after, on the 1st of May 2009, and the first death reported on the 14th of June 2009. Cases increased to a high of 15,000 in July, before falling in August. At the peak of the outbreak, the weekly incidence was highest in children under 5 years of age (44.4 per 10,000) and fell progressively with increasing age to 3.4 per 10,000 in adults over 70.

A mass public health campaign was rolled out on the 30th of April 2009, with the slogan 'Catch It, Bin It, Kill It' referring to washing hands regularly and using tissues when ill and disposing of them correctly. Additionally, the Health Protection Agency staff handed out advice at UK airports to passengers returning from Mexico, alerting them to the symptoms to watch for, and airlines were asked to keep passenger seating records for a longer period in the event cases were identified and contact tracing was needed to find anyone else on flights. Initially one school closed for seven days when one of the pupils was diagnosed with swine flu. The UK's Health Protection Agency advised schools to consider closing in response to a single case, because evidence from seasonal influenza suggested that swine flu could spread rapidly in a school setting. Face masks were not recommended, as they were seen as largely ineffective at stopping spread; as we have seen, this issue would repeat with COVID-19.

At first a containment approach was used to manage the pandemic in the UK. On the 2nd of July 2009 the Health Protection Agency announced a move from contain to mitigate: containment was no longer appropriate as a result of all the clusters throughout the UK and the overwhelming of the contact tracing capacity available. The focus then moved to hospitals. Anyone who presented with symptoms was eligible for antiviral treatment without needing a confirmed swine flu diagnosis. Daily reports of confirmed cases were no longer published (this was two months after the first case of person-to-person transmission in the UK). And on the 21st of October 2009 a vaccine became available in the UK.

Despite the dire predictions of more than 65,000 UK deaths, the 2009 swine flu pandemic resulted in 457 deaths in the UK and 284,000 worldwide. Many governments, including the British government, felt that WHO had overstated the pandemic's potential impact and rung the alarm bell too early. The Council of Europe and the

European Parliament both said that WHO moved to the pandemic phase too quickly, even when enough data was available to show that swine flu was milder than expected. They argued that overreacting was as much a problem as underreacting and that WHO should have been more cautious in evaluating the virus's potential. Among the conspiracy theories being shared was that the Emergency Committee members deciding on pandemic potential had financial ties with drug manufacturers, which meant that they would profit from the pandemic declaration.

Six months into the pandemic, a vaccine was available that built on existing influenza vaccines. But, as always with limited resources, fair allocation became a problem. While eventually 78 million doses were shipped to seventy-seven countries, they were sent after the pandemic peaked. The shortfall in vaccine production capacity was due to a reliance on viral egg cultures, a type of vaccine production process. It was agreed that for the future there needed to be greater production capacity. This lesson to prepare manufacturing capacity to mass produce vaccines at short notice was not learnt or acted upon, as became apparent in the COVID-19 pandemic. I return to this in Chapter 10.

In parallel with the vaccination campaign, swine flu became a manageable illness, and, while it continued to circulate, enough immunity had built up in the population to avoid large outbreaks. The then WHO Director-General Margaret Chan said, 'Although the new H1N1 is still here and will continue to cause disease, it has become much like any other flu strain, no longer causing the vast majority of flu cases nor triggering outbreaks during the summer.' She declared the end of the pandemic on the 10th of August 2010.

Swine flu fizzled out in a way that didn't match the global panic it set off. In the end most people around the world probably never heard about swine flu, and, if you did, it didn't affect your daily life.

## *The Scientific Advisory Group for Emergencies (SAGE)*

The experience of swine flu a decade earlier meant there was widespread complacency about the threat of a new pandemic among senior leaders and scientific advisers. Every month there are outbreaks across the world, and even one initially presumed to be as dangerous as the H1N1 flu did not prove overwhelming and slowly went away. Swine flu never shut down the UK economy or disrupted daily life. As noted above, the pandemic flu playbook continued to move quickly from 'contain', which involved finding cases, tracing contacts and ensuring they all isolated, to 'mitigate', which involved preparing hospitals for an influx of patients. A fast-moving respiratory pathogen was seen as unstoppable.

The scientific group responsible for advising the British government is called SAGE. Its job is to ensure that timely and coordinated scientific advice is made available to ministers to support key decisions. It doesn't have a long history, with its creation linked to the 1996 BSE (bovine spongiform encephalopathy) outbreak in Britain. BSE (or mad cow disease) is a severe neurological disease of cattle caused by misfolded proteins, first identified in 1984. In late 1994 a neurological disease – variant Creutzfeldt–Jakob disease (vCJD) in humans was identified; and in May 1995 the first known death occurred, when 19-year-old Stephen Churchill died. Suspicions were raised about the links to BSE. The government continued to downplay any links between cattle, beef and human infection, some think because of the major trade and economic consequences for the agricultural industry. Finally, on the 20th of March 1996, the then Secretary of State for Health announced that BSE-linked disease was caused by eating BSE-infected meat. The outbreak would result in 177 deaths and major criticism over how long it took the government to identify the link. In response, in 1997, the government published *Guidelines on the Use of Scientific Advice in Policy-making*.

The BSE crisis marked a significant turning point in the use of scientific advice, and in 1998 an independent enquiry was set up to review what had happened with BSE. The *Guidelines* were revised in 2010, and since then have determined how scientific and engineering advice should be sought and applied in order to enhance the ability of government officials to make better evidence-informed decisions regarding emergencies.

Since 2009 SAGE has been activated nine times for issues as broad as outbreaks of Ebola and Zika, the swine flu pandemic, the Nepal earthquake, the volcanic ash emergency and other flooding and nuclear incidents.

Until late into 2020 the membership of SAGE was confidential. It was clear that at the time of the COVID-19 pandemic it was chaired by the Government Chief Scientific Adviser (Sir Patrick Vallance) and had the Chief Medical Officer (Professor Chris Whitty) on it, but that was all that was known publicly. The lack of transparency about who sat on SAGE, and what their exact advice was, was incredibly problematic, especially when reading the minutes later. Difficult decisions were taken 'unanimously': one would have expected more disagreement even from one or two members, and it was largely men from similar educational institutions.

The group was composed mainly of infectious disease modellers with little practical experience in the logistics of outbreak management and who relied on their expertise in flu. As Dame Sally Davies, former UK Chief Medical Officer, said, 'One day we will certainly get another flu pandemic, so we prepared for that and I think we prepared well. But none of the experts seemed to think a coronavirus would be relevant.' This was an implicit bias within the members chosen to be on SAGE.

Modelling drove much of the scientific advice; Whitty was quoted as saying models are 'most useful when they identify impacts of policy decisions which are not predictable by common sense'. Models themselves are constructed using advanced statistics and mathematics as a technical tool to present different scenarios with considerable uncertainty intervals and underlying assumptions.

Unfortunately overreliance on modelling by SAGE led to major missteps and blind spots in the UK response. For example, early COVID-19 models did not factor in the effect of mass test/trace/isolate programmes, such as those implemented by South Korea, or potential staff shortages in hospital capacity due to illness. Including these might have led to an earlier focus on testing capacity and adequate PPE for frontline workers, both of which proved to be major problems in the UK's early response. SAGE members didn't seem to be tracking the policy responses of other countries in real time and learning from them. These omissions were a reflection of the composition of the group and the expertise present.

Modelling is of course useful as a way of projecting various scenarios for the future, but it should be used as one data source alongside other forms of information, such as looking at what other countries were doing, their policy documents and the effect these were having, talking to frontline staff in hospitals, and analysing what historically has worked in other outbreaks. Triangulation across these would have helped to create a bigger picture of the routes other countries, like South Korea, China, Greece and Senegal, were taking. The ramifications of not having public health experts or internationally focused expertise at the table, as well as adequate diversity, led to the poor early scientific advice on the part of SAGE to government from January to April 2020.

In short, UK experts were so used to telling poorer countries how to do global health that they completely forgot humility and to listen to what experts in those poorer countries were saying or doing. They tried to 'outsmart' the problem of the virus through complex models and maths, instead of doing the hard work of building the logistics of a response and using common sense to stop an infectious disease spreading. Sweden attempted something similar, as I explore in Chapter 7. But the initial stance on 'herd immunity' (to let the virus spread until enough people were infected to stop continual transmission) was indeed supported by SAGE, as I explain below. The government at the start was 'following the science'.

## *'Head in the Sand'*

There were four phases in the UK coronavirus action plan. Phase 1: containment that would 'prevent disease taking hold by detecting cases early and following up on close contacts'. Phase 2: delay aimed at slowing 'the spread of the illness by lowering peak impact and pushing it away from the winter season'. Phase 3: research that would help 'to better understand the virus and the actions that would lessen its effect on the population by identifying innovative responses including diagnostics, drugs and vaccines and the use of evidence to inform development of the most effective models of care'. And Phase 4: mitigate, which attempted 'to provide the best care possible for people who become ill, to support hospitals to maintain essential services and to ensure ongoing support for people ill in the community and to minimize the effect of the disease on society, public services and the economy'.

From January until the 12th of March 2020 the UK was in the containment phase of trying to catch cases early and to track all contacts to avoid the spread of the disease. Public Health England (an agency within the Department of Health and Social Care tasked with protecting public health) teams were on site at appropriate international ports and provided health advice and information, to prevent the disease from taking hold by detecting and isolating the first cases of COVID-19. Additionally, new regulations were introduced in the UK to provide medical and public health professionals with the power to detain individuals in quarantine if they were suspected of having the virus.

However, there was generally a sense of complacency and lack of leadership. Prime Minister Johnson missed five COBRA meetings in the build-up to the COVID-19 crisis while spending time at his official country retreat, Chequers. COBRA (Cabinet Office Briefing Rooms (COBR, but commonly referred to as COBRA)) meetings are the highest-level of emergency response in the UK, usually chaired by the Prime Minister, and they are convened so that fast and effective decisions can be made during a crisis. Johnson was distracted at work with Brexit and a complicated personal situation. Rebecca Long-Bailey, a Labour MP, complained at the time, 'The first duty of any prime minister is to protect people, but whether it's protecting the public from natural disasters like floods or public health emergencies like coronavirus, Boris Johnson is consistently AWOL. Our NHS is already at breaking point. The government has to come up with an immediate plan to reassure us that it can cope.'

Johnson did not hold an emergency COBRA meeting until the 2nd of March 2020, as he again headed to Chequers for the weekend, despite calls for an earlier meeting to be held. Like former US

President Donald Trump, who often took golfing holidays at times of crisis, Johnson preferred to be in elite country retreats relaxing rather than facing the crisis ahead. Countries like Greece, Senegal, South Korea, the Czech Republic and Singapore were racing ahead at this point, with their leaders understanding the severity of the situation.

On the 3rd of March 2020 SAGE said that the government should advise against handshakes and hugging; however, during a press conference that same day, Johnson said he was continuing to shake hands with people. 'I was at a hospital the other night where I think there were a few coronavirus patients and I shook hands with everybody, you will be pleased to know, and I continue to shake hands.' Johnson continued, 'We should all basically just go about our normal daily lives . . . The best thing you can do is to wash your hands with soap and hot water while singing Happy Birthday twice.' To sing happy birthday twice takes about twenty seconds, which is the amount of time needed to properly wash hands and clean them. It's a good tip for handwashing, but the overall message he was giving confused the public and made them complacent about COVID-19's ability to spread through the air. It ultimately cost lives.

## *'Herd Immunity' as a Strategy*

On the 12th of March 2020 when global health experts such as myself were expecting a lockdown similar to what Italy had just enforced, Johnson announced that all testing outside of hospitals and contact tracing would stop, and voluntary self-isolation would be introduced for those with symptoms, all part of a 'herd immunity' strategy supposedly endorsed by the 'best science'. This was a shift from 'contain' to 'mitigate'. Dame Sally Davies, the former Chief Medical Officer, noted, 'We didn't practise how to stop a coronavirus spreading because we were told by Public Health England that the next big [pandemic] would be influenza, and they didn't believe it could be stopped.' The preconceived notion about pandemic flu was that it would continue to spread, even with the harshest restrictions on mixing.

I found this decision shocking. After reflecting overnight, I made my concerns public on the 13th of March, tweeting,

> Part of my job is speaking truth to power. And the UK govt is (in my view) getting it wrong. Other countries have shown speed is crucial. There is a middle path between complete shutdown & carrying on as normal. 'Middle path is: 1. Increase testing & contact tracing (e.g., China, Singapore, S. Korea model). 2. Stop large public gatherings. 3. Stop non-essential travel. 4. Encourage employers to allow home working. 5. Over 70-s & those w/pre-existing conditions need clearer advice on risk.
>
> Why do we give up so easily on contact tracing & stop mild testing? Crucial is buying time for health services, also for treatment & vaccine. Unless we're accepting that many elderly & vulnerable people will die. Which I don't see any doctor or health professional agreeing to.

I also started to write a bi-weekly *Guardian* column and speaking out to the media on some of the concerns I had as an expert who had been tracking the virus from early January. My first article was titled 'Britain's gamble', which contrasted Britain's approach with the path East Asian countries had taken.

The British plan, as explained by England's Chief Scientific Adviser Vallance, was to work towards 'herd immunity', which is to have the majority of the population contract the virus (he estimated 60 per cent), develop antibodies and then become immune to it. This theory has been widely used by those advocating mass vaccination for measles, mumps and rubella. The thinking is that, if most of the population is vaccinated, a small percentage can go unvaccinated without cases emerging, because there are fewer opportunities for the virus to transmit. Those who are vaccinated (the herd) effectively block transmission and thus protect those vulnerable individuals who cannot be vaccinated.

The UK government was assuming that a vaccine or treatment would not be available any time soon, considering how long novel vaccines take to develop, and that the virus was 'unstoppable', in the light of how it spreads. Thus, working towards creating immunity within the UK population would ideally prevent widespread transmission and subsequent infection and illness for those most vulnerable, such as elderly people and those with pre-existing conditions. A key member of SAGE, Professor Graham Medley, said on the BBC on the 12th of March 2020 that, ideally, he would have liked to move all the vulnerable people to the north of Scotland, have a large epidemic in England and then have them return safely at the end.

Jeremy Hunt, a Conservative MP and a former Secretary of State for Health and Social Care, was worried that the government was too resigned to the virus spreading: 'I couldn't understand why they were so certain that nothing could be done to stop 60 per cent of our population becoming infected, when I had figures showing even in Wuhan, the centre of the outbreak in China, less than 1 per cent of the population actually became infected.'

The UK government was working to the swine flu and influenza playbook. Hunt himself was under fire too, because he had been the Health Secretary during one of the country's most important preparedness exercises. In 2016 Exercise Cygnus simulated a flu outbreak in the UK, aimed at war-gaming pandemic readiness: 950 officials from government, the NHS and various emergency response organizations took part, and it lasted three days. In this exercise it was imagined that a new virus had emerged in Thailand, with WHO soon declaring a PHEIC. This new pandemic could affect up to half the population and cause up to 400,000 excess deaths.

Cygnus found that the UK was not prepared – in terms of plans, policies and capability – to cope with the extreme demands of a severe pandemic because of a lack of PPE, not enough beds in hospitals, not strict enough infection control procedures in medical settings and poor coordination between ministries within government. However, almost nothing was done after this to prepare the UK for a pandemic, flu or otherwise. The UK still relied on its 2011 Influenza Pandemic Preparedness Strategy; it was the only one they had. As Sally Davies said, 'I did ask during a conversation in my office in around 2015, should we do SARS? But I was told no, because it wouldn't reach us properly. They said it would die out and would never travel this far.' This, again, was caused by the cognitive bias of experts, who were looking out for the next flu pandemic, and not for other types of viruses that could spread through respiratory ways.

## *SAGE Minutes*

The UK government's belief that 'herd immunity' through mass infection was the only way forward meant that other measures undertaken by East Asian countries like mass testing, tracing, banning large gatherings, screening incoming passengers at airports and cancelling non-essential travel were largely ignored. The only advice at that point was for those over the age of seventy to avoid cruise ships.

This planning seemed largely aligned with what SAGE itself advised the government, following a careful review of their meeting minutes. The minutes for the seventh SAGE meeting, in February 2020, noted, 'SAGE should continue to work on the assumption that China will be unable to contain the epidemic' and 'SAGE concluded that neither travel restrictions within the UK nor prevention of mass gatherings would be effective in limiting transmission.' While against common sense and baffling, this was indeed what theoretical mathematical models were showing, and several SAGE members continued to argue this in public interviews.

The minutes for the fourteenth SAGE meeting, on the 10th of March 2020, noted, 'the UK is considered to be 4–5 weeks behind Italy but on a similar curve (6–8 weeks behind if interventions are applied)' and 'SAGE noted that public gatherings [such as football matches or music concerts] pose a relatively low but not zero public risk.'

At the 15th meeting, on the 13th of March, SAGE advised against suppressing the outbreak, as this could cause a second peak during the winter months when health services would be strained and there would be less capacity to care for COVID-19 patients. This decision was made unanimously: 'SAGE was unanimous that measures seeking to completely suppress spread of COVID-19 will cause a second peak . . . Community testing is ending today . . . The science suggests household isolation . . . of the elderly and vulnerable should be implemented soon.' At the 16th SAGE meeting it was reiterated that 'The objective is to avoid critical cases exceeding NHS intensive care and other respiratory support bed capacity.'

Scientists like myself struggled to understand why these policy decisions were being made and why time was being discounted. Every infection averted until a vaccine or therapeutic was developed could mean a life saved. I wrote an email to a senior government adviser on the 15th of March 2020, noting, 'I take on board the worries about a second peak, but we need to buy time to get NHS staff protected properly, more equipment, and get more data in on the virus itself. Plus there are vaccine trials ongoing and antivirals as well. We need to maintain testing, contact tracing, appropriate social distancing and protect our health workforce, who are limited and precious.'

The language of unanimity in the SAGE minutes is concerning, in that it reflects a lack of dissent from the consensus that complete suppression was not the right short- or long-term strategy. There seemed little reflection that buying 6–12 months' time might help science deliver solutions. This also indicates why diversity and disagreement are healthy and necessary in scientific advisory groups, especially when there is considerable uncertainty attached to the topic and no 'correct' way forward. I'll return to this in the final chapter.

## *Shielding the Vulnerable*

Behind the idea of letting the virus spread and herd immunity was the core idea of 'shielding the vulnerable'. The basic idea was explained by David Halpern, the Chief Executive of the Behavioural Insights Team, which is partly owned by the UK Cabinet Office: 'There's going to be a point, assuming the epidemic flows and grows, as we think it probably will do, where you'll want to cocoon, you'll want to protect those at-risk groups so that they basically don't catch the disease – and by the time they come out of their cocooning, herd immunity's been achieved in the rest of the population.'

In a UK government report on coronavirus policy from July 2020, just after the first hard lockdown, shielding was mentioned thirty-six times. Shielding meant asking those who fell into risk categories such as those with comorbidities (cancer survivors, immunocompromised) or of a certain age (above seventy) to avoid any face-to-face interaction and stay at home.

Two assumptions were behind this approach in the UK. First, that it was possible to identify and separate those vulnerable to severe infection from COVID-19 from those who would only have mild illness. And, second, that immunity would last longer than it did with common cold coronaviruses, which can reinfect the same person over time. There continues to be a large question over whether 'herd immunity' could ever be reached with a virus that could continue to infect those who had built up some immunity but just not enough to stop transmission.

As explained in my *Guardian* column on the 29th of May 2020, the first assumption falls apart with a cursory look at the evidence. A study in China showed that 80 per cent of transmission occurs within households, so how exactly are elderly and vulnerable members supposed to isolate from their own families? Especially if children in these households need to go to school, where we know children can indeed transmit and become infected.

The most vulnerable rely on others for assistance, either paid or unpaid carers, who provide medical care, food, transportation and deliveries. And attempts to shield care homes, which are full of vulnerable people, in Sweden and the UK failed. Sweden also attempted to 'live with the virus' and moved to mitigate, as I'll discuss in detail in Chapter 7. Sweden's State Epidemiologist, Dr Anders Tegnell of the Swedish Public Health Agency, noted that at least 50 per cent of Sweden's coronavirus deaths in the first wave occurred within care homes. A similar death toll is evident in the UK, where care homes became hotbeds for infection.

In addition, shielding vulnerable individuals puts them at risk of further isolation and depression. Humans need social contact and mixing for mental wellbeing, especially those suffering from dementia. And, finally, age is not the only element of risk. Other factors for death in the first wave of infection in the UK were uncontrolled diabetes, severe asthma, obesity, poverty and ethnic minority status. Should all these people also be shielding at home?

In some strange way the idea of shielding creates an Orwellian society in which only the young, healthy, white and fit can circulate and interact, while the rest of society must hide. It's as if only those who would fit into the cast of the 1990s show *Friends* could stay part of mainstream society. This is unrealistic as well, given the numbers involved: one study estimated that upwards of 40 per cent of the population would have had to shield, including people working as NHS medical and support staff, cleaners, security staff, supermarket staff and other essential workers who were needed to keep basic services running.

The idea of 'shielding the vulnerable' (also called 'segmentation' and 'focused protection') would continue to be pushed by extreme right-wing sources as a 'solution' to the COVID-19 crisis; they even managed to find several fringe scientists to form the face of a lobbying effort called the 'Great Barrington Declaration' to remove all restrictions.

While fringe among mainstream public health and infectious disease researchers, and called unethical and immoral by the head of WHO, certain political leaders were keen on this approach. Specifically, former US President Donald Trump, Brazilian President Jair Bolsonaro and UK Prime Minister Boris Johnson: populist leaders who wanted to do nothing rather than act to protect their people. They sought and found scientists who were willing to tell them what they wanted to hear, rather than listen to the scientists providing a realistic picture of the crisis and the steps needed to solve it. But even these leaders couldn't hide behind spin: it's not easy to hide hospitals collapsing and bodies piling up.

## *'Stay Home, Protect the NHS, Save Lives'*

When it became clear that it was impossible to separate the 80 per cent who would largely be fine after contracting the virus from the

20 per cent who would suffer severely, and the NHS looked like it would collapse under the burden of too many COVID-19 patients, the government changed its message. On the 23rd of March 2020 measures were introduced to delay the spread of the virus. These were similar to what other countries had done: closing schools except for the children of key workers, closing pubs and other gathering places, asking households to self-isolate for fourteen days and focusing on scaling up testing from 25,000 per day to 100,000 over the next months. The message was: 'Stay Home, Protect the NHS, Save Lives'.

Patient numbers climbed fast, and the UK found itself just two weeks behind Italy instead of four. But the government had done little in those four weeks to prepare. Jeremy Hunt, Member of Parliament and Chair of the Health and Social Care Select Committee, said, 'I think it is surprising that we're not doing any of it at all when we have just four weeks before we get to the stage Italy is at. You would have thought every single thing we do in that four weeks would be designed to slow the spread of people catching the virus.' Testing was not ready to go at the level needed and PPE supply chains were strained.

Whitty defended the decision to lock down late, arguing that people would tire of restrictions if imposed too early, and it was important to time them: 'There is a risk if we go too early people will understandably get fatigued and it will be difficult to sustain this over time.' Whitty's view was that the real problem was how to control the spread of the virus through the population over time, not how to stop the spread. He continued to compare it to seasonal flu, noting that we accept deaths from flu and live with them, and so we should accept deaths from COVID-19 and live with them. In a bad flu year roughly 30,000 people die in the winter, largely those who are elderly and frail.

Challenging this, five members of SPI-B (Scientific Pandemic Insights Group on Behaviours, a SAGE subcommittee on behavioural science) noted that they had never asserted that people would tire of restrictive measures, which was confirmed by publicly available minutes from the group.

Testing capacity was a particular problem, with limited ability even to identify who had the virus in the community. After the 12th of March 2020 testing was restricted to people in hospital, with no community testing. At this point roughly 28,000 people were tested and 590 had tested positive for COVID-19.

While rich and famous people like actors Idris Elba, the magician Dynamo and British football manager Mikel Arteta could access testing at private clinics (London clinics charged £375 per test), testing was not available for the medical staff working on COVID-19 wards who felt unwell. I wrote a piece for the journal *Undark* on the 19th of March 2020 titled 'For the rich, COVID-19 protections. For health workers, a shrug'. My message was simple: 'Test and protect our health care workers and frontline responders. They are society's most precious defence line against COVID-19.' A week or so later, on the 27th of March 2020, frontline health workers got access to COVID-19 testing in the UK.

However, the government was reluctant to acknowledge that testing was constrained because of capacity. Instead it said this was a strategic decision. Dr Jenny Harries, England's Deputy Chief Medical Officer, defended the policy in a press briefing. 'There comes a point in the pandemic when that is not an appropriate intervention' and argued that WHO's advice to 'test, test, test' was intended for poor countries, not the UK. This arrogance would come back to haunt Britain as hospitals filled up.

Mixed messages emanated from No. 10 once lockdown began on the 23rd of March 2020. While the population largely complied with the stay-home message, key scientists within government continued to argue against community testing, testing of health workers and aggressively going after the virus – all part of 'following the science'.

Scientists like myself struggled to understand why decisions were being taken that ran completely counter to WHO advice. It was unclear who exactly was advising the government, who sat on SAGE, and what factors Johnson and his colleagues were taking into account in their decision-making. The lack of transparency became a problem for scientists, who wrote to the UK government several times asking to see the evidence behind key decisions and asking for full membership of SAGE to be made public. It took weeks for the government to release the SAGE papers and a partial list of members, and at that point it was clear that, while meetings revolved around what members thought was the best data and evidence at the time, the government was choosing which aspects of that advice to listen to. Professor Neil Ferguson, of Imperial College, and Sir Jeremy Farrar, of the Wellcome Trust, both spoke afterwards about their frustration with minutes not being detailed enough, meetings not being recorded and with feeling that their views on how to manage the pandemic were not being listened to.

In a surreal few weeks Prime Minister Johnson and his Health Secretary, Matt Hancock, were both diagnosed with COVID-19, with Johnson becoming increasingly ill and needing admission to hospital on the 5th of April 2020. He was administered oxygen through a face mask, and, as a precaution, was moved to ICU on the 7th of April, but did not need mechanical ventilation. Dominic Raab, the Foreign Secretary at the time, took over Johnson's duties, until the Prime Minister had fully recovered and could return to work on the 27th of April 2020.

## *'We're being sent out to die'*

At the same time doctors and nurses continued to show up for work but were not offered adequate PPE to protect themselves. The same situation was playing itself out in the social care sector with care home workers. As the number of infections increased dramatically from dozens to hundreds to thousands, over a hundred health care workers became infected with COVID-19 and died from it, often denied a test for the virus until at quite an advanced stage of illness. Thousands of others became ill with 'long COVID', a post-COVID-19 syndrome that causes fatigue and recurrent illness, which I will discuss in detail in Chapter 7.

On the 14th of March 2020 one of my former medical students sent me an email from England.

> Hi Devi, we met many years ago at the Global Health Governance Programme in Oxford. We are being asked to reduce the amount of PPE we use to the same precautions as normal flu (surgical mask, flimsy apron, short gloves) as there is some belief SARS-CoV-2 is not dangerous. I am treating a ward full of positive patients. I am unsure how I can do my job when I could be transmitting the virus as I care for patients around the hospital and if I am exposed in this way and fall ill. I do not have testing if I fall ill and have mild symptoms. I have already been exposed and told to continue working. We will lose the battle against this virus if we lose our health care workers this early in the outbreak. We don't know enough yet about this virus to downgrade it, but we are running out of PPE.

The question was how could the UK government be taking such a gamble with the frontline workforce, expecting them to shoulder the burden of the pandemic? It's like leaving your goalie (health services) alone on the pitch while multiple footballs are being shot at them. Where are the lines of defence on the pitch to help them? Another former student wrote to me about PPE in hospitals. He said:

> Doctors, nurses and support workers are unprotected. We have been asked to perform procedures with only a face mask, gloves and an apron. This falls short of WHO guidance. Wards are being told they can't have visors because there aren't enough to go around. FFP2 masks are being rationed – we have been told there aren't enough in the hospital. This is starting to become the new normal. Last night at the end of my shift we had a handover meeting. The new policy is to stop testing patients that were well enough to go home. We are to only test patients who need hospital admission. We have missed the window of opportunity to contact, trace and isolate. If we fall ill with mild symptoms we will not get tested. Our patients will not know if they were exposed because of us.
>
> Doctors and nurses I've spoken to, colleagues and friends, are worried that we are being pressured to treat patients without proper protection. Some are in their fifties. Some have grandparents. Some are fully aware that they could catch COVID-19 and die because of the lack of PPE. They say, 'It's like we're being sent out to die.' Some,

> myself included, have echoed this sentiment. The worry is growing. Forget the UK becoming the 'next Italy'. At the risk of sounding sensationalist I worry in a few months' time countries will worry about becoming the 'next UK'. We don't have confidence that the NHS will be able to take care of its health care workers if we get sick in the line of duty. And this worries me immensely. With staff vacancies at an all-time high, we have no reserves to call on.

## *The Lost Summer of 2020*

The first lockdown, which lasted from March to June 2020, managed to bring cases down to a low level, but this was followed by a push by the UK government to return to 'normal' in the summer with schemes such as 'Eat Out to Help Out' (subsidized meals in indoor hospitality) and messaging such as 'Return to the Office or Lose Your Job'. Travel corridors throughout Europe without any testing or other measures also resulted in a new strain being imported. The second wave in January 2021 would be even more deadly than the first, with the criticism being that the lessons from the first wave had been squandered.

The tension within the UK turned on an internal struggle between two opposing camps. One seemed to think the government should attempt to get over the worst of the pandemic by allowing the virus to spread through the population, albeit at a slower pace to ease the strain on the NHS, and by creating more hospital and mortuary capacity to cope with the inevitable spike in deaths. The other camp wanted to drive down the number of coronavirus cases and reduce the rate of infection – or R – to as close to zero as possible. Those were two different goals, with two different strategies behind them. England clearly took the first route, trying to keep the virus simmering at a certain level within NHS capacity, and the economy running alongside. This path took them to multiple lockdowns and a very high death toll.

By contrast, Scotland attempted to chart an 'elimination' path, which was to drive the virus out of the country, and hold and wait for mass vaccination. In July 2020, after the first lockdown, this seemed extremely possible. Deaths and hospitalizations were absent, and cases were down to a handful. In fact, genetic sequencing work, which tracks strains of the virus, showed that the first strains were indeed eliminated, and that Scotland's second wave in the autumn was driven by new viral strains imported during the August 2020 tourism season.

## *Scotland Charts Its Own Course*

These contrary goals created political tensions within the four nations of the UK. Scotland and First Minister (the leader of Scotland) Nicola Sturgeon committed to the second goal of maximum suppression and elimination, while England silently adopted the first of allowing a certain level of spread of the virus. I'm going to spend some time on Scotland, given my close role in advising the government there and living in Edinburgh over the course of the pandemic. I sat on the Scottish Government COVID-19 Advisory Group as well as on the Advisory Subgroup on Education and Children's Issues and the Advisory Subgroup on Universities and Higher Education.

I also spoke regularly with Sturgeon, offering impartial advice, particularly on what challenges might lie ahead and what best practice from other countries seemed to be at the time. We developed a close working relationship. I was also studying to become a personal fitness trainer, and Sturgeon even agreed to become my first client. I should say clearly that she never asked me to change what I said publicly; she listened carefully, asked thoughtful questions, and tried to understand the best data and evidence. I never felt any political pressure to say what she wanted to hear: she wanted the blunt truth from me, and I gave it without fear or favour, in my typically American direct way. I have no ambition to go into politics or into government, and just wanted to bring what expertise I could to help support her in making extremely difficult leadership decisions. If anything, the COVID-19 crisis reignited my passion for physical fitness and my bucket list goal of having visible abs and a six-pack.

Before delving into what Scotland tried to do differently, it is important for those unfamiliar, like I was pre-COVID, with how the UK political system works (I came to realize this after stepping into a few minefields). I apologize now for how basic it will seem to British readers.

The United Kingdom is made up of four nations: England, Scotland, Wales and Northern Ireland. In 1707 the Acts of Union merged the Scottish and English parliaments into the single Parliament of Great Britain. However, tension remained between those who thought Scotland should be its own country and have its own government and those who wanted one government in London. Various Scottish independence parties were rolled into a single political party, the Scottish National Party (SNP), in 1934. In May 1997 Tony Blair was elected Prime Minister of the UK with a promise of creating devolved institutions in Scotland, including a Scottish Parliament that could cover areas that were devolved (such as health, the NHS and education) but not areas that were reserved (such as science, economic power and border policy).

Even with this conciliatory move, support for Scottish independence continued to grow, and the SNP continued to win more and more seats. In September 2014 a referendum on Scottish independence was held in which 55.3 per cent voted against and 44.7 per cent voted for. The Leader of the SNP party at the time, Alex Salmond, was a divisive personality who in the end couldn't convince enough voters of the economic viability and practicalities of an independent Scotland.

In 2014 Nicola Sturgeon took over as Leader and aimed to unite the country and work with all parties: Conservative, Labour, Green and her own SNP. Sturgeon is a progressive politician in the same vein as Angela Merkel, Jacinda Ardern and Hillary Clinton, aiming to take an international view of Scotland's role both within Europe and in the wider world. A keen reader, careful thinker and lawyer by training, she wants to understand the detail and evidence of all policy matters and has struggled against misogyny in a largely male-dominated senior political world.

Tory (Conservative) leadership in No.10 continued to drive a wedge between Scotland and the rest of the UK, especially after the 2016 referendum on EU membership. Scotland voted to remain part of the EU, yet, as part of the UK, it was forced to accept the terms of a Brexit deal. This aggravated voters, who felt their voices were not being heard in London. In 2021 the SNP won a fourth consecutive term in government, and, along with the Greens (which also support Scottish independence), won 72 of the 129 seats in the Parliament. Sturgeon was overall seen as providing good direction to the country during the COVID-19 crisis, with polls showing she was more popular than UK Prime Minister Johnson not only in Scotland but in the entirety of the UK.

## *Scottish SAGE*

The context just described is vital to an understanding of how the pandemic played out in the UK. From January until mid-March 2020 all four nations were in lockstep, following the advice from SAGE to treat the virus much like flu. However, soon after going into lockdown, serious questions by Sturgeon were raised over both the transparency and the advice coming from SAGE. She decided to set up the Scottish Government COVID-19 Advisory Group, chaired by Professor Andrew Morris of Edinburgh University. Its minutes and membership would be published, and it would brief Sturgeon directly.

Referred to colloquially as 'Scottish SAGE', this group (which I was invited to join on the 2nd of April 2020) would have access to SAGE papers but would tailor the advice and recommendations to the Scottish context. It also had various subgroups – on the public health threat assessment, hospital associated and onset transmission of COVID-19, education and children, and university and higher education, which SAGE did not have. Scottish SAGE advised that a mitigation strategy (just to stay within hospital capacity) would result in tens of thousands of deaths as well as recurrent lockdown cycles. Instead, the advice was containment and maximum suppression, until mass vaccination. On the 23rd of September 2020 I wrote an article for the *Scotsman* newspaper titled 'Herd immunity strategy is flawed until we have a coronavirus vaccine'.

In April 2020 the Scottish government adopted a strategy to get cases down to the lowest possible level and hold them there, with the explicit aim of having no one intentionally exposed to the virus. In June 2020 the advice from several group members was to push to eliminate the virus with the first lockdown and then seal off travel to the rest of the world until scientific solutions could be rolled out, such as a vaccine. While recognizing the land border with England, the aim was to stop community transmission and to focus on catching all imported cases. An article in the *New Scientist* on the 30th of June 2020 was titled 'Scotland could eliminate the coronavirus – if it weren't for England', with the subtitle 'Scotland may be only weeks away from no new daily cases of coronavirus. As the nation gets close, cases from over the border will become a big problem'.

Aside from strategy, crucial differences between England's approach and Scotland's were around test and trace, easing of restrictions and messaging, and finally leadership. England decided to outsource its test and trace to large private companies for billions of pounds – reportedly £37 billion. These private firms, like Serco, then outsourced tracing to dozens of smaller firms – which rented empty office floors, hired largely untrained staff and attempted to do contact tracing by phone with a centralized system. It's generally agreed this model did not work and created an additional and expensive level of bureaucracy that was an impediment to the work of the local public health authorities.

Instead, Scotland built up 'test, trace, isolate, support' within local NHS public health boards, referred to as 'Test and Protect'. This system involved testing symptomatic individuals, tracing contacts, isolating those who were carrying or had been exposed to the virus, and providing them with the necessary support to meet their needs. The political decision was made to reinforce the existing public health systems, rather than to outsource to private firms.

On the 10th of May 2020 Johnson announced the easing of restrictions and continued to lift them quickly to get economic activity moving, changing the messaging from 'Stay Home' to 'Stay Alert'.

That same day Sturgeon told the people of Scotland to follow her advice instead and to continue to stay home. The idea was to ease lockdown measures more cautiously so that case numbers would be low enough for 'Test and Protect' to operate effectively. Clear messaging was vital: people in general seemed to want to follow the guidance but needed it to be clear on what was and was not permissible.

Finally, there was a disparity in the extent to which each government was trusted by the public. Not only did Johnson often hide from the media (e.g., in December 2019 he retreated into an industrial refrigerator packed with milk bottles to avoid an unplanned interview with *Good Morning Britain*'s Piers Morgan), he supported his Chief Adviser and right-hand man Dominic Cummings, even when it was clear Cummings had broken lockdown rules and refused to apologize.

In Cummings's initial version of the story (which he subsequently changed), he noted that he and his wife were feeling unwell and frightened they had COVID, and so drove 425 kilometres from London, where they lived, to Durham, where they would stay near his parents. This occurred in early April 2020, when the country was under a strict 'Stay at Home' lockdown and told not to visit family members outside the main home. When the media revealed he had been spotted not only in Durham but also walking around tourist destination Barnard Castle (on the day of his wife's birthday), Cummings defended this, saying he had driven to Barnard Castle to test his eyesight before the drive to London. (To readers not from the UK: No, I'm not making that up.)

The Cummings episode was covered extensively by the press and led to public outrage that there could be one set of rules for 'elites' to do as they please and another for ordinary people. If the police had pulled over someone else doing this, they would have been reported and fined. Compliance was eroded from this point. The young, especially, who felt they had been making sacrifices, wondered why they were giving up so much, when others were not.

A study in the *Lancet* tracked more than 40,000 people's views and found that, over the three weeks after the Cummings headlines, willingness to comply with restrictions dropped steeply in England. There was no similar drop in Scotland or Wales. The former Chief Constable of Durham, Mike Barton, said in July 2020, 'People were actually using the word "Cummings" in encounters with the police to justify anti-social behaviour . . . People who make the rules shouldn't break them. Otherwise, you can't expect the little people to do it.'

## *Scotland's Limits*

Just a few weeks earlier, Sturgeon's Chief Medical Officer, Catherine Calderwood, had also broken lockdown to visit her holiday home on three weekends. She had to swiftly resign when newspapers reported on it, in the light of government concerns that her actions, if left unpunished, would undermine compliance with restrictions. Calderwood was replaced by her Deputy, Dr Gregor Smith, a serious and reserved GP who had risen through the ranks by virtue of hard work and strong management skills. Smith, a talented guitarist, Iron Maiden fan and fitness enthusiast, would become the key behind-the-scenes adviser to Sturgeon on how to manage the pandemic. Given his difficult childhood from a working-class background in the West of Scotland, he focused on COVID-19's impact on poorer communities and deprived areas.

Sturgeon's other key adviser was Professor Jason Leitch, a brilliant communicator and a favourite of the public, who managed to keep key external partners in line with government strategy. Leitch would become the government's 'face' of COVID-19 in all their communications, on top of his core job of NHS quality and delivery as well as his side project of supporting a children's home, school and college in south-east India. Smith and Leitch would come to flank Sturgeon in the regular press briefings and form the inner circle of decision-makers.

One of the frustrating aspects of advising on the Scottish response was how little power to take the necessary public health measures resided in the Scottish government. This is my opinion and analysis after observing the response over two years, close up. Any kind of restrictions on economic activity required 'furlough', an economic package that provided 80 per cent of a claimant's usual income in the event they couldn't work because of legal restrictions; but the power to grant furloughs was invested in the UK Chancellor. Limitations on travel required the ability to stop flights arriving and departing, and tight screening, but border policies are reserved to the UK government. And a large land border with England meant that cross-infection (both ways) could occur. It would have been easier if Scotland had been an island in the North Sea that could manage its own affairs tightly. On the flip side, being part of the UK gave Scotland access to the vaccination programme negotiated by No. 10.

However, an immediate public backlash in England awaited anyone who formally advised the Scottish government and Sturgeon, or who said anything positive about Scotland's approach to the pandemic. Many of my comments or observations were deliberately misquoted or misrepresented by right-wing outlets such as the *Mail* and the *Spectator*, which made repeated attempts to discredit my credentials and reputation as part of a larger attempt to discredit the majority Scottish government and an increasingly popular Sturgeon. It took me a while to understand this; now I know that this is all part of an overarching political agenda.

## *Scientists under Attack*

My experience was far from unique. The anger, frustration and sense of loss that certain people felt started to be directed at those seen as most responsible for lockdown. The focus for this became the scientists in advisory roles who had increasing visibility on television, as they tried to explain the reasons why decisions had been taken; and the anger levied at them became more pronounced as governments pointed the finger at the scientists who had advised on the emergency rules and the loss of freedoms that followed. A typical refrain transpired: 'We were just following the science.' This was a phenomenon repeated across the globe.

Few experiences are as extreme as that of Professor Marc Van Ranst, Professor of Virology at the Katholieke Universiteit Leuven and a top adviser to the Belgian government on COVID-19. On the 18th of May 2021 he and his family had to go into hiding after Jürgen Conings, a serving officer with the Belgian Air Force and a former Special Forces soldier with far-right beliefs, made threats against Van Ranst and then disappeared after taking a bulletproof vest, a sub-machine gun and four anti-tank missile launchers from the barracks where he was stationed.

Before disappearing, Conings wrote a letter to his girlfriend, in which he mentioned that he 'could no longer live in a society where politicians and virologists have taken everything away from us'. The Belgian Minister of Justice, Vincent van Quickenborne, said, '[Conings] is the only soldier on the OCAM list, and that's a terrorist list. People are not placed on such a list lightly. This is someone who wants to use violence because of his extremist ideologies.'

Van Ranst said in an interview, 'The threat was very real. The ex-soldier, heavily armed, was on my street for three hours, right in front of my house, waiting for me to arrive home from work.' The government moved Van Ranst, his wife and their twelve-year-old son into a safe house. The family tried to make it feel like an adventure and as pleasant as possible.

Van Ranst said, 'We're not scared, we're just being careful. And my twelve-year-old son, Milo, he's pretty brave about it.' But of course it was challenging for life to change so radically. He said, 'Moreover he finds this situation very unpleasant because he cannot go to school. He misses classes, but especially his friends . . . It is pretty surreal, but knowing is better than not knowing, because at least I can take these precautions. The thing that makes me mad is that my son has been inside for almost three weeks. That I really hate.'

The Dutch leader of the anti-lockdown/anti-vax group Viruswaarheid, Willem Engel, a dance teacher, subsequently mentioned that Van Ranst had it coming to him, as he'd been complicit in creating a 'pandemic of fear'. Van Ranst responded with a tweet that went viral: 'When there's a salsa pandemic, I'll listen to you with great pleasure. But at this moment, I don't give a flying f— what you have to say, and nobody in the Netherlands should either.'

This Twitter exchange was shown on 2,000 advertising screens across the Netherlands, including those on motorways and at petrol stations, bus stops and Schiphol Airport. The campaign was organized by Dutch businessman Alexander Klöpping, who raised €25,000 in less than a day through crowd-funding.

Coning was found dead in the remote area of Dilserbos in Belgium on the 21st of June 2021, after taking his own life. Van Ranst then came out of hiding and reflected on the attacks he had experienced: 'If you're on television a couple of times every day for months on end, people get sick and tired of you. That's unavoidable. There are a group of people that hate science and hate scientists. Very often they are scared and uncertain.'

It wasn't only extreme members of the public attacking scientists: governments also put on the pressure when the message they were receiving didn't fit with what they wanted to hear. For example, Rebekah Jones, a data scientist with degrees in geography and journalism, was Florida's Geographic Information Systems Manager for the state's COVID-19 database. She was sacked on the day Florida opened from lockdown measures because of her refusal to hide COVID-19 data. Jones said in an interview that she had been asked to alter the numbers to show that test positivity had reduced from 18 per cent to 10 per cent, meaning the state was on target for reopening – the lower the test positivity, the better in terms of having COVID-19 under control. She said, 'To me, it did not read like some kind of political conspiracy or some higher directive. It seemed like people who expected when I brought in those results, the results to support the plan they had written, and they did not, they seemed panicked and like they had to figure out a way to make the results match the plan.'

Governor Ron DeSantis (a Republican) denied this allegation and insisted that Jones was let go for disruptive behaviour and insubordination. However, Terrie Rizzo, the Chair of the Florida Democratic Party, said, 'Allegations that Florida's government may have tried to manipulate or alter data to make reopening appear safer is outrageous. These kinds of actions are dangerous, and frankly should be criminal. An independent investigation is needed immediately.'

A month after being fired, Jones opened an independent dashboard for COVID-19 data, which was similar to the official data systems but showed which counties were ready for reopening based on set targets. While Florida's Department of Health said that Jones's new data portal was using unreliable data, academics like Professor Cindy Prins of the University of Florida believed Jones's dashboard was more comprehensive and accurate. Jones noted: 'When I went to show them what the report card would say for each county, among other things, they asked me to delete the report card because it showed that no counties, pretty much, were ready for reopening. And they didn't want to draw attention to that.'

Soon afterwards Jones's house was raided by armed police who confiscated her computers while her husband and children were also there. The search warrant was executed after Jones allegedly accessed a Department of Health-run emergency alert platform and sent a group text urging people to speak out about Florida's COVID-19 strategies, warning users that 'It was time to speak up before another 17,000 are dead.'

The Florida Department of Law Enforcement called the action a computer hack. After the raid Jones posted videos online showing police escorting her out of her Tallahassee home. 'He just pointed a gun at my children.' Jones complained, 'This is what happens to scientists who do their job honestly. This is what happens to people who speak truth to power.'

Jones claimed the raid was the work of Governor DeSantis and filed a lawsuit saying that the raid was an act of retaliation against her, and that Florida Department of Law Enforcement officials violated her right to free speech and deprived her of due process when they unlawfully seized her computers, mobile phone and storage media. Jones said in a TV interview, 'DeSantis needs to worry less about what I'm writing about and more about the people who are sick and dying in his state. Doing this to me will not stop me from reporting the data, ever.'

After the raid, Jones was charged with several offences and turned herself in to Leon County Detention Facility in Tallahassee: 'To protect my family from continued police violence and to show that I'm ready to fight whatever they throw at me, I'm turning myself in to police in Florida.' She later moved to Washington DC, as she did not feel her family was safe in Florida.

On the 7th of June 2020 an official statement from DeSantis's office called Jones a 'super-spreader of COVID-19 disinformation'. In July 2020 Jones filed a whistle-blower complaint against the state of Florida, and on the 29th of May 2021 she officially became a whistle-blower under Florida law, which offered her protection from retaliation for disclosing information related to neglect of duty by her employer. A letter from Inspector General Michael J. Bennett mentioned that Jones's complaint demonstrated 'reasonable cause to suspect that an employee or agent of an agency or independent contractor has violated any federal state, or local law, rule or regulation'. Again, an example of a scientist trying to do their job under intense political pressure.

A third example of scientists under attack is from India, and this time not from an extreme member of the public, nor government pressure, but from a professional association. In May 2021 Dr Soumya Swaminathan, the Chief Scientist of WHO and a highly respected paediatrician, was served a court notice when the Indian Bar Association (IBA) filed a lawsuit against her for allegedly distributing false information on the use of ivermectin to treat COVID. The association said Swaminathan was 'spreading disinformation and misguiding the people of India in order to fulfil her agenda' and that she was causing 'further damage'. The IBA pointed to a tweet by Swaminathan which stated, 'Safety and efficacy are important when using any drug for a new indication. @WHO recommends against the use of Ivermectin for #COVID19 except within clinical trials.'

The IBA claimed that Swaminathan had overlooked research and clinical studies from various alternative groups that presented data demonstrating ivermectin prevents and treats COVID. Based on this, the IBA called for actions under sections 302 (punishment for murder), 304 (culpable homicide not amounting to murder), 88 (act not intended to cause death), 120 (party to criminal conspiracy) and 34 (acts done by several persons in furtherance of a common intention).

Swaminathan deleted her tweet but still faced a barrage of attacks. The court claimed that she had abused her position as Chief Scientist at WHO to adversely influence people, including medical scientists and medical doctors, by trying to impose upon them the fact that WHO does not support the use of ivermectin. And it even claimed that she deliberately opted for the deaths of people to achieve her ulterior goal.

While this might seem far-fetched to those reading the story, it is easy to imagine the mental toll this situation would have taken on a scientist who had devoted her life to helping people and had worked non-stop for eighteen months to help steer countries through a pandemic. It also reflected how social media, particularly Twitter and Facebook, could trigger an onslaught of abuse by cult-like accounts with large followings. Scientists and medical doctors are not used to the sort of public scrutiny usually applied to politicians and actors; but throughout the COVID-19 pandemic they faced the same intrusions into their personal lives, and blame for all that was lost during the pandemic.

In fact, someone went ahead on social media and did blame me for causing the pandemic, pointing to my Hay Festival talk on emerging infections as evidence that this had been planned by 'the deep state'. I have a sense of humour and tried to point out that this was just expertise and science, and not witchcraft. If I had caused COVID-19 through casting spells, wouldn't I have first bought shares in Zoom video-conferencing, Pfizer and Moderna pharmaceutical companies, and Andrex, the toilet roll company? Regrettably, I have shares in none of these.

This was only the tip of the iceberg of social media attacks – whether through tweets, YouTube videos, blogs, viral Facebook postings, or even daily mal-intentioned edits to my academic record on Wikipedia. It was so much, and so absurd, with the various groups blending together in bizarre ways: the anti-vaxxers, with the anti-maskers, with the anti-lockdowners, and even distant colleagues who felt angry about swimming pools being closed or not being able to go abroad on holiday.

And there was also the reassuring consistency and dedication of Scottish Unionists (those against the pro-independence SNP), who would show up daily on my timeline to misquote something I had said or criticize my credentials as a scientist. One Conservative MP even commented that England had three senior and distinguished men as advisers (Sir Patrick Vallance, Professor Chris Whitty and Professor Jonathan Van-Tam), while Scotland had . . . me. The *Spectator* ran regular hit jobs, including a story in August 2021 calling me 'the pin-up girl' of Scottish Nationalists. When male professors defended my qualifications, trolls would reply, 'You just want to shag her', which was a highly effective strategy at ensuring they wouldn't speak up again in the future.

Observing months of attacks, a journalist noted, 'The lingering obsession Scottish Tory MSPs continue to show towards Professor Devi Sridhar is clearly unhealthy.' I learnt to cope with it, but it was definitely unhealthy for them, as any obsession is. I hope post-pandemic they find another hobby, such as gardening or yin yoga.

Scotland is the only place I have lived where saying anything good about it, in this instance Edinburgh, triggers anger among a tiny minority. When I lived in Miami, I was proud to live there, and the same thing applied to Oxford, Munich, Seoul and Chennai; but when it comes to Scotland, this seemed to upset those who see anything positive about Scotland as a soundbite for Scottish independence.

Yet, in person, everyone I have met has been lovely and kind, whether taxi drivers, supermarket staff, health workers or random people in the park. Social media amplifies the extreme voices, which are often bots or those looking for attention, rather than reflecting the vast goodness of humanity and the care and consideration most people show to their fellow humans and community members.

Scientists are not heroes. We are not villains. We are people who tried to offer our expertise and knowledge during a fast-changing and complex situation where there was no good path forward, only various suboptimal options with trade-offs attached to each. Yes, indeed, governments turned to scientific experts to become the oracles and predict the future; first to mathematical modellers and behavioural scientists, then to virologists and vaccinologists, and finally to public health experts and economists. And when these experts were not always able to provide the answers that appeased the need for concrete predictions with immovable dates, a backlash followed against their frustratingly balanced, contextualized and nuanced statements.

This gap was exploited by swathes of Facebook and Twitter pseudo-scientist celebrities, who predicted the future and offered comforting lies on what was to come. These fake gods would make promises of the virus disappearing in the summer of 2020, or promote miracle cures like ivermectin, or sell clickbait to ensure they got the most retweets and likes on Facebook and Twitter. Misinformation on social media has caused huge amounts of harm during this pandemic: whether it was people thinking COVID-19 was a hoax, reading scare stories about the side effects of vaccination or viral posts on miracle cures. How is one supposed to decipher what is real information and what is fake news on the internet? Deep Fake technology is particularly scary: someone sent me a three-minute video from the Disney live-action movie *Aladdin*: but in this version they had put my face on Jasmine's body, and it looked real. The confusion that results in the brain when watching yourself in a scene you've never been in is palpable.

Entire books will be written on the harm Facebook has caused. American public health expert Dr Tony Fauci said in July 2021, 'If we had had the pushback for vaccines the way we're seeing on certain media, I don't think it would've been possible at all to not only eradicate smallpox, we probably would still have smallpox, and we probably would still have polio in this country if we had the kind of false information that's being spread now.'

And to those who accused me of being in the pocket of the Scottish government, or Big Pharma, or vaccine manufacturers, I can only clarify – as I have done throughout the pandemic – that I have never received any funding from those sources. I turned down memberships on the advisory boards of J&J, GSK and Pfizer, because of the perceived conflict of interest and because of how important I felt it was to be a neutral and impartial voice. And, as for the 150-plus hours I sat in government and advisory meetings: all of that was unpaid and voluntary. Part of our job as academics is to serve in public roles for the public good. The joke was that when our advisory group finally met in person for dinner, we would have to pay for it ourselves. This didn't stop Freedom of Information (FOI) requests on how much we were each being paid. When hearing it was nothing, the reaction was 'But why do it?', with the answer being 'Because we care.'

And we continued to do our jobs alongside the associated strain of the pandemic on our personal lives. Science for most is a vocation, a calling, driven by internal motivation to address unanswered questions about the world and to improve the human experience. I went into public health because of the experience of my father falling seriously ill and eventually dying of cancer, when I was a teenager. I do wonder, given their experience of COVID-19, how many scientists will step up again for the next pandemic, and how many will prefer to remain hidden.

### *The Hunt for a Treatment for COVID-19*

While the politics were indeed messy, Britain shone when it came to cutting-edge science. A treatment for COVID, to stop people dying, or to stop hospitalization, or to stop critical disease, would have been transformational. Doctors were dealing with a novel virus and had to learn how best to treat it in real-time.

Developing a new drug takes too long. With an emergency like COVID-19, it made sense to look at our existing medicines to see which of them could have an impact. Repurposing essential medicines means examining existing therapeutics that have already been tested for safety in humans and attempting to use these in real time studies in hospitalized patients to assess if they affect clinical outcomes.

In China doctors tried all kinds of cocktails of antivirals. But how to assess and conduct science into what works and doesn't in an emergency situation? The most robust way to do this is through the double-blind randomized clinical trials (RCTs) mentioned in the previous chapter – that is, enrolling patients with similar profiles into two different groups, one of which receives a treatment and one of which receives a placebo. 'Double blind' means neither patients nor doctors know who gets which treatment. With enough patients and time, outcomes can be compared to assess the impact of various treatments.

Giving placebos to those in hospital carries ethical considerations. American public health adviser Dr Fauci said, 'Whenever you have clear-cut evidence that a drug works, you have an ethical obligation to immediately let people know who are in the placebo group, so they can have access.' At the same time the only way to study what interventions work is to have a placebo group until clear evidence is available that an intervention is effective.

In March 2020, just at the start of the first wave in Britain, two large trials started to recruit patients in order to evaluate the effect of potential treatments for COVID. In the UK the RECOVERY (Randomized Evaluation of COVid-19 thERapY) Trial was led by researchers at Oxford University and performed in 181 NHS hospitals. The RECOVERY research protocol was drafted on the 10th of March 2020, the first patients were recruited nine days later, and, by the 14th of May 2020, 10,000 patients were involved. Thousands of participants continued to be included in the trial to test the effects of azithromycin, convalescent plasma, dexamethasone, hydroxychloroquine, lopinavir–ritonavir and tocilizumab.

The Solidarity Therapeutics Trial was set up by WHO in March 2020 to evaluate the effects of four drugs – hydroxychloroquine, interferon beta-1a, lopinavir and remdesivir – on the in-hospital mortality of COVID-19 patients. The Solidarity Trial ran in 405 hospitals in 30 countries and 11,330 patients were assigned to random groups.

Dr Jeremy Farrar, Director of the Wellcome Trust, was involved with setting up both trials. He noted, 'I think that RECOVERY and the Solidarity Trial between them have set the standard of the scale that's required in order to give you clear answers.' Unfortunately, none of the investigated drugs in the Solidarity Trial showed a statistically significant impact in the treatment of COVID-19 patients. Farrar noted, 'It's disappointing that none of the four [drugs] have come out and shown a difference in mortality, but it does show why you need big trials.' Dr Soumya Swaminathan, WHO Chief Scientist, agreed with this assessment: 'We would love to have a drug that works, but it's better to know if a drug works or not than not to know and continue to use it.'

Three of the most important drugs identified early on were dexamethasone (a corticosteroid used to suppress an overreactive immune system); hydroxychloroquine (typically used as an anti-malarial and as a maintenance treatment for lupus); and remdesivir (an experimental drug against Ebola).

Dexamethasone is used for a wide range of health conditions including inflammatory and allergic disorders, severe skin conditions, croup (a viral infection that causes a nasty cough that sounds like a seal barking), swelling in the eye and autoimmune diseases. Additionally, dexamethasone can help reduce the nausea and vomiting associated with chemotherapy and is also used to reduce pain for patients receiving palliative care. The RECOVERY Trial showed that dying within twenty-eight days of diagnosis was significantly lower for patients treated with dexamethasone compared with patients receiving usual care.

Sir Patrick Vallance, the UK Government's Chief Scientific Adviser, released a statement noting, 'This is tremendous news today from the RECOVERY Trial showing that dexamethasone is the first drug to reduce mortality from COVID-19. It is particularly exciting as this is an inexpensive widely available medicine.' Professor Peter Horby, of Oxford University and the Chief Investigator in the RECOVERY Trial, agreed: 'Dexamethasone is the first drug to be shown to improve survival in COVID-19. This is an extremely welcome result. The survival benefit is clear and large in those patients who are sick enough to require oxygen treatment, so dexamethasone should now become standard of care in these patients. Dexamethasone is inexpensive [£5 per patient], on the shelf and can be used immediately to save lives worldwide.'

Hydroxychloroquine was also investigated, because of its role in preventing and treating malaria, as well as treating rheumatoid arthritis, lupus and several dermatological conditions. Unfortunately both the RECOVERY and Solidarity trials showed it to be ineffective in combating COVID-19. In the WHO Solidarity Trial, 954 patients were randomly assigned to the hydroxychloroquine group and 104 of them died during the trial, compared with 84 of 906 in the control group. In the RECOVERY Trial 1,561 patients were chosen at random and given hydroxychloroquine and 3,155 were given usual care. After twenty-eight days, 27 per cent of patients in the first group died, and 25 per cent in the second group. It was also found that patients in the hydroxychloroquine group had a longer duration of hospitalization than those in the usual care group.

Horby noted, 'Hydroxychloroquine and chloroquine have received a lot of attention and have been used very widely to treat COVID-19 patients despite the absence of any good evidence. The RECOVERY Trial has shown that hydroxychloroquine is not an effective treatment in patients hospitalized with COVID-19. Although it is disappointing that this treatment has been shown to be ineffective, it does allow us to focus care and research on more promising drugs.'

The third drug of focus was remdesivir, which was originally developed to treat Ebola but was never approved by regulators. It is an antiviral drug that tries to stop viruses reproducing by targeting their replication mechanisms. The trial results were mixed: one trial (ACTT-1) showed a clearly shorter time to recovery compared with patients in the placebo group, but there was no significant difference in terms of mortality. The Principal Investigator of the trial, Andre Kalil, said, 'The ACTT-1 trial results demonstrate that in hospitalized patients with COVID-19 pneumonia, remdesivir is the first antiviral medication significantly associated with a shorter time to recovery, in combination with a lower progression to mechanical ventilation.'

While the trials have been lauded for their speedy establishment, rigour and findings, they haven't escaped criticism. For example, the first results of the dexamethasone trial were reported in a press release and in a press briefing by the UK government, without methods and data analysis being made available. Some scientists have criticized this as 'science by press release', with findings being accepted as a given without the usual level of independent scrutiny of clinical trial data and results.

## *Judgement Day*

How will the UK's response be judged by history? Probably in quite a mixed way. The initial plan to move towards mitigation and 'herd immunity' was clearly a mistake, considering the loss of life and the multiple lockdowns it entailed. As the virus spread, care homes were particularly hit, as efforts to bubble them became undermined by patients being released from hospital into care homes without having been tested and by social care workers moving between care homes without adequate PPE or testing. A comprehensive, cross-party House of Commons Joint Inquiry, co-led by Tory MPs Jeremy Hunt and Greg Clark, bluntly reported that the UK government had made serious mistakes in its response along the lines discussed in this chapter. Their findings were based on days of testimony to the Health and Social Care Committee (which I also gave evidence to).

As frustration with Johnson's leadership grew, the various nations of the UK each started to try to chart their own course. But, while they would try their best to develop a tailored and forward-looking response, the UK government overall held the purse strings and the main powers for furlough and border control, meaning that the fates of all four nations were inextricably tied together.

On the other hand the UK government did invest in science, including RECOVERY, which resulted in the first approved drug to treat COVID-19 and better data on how best to clinically treat COVID-19 in hospital. One major success of Britain would be in its science and university research, which was, and continues to be, world-leading. In parallel with research into treatments, Oxford University was steaming ahead with vaccine research, as detailed in Chapter 9.

Another clear success was the vaccine rollout that started in January 2021. The UK government had made advanced preparations and negotiated arrangements to have preferential access to vaccines. Relying on the strong universal health care system that is the NHS, the rollout happened quickly, effectively and aided in the lifting of restrictions.

In the end the public just wanted the COVID-19 crisis to be over. The successful vaccination programme administered through the NHS enabled restrictions largely to be lifted in the summer of 2021. But we must not forget the policy decisions made pre-vaccine, and how many lives could have been saved had the advice and the leadership been different.

# 6. Trump's Divided America and the Children

No one could have predicted how badly hit the US would be, and how the country would become the world's epicentre of the pandemic for much of 2020. By February 2021 the US was responsible for a quarter of worldwide cases (28 million) and 20 per cent of all deaths (500,000). Life expectancy there was reduced for all Americans by a full year in 2020. But behind these numbers are inequalities that reveal how fractured American society is. Life expectancy dropped by over two years for Black Americans and three years for Hispanic Americans. Black and Hispanic Americans died at over 2.6 times the rate of white Americans. And it wasn't only in deaths that the US had major losses. Unemployment rose from 3.8 per cent in February 2020 to 13 per cent in May 2020; homelessness increased by 15 per cent due to the pandemic; and millions of children were out of school.

Why did the US suffer so badly, despite being ranked as top for pandemic preparedness and being one of the wealthiest countries in the world? It would be too simple an answer to say former US President Donald Trump, although his leadership clearly played a role. The crisis revealed a health system inaccessible to those who would need testing and treatment the most, an inadequate social care system for those who needed time off work and sick pay, and huge inequalities between racial groups. The Black Lives Matter protests, as I describe below, underscored that racism was indeed a 'pandemic within a pandemic'.

Children also became a political issue, with few subjects as controversial as school closures. While children themselves were rarely severely ill with COVID-19, their lives were nonetheless affected in an irreversible way. Some states implemented 'Zoom school' for over a year, playgrounds shut, and divisive debates ensued within the scientific community about the role of children in transmission. This was the case not only in the US but across the globe: UNICEF warned that the world was facing a global educational emergency.

## *The US Response*

In terms of the ability to respond to a pandemic, such as lab capacity and surveillance (as specified by the International Health Regulations), the US was the best prepared country in the world – in theory. In practice, as noted earlier, the country was hit badly with a staggering death toll. Gregory Treverton, the former Chairman of the National Intelligence Council, said, 'This has been a real blow to the sense that America was competent. That was part of our global role. Traditional friends and allies looked at us because they thought we could be competently called upon to work with them in a crisis. This has been the opposite of that.'

On the 3rd of January 2020 the Trump administration received the first notice of the outbreak in Wuhan. It took former President Trump more than seventy days to treat it like a serious public health emergency and not a distant threat or supposedly harmless flu strain. He liked to call it 'the China virus'. Trump himself kept underplaying the crisis, claiming on the 22nd of January 2020, 'We have it totally under control. It's one person coming in from China. We have it under control. It's going to be just fine.'

In February 2020 Mike Pence, Trump's Vice President, was made head of the White House Coronavirus Task Force, which was charged with coordinating the federal response on a day-to-day basis. Pence was a controversial choice. He was a born-again evangelical Christian who had refused to allow needle exchanges during his time as Indiana Governor, leading to spikes of HIV infections. The epidemic subsided only once Pence gave in and needle exchanges resumed. He had also written an op/ed in March 2001 saying that 'Smoking doesn't kill' and another one in June 2020 asserting that 'There isn't a coronavirus second wave', even as cases quickly rose. Pence's strategy when guiding the taskforce involved downplaying the pandemic and urging the country to reopen the economy, rather than containing the disease and minimizing loss of life.

## *What Went Wrong?*

Several clear failings in response were evident. First, the US failed to quickly develop a diagnostic test that could be mass produced and delivered across the country, enabling case finding and quarantine. In late January 2020 scientists in the US started vetting and developing tests that could detect the genetic sequence of SARS-CoV-2. This vetting included reviewing a protocol developed in Germany and validated by WHO. However, the efforts of American scientists hit a wall, because the Food and Drug Administration (FDA), the government's regulatory agency, did not approve the German test, and the Centers for Disease Control and Prevention (CDC) insisted its labs could develop an 'American' one.

On the 6th of February 2020 the CDC started to ship out tests to public health departments. However, three days later, it turned out these tests didn't work. The FDA finally approved the CDC's modified tests on the 29th of February 2020. The wish by senior government officials to have American tests, rather than use foreign, German ones, resulted in a delay of several weeks in getting testing going.

In March 2020 former Vice President Pence claimed that every American could get tested, but he quickly had to pull back from that because of the high demand from the public. Experts believed that the US needed the capacity for millions of tests each day, yet the federal government didn't want to pay the cost of these, and Trump wanted to slow down testing, as it made the US case rates look higher. He argued that more testing, and not an actual increase in cases, was responsible for the documented increase in confirmed cases. Or, as became widely recognized: test more, find more cases; test less, find fewer cases.

Whether intentionally (as with Trump) or unintentionally (as in certain countries without adequate diagnostic capacity), confirmed cases reflected only a certain percentage of actual cases. This is when testing positivity became important, or the number of people testing positive out of all people tested, because it indicated whether enough testing was taking place to catch most infections. Throughout the pandemic, WHO specified that under 5 per cent positivity was a good range to be in, particularly when easing restrictions and reopening the economy.

Second, PPE was limited and not well stockpiled – meaning that health workers had to go on to COVID-19 wards in bin bags and self-assembled kit. In February 2020 the US Health and Human Services Secretary testified before the Senate that the national stockpile for PPE had only 30 million surgical masks and 12 million respirators. Considering the US's population size and range of medical settings, it was estimated that the number needed would be much larger: 2–7 billion respirators and face masks to cover all health care workers.

Third, even before COVID, Trump slashed several key pandemic response programmes. He shut down one called Predict (set up by the Bush administration and continued by Obama) aimed at identifying and researching infectious disease in animal populations in early 2019; and he made budget cuts to the CDC of $1.2 billion, or 17 per cent of its budget. The Trump administration also sacked the government's entire pandemic response chain in 2018, including the position of Global Health 'Czar'.

Concerned about this lack of preparedness, daughter of former US President Bill Clinton and Columbia adjunct Professor Chelsea Clinton and I wrote an op/ed for CNN in February 2020, saying, 'To put it simply, the Trump administration, by seeking to cut funding and rejecting proven solutions, is not sufficiently prepared to face a major threat like a novel coronavirus outbreak.' We continued, 'The dissolution of the entire global health security unit and removal of global health security expert Timothy Ziemer from the National Security Council was the final step in a process of undermining one of the Obama administration's key decisions on global health. The unit monitored epidemics and ensured that public health planning was coordinated with the more traditional security infrastructure.'

Fourth, Trump rejected various evidence-supported measures to combat COVID-19, including distancing, test and trace, and masking, and instead encouraged people to protest against lockdown rules. Professor Jeffrey Shaman of Columbia University deplored Trump's actions: 'This is not just ineptitude, it's sabotage. [Trump] has sabotaged efforts to keep people safe.' Trump continued to compare COVID-19 to flu and told his supporters not to worry, acknowledging in February 2020, 'I still like playing it down because I don't want to create panic.' He even continued to hold in-person large indoor rallies ahead of the November 2020 presidential election.

One clear example of Trump's rejection of science is his attitude to face masks, which turned into a partisan issue: Republicans refused to wear masks and Democrats wore masks. When the CDC recommended the wearing of face masks, Trump claimed that face masks were a personal choice and refused to wear one, saying that people who wear masks do so to spite him. He even said, 'I don't wear masks like [Biden]. Every time you see him, he's got a mask. [Biden] could be speaking 200 feet away [and then] shows up with the biggest masks I've ever seen.'

Trump clashed with American expert Dr Tony Fauci, who had worked on infectious disease for the government for fifty years and led the National Institute of Allergy and Infectious Diseases. Fauci became a hero to those looking for science-based leadership and for his fearlessness in standing up to Trump. Trump threatened to sack Fauci if he were re-elected. He even issued an executive order that chipped away at the laws that prevented him from firing Fauci, arguing that agencies should be able to fire 'poorly performing employees'. He also blocked Fauci from testifying about the US pandemic response in front of the Democrat-led House Committee on Appropriations.

Fauci continued to speak to the American people through the media, to give public talks at universities about the risks that COVID-19 posed and to offer clear public health guidance. Unfortunately, he and his family, including his daughters, became a target of Trump supporters and they received death threats. Fauci noted, 'I wouldn't have imagined in my wildest dreams that people who object to things that are pure public health principles are so set against it, and don't like what you and I say, namely in the word of science, that they actually threaten you. I mean, that to me is just strange . . . Getting death threats for me and my family and harassing my daughters to the point where I have to get security is just, I mean, it's amazing.'

## *Political Pressure on Science*

Trump also put pressure on the FDA, usually an independent and strongly respected agency, to approve various new treatments, such as hydroxychloroquine and antibody-laden blood plasma treatments. He managed to pressure the FDA into approving the latter for emergency use despite no clear evidence on its effectiveness. This started to erode trust and public confidence in the FDA. Dr Ezekiel Emanuel, of the University of Pennsylvania, noted, 'Everyone is wondering: "Am I going to trust the Food and Drug Administration's decision on the vaccine?" The fact that people are even asking that question is evidence that Trump has already undermined the agency.'

The CDC, one of America's greatest assets for its deep experience and expertise in infectious disease management, was also sidelined by Trump. For the H1N1 (swine flu) crisis in 2009, the CDC held 32 out of 35 press conferences in the first thirteen weeks of the pandemic. By contrast, Trump led 75 per cent of the 69 press conferences held in the first thirteen weeks of COVID-19. He was particularly incensed that the CDC tried to get Americans to take COVID-19 seriously, tweeting, 'The number of cases and deaths of the China Virus is exaggerated in the United States because of @CDCgov's ridiculous method of determination compared to other countries, many of whom report, purposely, very inaccurately and low.'

According to an article in the prestigious journal *Nature*, Michael Caputo, a vocal Trump ally and spokesperson for the US Department of Health and Human Services, tried to meddle with the peer-reviewed *Morbidity and Mortality Weekly Report*, a digest published by the CDC. Caputo and his team started to interfere politically with reports that they felt were too critical of the Trump administration. CDC officials pushed back by asserting that the *Report* 'was merely recounting the state of affairs and not rendering judgement on the response'.

In July 2020 four former directors of the CDC wrote an editorial for the *Washington Post* in which they described how the CDC guidelines on school reopenings – in terms of how mitigation efforts might reduce transmission – were undermined by the Trump administration, which continued to cast doubt on the agency's recommendations and role in informing and guiding the nation's pandemic response. The directors mentioned how science is challenged by partisan politics, sowing confusion and mistrust when leadership, expertise and clarity are essential.

## *Chaos in the White House*

In an eerie parallel with UK Prime Minister Johnson, Trump was diagnosed with COVID-19 on the 2nd of October 2020. His administration downplayed the severity of the disease, saying that he was going to hospital not because of low oxygen or fever but 'out of an abundance of caution'. On the 4th of October he was driven outside Walter Reed Hospital, where he was being treated, for a photo op, to show he wasn't that ill and COVID-19 wasn't that bad an illness. He exposed the driver and another passenger to COVID-19 for the sake of publicity. James Phillips, a physician at Walter Reed, noted, 'That Presidential SUV is not only bulletproof, but hermetically sealed against chemical attack. The risk of COVID-19 transmission inside is as high as it gets outside of medical procedures. The irresponsibility is astounding.'

On the 5th of October 2020 Trump tweeted that he would be leaving the hospital and said, 'Don't be afraid of Covid. Don't let it dominate your life.' His callousness with regard to how different people might experience COVID, including those without health insurance and access to top medical care, led some critics to suggest that he had faked getting the disease in order to prove it wasn't as serious a threat as scientists had claimed.

At least twenty people working in the White House tested positive soon afterwards, including Trump's press secretary Kayleigh McEnany, who, following contact with an infected Trump, gave two press briefings and continued her daily routine, until she tested positive. This was despite the CDC advice for her to isolate. Others who didn't isolate were Senator Chuck Grassley, Attorney General Bill Barr and Vice President Mike Pence. Most of the White House cluster did not wear face masks or take precautions. On the 6th of October 2020 I did a TV interview with CNN journalist Christiane Amanpour and said, 'The White House has more cases in this cluster than Taiwan, Vietnam and New Zealand all put together.' At that moment it felt like the US was the laughing stock of the world.

Overall there was chaos emanating from the federal response led by a president uninterested in global health, sceptical of experts, intent on downplaying the crisis and distracted by impeachment and re-election concerns. Columbia University tracked more than 400 cases of the Trump administration's efforts to restrict or dismiss scientific research or evidence. This led to Trump's being removed from Twitter on the basis of his multiple tweets spreading pseudo-science and disinformation about COVID-19, as well as his role in inflaming violence against scientists. Trump was absolutely outraged and tried to establish his own separate media platform, but the damage had been done in terms of spreading false information during his presidency and creating a huge divide among the American people.

At the close of Trump's 2020 presidential campaign he argued that, 'We got hit with the China Virus. We've done an incredible job with respect to that – other than public relations.' He also noted that, based on initial modelling, 2.2 million people should have died, and until that point 'only' 230,000 Americans had done so. Trump also put the blame on China: it was 'China's fault' that Americans had died, and he claimed that the virus had originated in a Wuhan lab. Additionally, he argued that WHO had helped China to cover up the outbreak, leading to his pulling the US out of WHO by presidential executive order, an issue that will be covered in detail in Chapter 10.

## *President Biden*

Fortunately for public health and the COVID-19 response, Joe Biden was elected US President and took office in January 2021. He made it the top priority for his government to stop COVID-19 and linked this goal to getting the economy in order. As his Vice President, Kamala Harris tweeted on the 10th of December 2020, 'The scale of this pandemic is heart-breaking. Over 15,000,000 cases. More than 3,000 deaths in a single day. The economic devastation. Getting COVID-19 under control and safely reopening our economy starts with listening to experts. @JoeBiden and I will do just that.'

On the 6th of April 2021 the US Secretary of State Antony Blinken said, 'Stopping COVID-19 is the Biden–Harris Administration's number one priority. Otherwise, the coronavirus will keep circulating in our communities, threatening people's lives and livelihoods, holding our economy back.' This was finally an acknowledgement by a senior government minister that it was not a choice between the economy or health, but that it was necessary to protect public health in order to protect the economy. I will delve into that debate in Chapter 11.

But, while the federal government finally brought in senior experts to move forward with a science-based approach, including Dr Fauci and Dr Vivek Murthy as Surgeon-General (a childhood friend of mine from Miami), states continue to diverge in their public health policies based on whether they had Republican governors, like Florida and Texas, or Democratic governors, like Connecticut and Minnesota. The result was major political battles between the White House and individual states over issues such as masks, school return and vaccine mandates for events and employment. Clear divisions in the population remained over how seriously to take COVID.

## *Black Lives Matter*

While the first wave was taking off in the US, a man called George Floyd died after being arrested by the police outside a shop in Minneapolis, Minnesota, on the 25th of May 2020. Video footage on social media showed the policeman Derek Chauvin kneeling on Floyd's neck for eight minutes and forty-six seconds until he died, while Floyd kept pleading, 'I can't breathe.' Darnella Frazier, a witness to the scene, cried, 'They was pinning him down by his neck and he was crying. They wasn't trying to take him serious. The police killed him, bro, right in front of everybody.'

A day later, on the 26th of May, hundreds of demonstrators took to the streets in Minneapolis asking for justice. Jacob Frey, the Minneapolis Mayor, expressed the outrage: 'For five minutes we watched as a white officer pressed his knee to the neck of a Black man. For five minutes. The officer failed in the most basic human sense. [Floyd] should not have died.'

These protests under the banner of 'Black Lives Matter' spread across the US. Memphis, Chicago, Indianapolis, Los Angeles and Portland all saw protests, and demonstrations eventually occurred in more than seventy-five cities with tens of thousands of protesters. Sarina LeCroy, a protester from Maryland, said, 'There are hundreds of deaths that aren't caught on video, but I think the gruesomeness and obvious hatred of the video woke people up.' It soon became a global protest and extended to Australia, France, Germany, Spain and Britain.

Why did Floyd's death capture global interest, even though there had been numerous deaths of Black people at the hands of the police, including Tamir Rice, Michael Brown and Eric Gardner? Frank Leon Roberts, one of the activists, felt it linked to the deep inequalities revealed by the COVID-19 pandemic that had become impossible to ignore: 'You have a situation where the entire country is on lockdown, and more people are inside watching TV . . . more people are forced to pay attention – they're less able to look away, less distracted.'

Protesters faced criticism for gatherings in the streets during a 'Stay at Home' lockdown. Some protesters travelled from outside their local area and marched for hours, with much shouting and chanting, all factors that usually increase the risk of spreading COVID-19. Some US cities, like Washington DC and New York, urged protesters to get tested for COVID-19 and to wear masks.

And, predictably, some politicians jumped to blame increases in cases on the protests. Carlos Giménez, then Miami-Dade's Republican Mayor, noted 'It wasn't a coincidence that, about two weeks after these demonstrations started, we started seeing these spikes. That probably was the main cause of why this virus went up.'

Others pushed back on this narrative and pointed to Florida's lax policies towards COVID-19 spread. Dwight Bullard, a former Florida Senator, said,

> Everyone knows that we're having a surge because Florida never properly shut down in the first place. Time and again, state and local leaders decided to ignore the advice of scientists and experts and instead blindly followed the baseless claims from the Trump administration. This is just another example of why we must continue to say Black Lives Matter. Because in the same way that police officers kill us and corporations abuse us, politicians are quick to blame us for almost anything including a global pandemic.

And, surprisingly, evidence suggests that the BLM protests in the US did not lead to significant surges in cases. An infectious disease epidemiologist at the University of Arizona said, 'In general, the outdoor protests, widespread mask usage, distancing, quarantine after protests and focused communication/testing appears to have helped.'

In an interview with Sky News about the BLM protests in late May 2020, I was asked whether protesters were wrong to take to the streets. I replied that racism is also a pandemic, and one that Black Americans feel can't be swept under the carpet any longer. While clearly mass gatherings during a pandemic are risky, I could understand that people were willing to take this risk in order to effect change for their children and the children of their children. This is how the civil rights movement has attempted to progress racial equality over decades.

## *Why Did COVID-19 Kill So Many Black Americans?*

The BLM protests focused attention on the underlying institutional racism in the US and how the COVID-19 pandemic was accentuating this problem. Black people were disproportionally affected by COVID-19: a cohort study of people living in ninety-nine counties in fourteen states, conducted in March 2020 by the CDC, showed that, while Black people constituted 18 per cent of the population of this cohort, they represented 33 per cent of those who were hospitalized for COVID-19.

Apart from higher hospitalization rates, COVID-19 also seemed to be killing Black Americans at a disproportionately high rate: while 32 per cent of Louisiana's population is Black, they made up 70 per cent of COVID-19 deaths, a trend also seen in cities like Milwaukee in Wisconsin. Other examples include Illinois, where 14 per cent of the state's population is Black but 42 per cent of COVID-19 deaths were among Black Americans, a disparity that was even greater in Chicago, Illinois's biggest city, where 32 per cent of the population is Black, and 67 per cent of deaths from COVID-19 were among Black people.

Dr Fauci addressed this directly. He said, 'Yet again, when you have a situation like the coronavirus [minorities] are suffering disproportionately.' What was causing this unequal burden? An academic physician at a hospital in Washington DC noted, 'The virus is not really the problem. It's actually the systems which are killing people.' Unpacking this is complex.

First, certain comorbidities increase the risk of severe COVID, and, in the US, Black people suffer from higher rates of diabetes, hypertension, asthma, HIV and obesity than white people. Dr Jerome Adams, the US Surgeon General under Trump, noted, 'I've shared myself personally that I have high blood pressure, that I have heart disease and spent a week in [the intensive care unit] due to a heart condition, that I actually have asthma and I'm pre-diabetic, and so I represent that legacy of growing up poor and Black in America.'

Second, certain key worker frontline jobs, often low paid and with no health or sick pay benefits, have clear occupational exposure to COVID, and people from minority communities are more likely to do these jobs. While 14.8 per cent of the US population is Black, of the 7.5 million low-income families, 20.8 per cent are Black. Nearly 40 per cent of Black people in the US are low-wage workers, with no sick pay, making it more likely that they go in to work despite showing symptoms. Quite a few of these low-wage jobs are in-person employment (not able to work from home) with almost 25 per cent of Black Americans being employed in service industry jobs, compared with 16 per cent of white Americans. Finally, many (roughly 45 per cent) low-wage workers rely on their employer for health insurance, forcing them to continue to go to work.

The reality in America is that access to health care or to sick pay is still enjoyed only by the privileged; it is not regarded as a right or part of government's basic duty to its citizens.

This story reflects the real experience of those unable to take time off:

> Sarah called the urgent care paediatrician in tears. Her two-year-old son Eddie had been diagnosed with COVID-19 during an emergency department visit the previous day. She simply couldn't get his fever down and he wouldn't drink. Sarah, a Latina waitress earning a minimum wage, had no paid sick leave or employment protections. She was exposed to COVID-19 by a coworker who could not afford to isolate and came to work infected. Sarah also became ill, along with many of her coworkers.
>
> Unable to isolate from her large family, the virus spread rapidly through her household of eleven, including her three children, cousin, elderly parents and sister's family. Her cousin, aged thirty-four, was now in the intensive care unit with severe COVID-19 pneumonia. Her elderly mother with heart disease had started coughing. She sobbed questions over the phone: Would Eddie recover? Would her cousin live? Would her mother die from a virus she had brought home? Who would bring them groceries or pick-up Eddie's medicine if she isolated? Her husband, the only person in the household without symptoms, knew he should quarantine but couldn't because they needed his paycheck to survive.

Third, Black Americans are more likely to have living conditions that put them at risk of COVID-19. For example, 40 per cent of the US homeless population is Black and people experiencing homelessness often live in close quarters, have a compromised immune system and are vulnerable to communicable diseases. Black Americans are more likely to live in neighbourhoods with concentrated poverty, pollution, lead exposure, higher rates of incarceration and higher rates of violence. Even handwashing and basic hygiene measures can be difficult for those in prison, or those living in highly dense communities with bad housing, poor sanitation and limited access to clean water.

Finally, Black Americans are less likely to have health insurance and more likely to experience barriers to accessing health services. This could be in testing or in getting access to health care at an earlier stage of COVID-19. Without health insurance, seeing a doctor or having huge hospital bills can result in major debt and bankruptcy. This makes engaging with the healthcare system impossible, because of what the bill will be like at the end. This is in stark contrast with the UK, where the NHS provides free and accessible high-quality health care to all.

All these factors meant that, as of the 6th of March 2021, 178 per 100,000 Black or African Americans had died of COVID-19, in comparison with 172 American Indians or Alaska Natives, 154 Hispanic or Latino Americans and 124 white Americans. This is not just true for COVID-19. Overall Black Americans have substantially worse health and shorter life expectancies than white Americans.

COVID-19 was just the latest example of racism in health care and health inequalities between Black and white Americans. In the nineteenth century Black slaves were exploited for the development of some aspects of US medical education, as medical schools relied on their bodies for anatomical material. The twentieth century witnessed the Tuskegee Syphilis Study: between 1932 and 1972 the effects of syphilis on Black men were studied but left untreated, so the natural course of the disease could be observed, even though treatment was available by the end of the experiment. The participants were unapprised of their diagnosis and more than a hundred died of the disease.

There are also examples like Henrietta Lacks, who was diagnosed with cancer in 1951 and whose bloods were taken and, without her knowledge, used for research purposes. The HeLa cells, as they were called, were subsequently mass-produced and used in various research projects worldwide: Jonas Salk, for example used them to test the first polio vaccine. It was not recognized until the 1970s that the cells were from Henrietta Lacks, and none of her family members were given any compensation or profits from their commercialized use. Furthermore many Black women were involuntarily sterilized or given hysterectomies without informed consent in the twentieth century. Between 1930 and 1975, 65 per cent of the sterilizations carried out by North Carolina were on Black women.

There is also systemic racism in all aspects of health care, from access to the top medical treatments to attention from nurses. Dr Randi Abramson, the Medical Director for Bread for the City (a medical clinic in Washington DC offering care regardless of ability to pay), said, 'The stress and racism in this country, the lack of access to other resources. It's not just their health but also their access to what you need to survive. And the fact that we don't have all that equal access in the city is really driving up morbidity and mortality within this community and within peoples of colour.'

Lack of trust in the government would become a factor when Black communities were more hesitant to take up the vaccine than their white counterparts. Professor Sandra Quinn, of the University of Maryland, said, 'We know that because of the history of segregation, discrimination and racism there are reasons people don't trust the government. The government doesn't always have their best interest at heart.'

The anger about George Floyd's murder reflected the ongoing unfairness and daily racism faced by Black Americans, which COVID-19 put into high relief. The US handling of the virus tells us how interlinked public health issues are with issues of social and economic inequality, and how any public health strategy can't afford to ignore these realities. This is true not only of the US but of every country across the world. For example, in the UK, risk factors for dying of COVID-19 included being from a deprived background and being from an ethnic minority group (such as Black or South Asian). *The Times* ran an article on the 12th of April 2020 titled 'Ethnic minorities bear the brunt of Sweden's coronavirus deaths', highlighting the higher morbidity and mortality in certain racial and ethnic groups.

## *Children and COVID*

Few issues provoke as wide a range of opinion and as much emotion as children and COVID-19. When the virus emerged in China, and spread throughout East Asia, the message from China to governments across the world (and concerned parents) was that children were not badly affected by COVID-19. The concern was about adults, especially those over sixty. But the picture was far from simple as the virus spread through the US and Europe. The two key concerns soon became how to manage infection in children and the role of children in transmission. Both would defy easy answers.

On the first, most children seemed to have mild or asymptomatic illness. Early on paediatricians weren't concerned about COVID-19, given all the other causes of severe illness in children such as other respiratory viruses and bugs. But red flags started going up among doctors in late April 2020 in the UK. Dr Patrick Davies, a consultant in paediatric intensive care at Nottingham University Hospital, shared details with his doctors' WhatsApp group about a rising number of children in intensive care with symptoms indicative of a rare and dangerous inflammatory disorder. Davies noted in an interview, 'This was something we'd never seen. They were so unstable. If you took one wrong turn they would become even more unwell.'

An increasing number of children in the UK started to present several weeks after exposure to COVID-19 with a strange clinical disorder, which would be termed Paediatric Inflammatory Multi-Syndrome Disorder (PIMS). Within six weeks, in April and May 2020, seventy-eight children were in ICU with PIMS, most requiring life support. Paediatricians were not expecting this at all. As Davies said, 'We were gearing up to take adults. This whole issue completely blindsided us. Suddenly all these patients [were] popping up with a hugely complex illness. The clinical variability frightened us.'

PIMS also emerged in Lombardy, Italy, where scientists reported a thirty-fold increase in cases. It presented like Kawasaki disease, meaning rashes, prolonged fever and abdominal pain, and hit those children of African or Asian descent considerably harder. It would also be seen in the US in larger numbers, with over 4,000 cases by July 2021. These were children testing negative on PCR tests (they didn't have the virus at the time of hospitalization) but positive on antibody tests (they had had COVID-19 at some point in the previous weeks).

Several key scientific questions didn't have good answers then: what was the prevalence of PIMS in all those infected with COVID? Why was this even occurring several weeks after infection? Answers to the second question would help to develop better clinical treatment, but the answers to the first were essential to communicating risk to parents and to political leaders, and in giving scientific advice on the risks of keeping schools open in the spring of 2020.

Eventually it would become clear that PIMS is rare, but this took several months of research to discover and it revealed the dangers of letting a new virus go uncontrolled even in a paediatric population. Exact incidence was difficult to assess, but ranged from 0.1 per 100,000 children and adolescents in Poland to 20 times that in New York State. While the percentage of children affected may be small, if the denominator of those infected grew, absolute numbers of children getting severely ill would become high.

The second main concern with children was their role in transmission. Usually with flu and other infectious diseases children play a super-spreading role. As they play together, their naive immune system and their general lack of hygiene causes them to pick up every infection. But, strangely, this did not seem to be the case with children and COVID-19. Data from Public Health England in late 2020 revealed that nurseries had fewer outbreaks than primary schools, which had fewer outbreaks than secondary schools, which had fewer outbreaks than universities. It seemed to form a gradient – from the youngest children, who didn't appear to transmit much, to the older children, who biologically resembled adults. Children could transmit the virus in some instances, but their role just didn't seem as central to spread. This knowledge became helpful in reopening schools and was supported by both WHO and UNICEF.

The picture changed slightly with the emergence of the Alpha and Delta variants in December 2020 and June 2021 respectively; as they were more transmissible, they also made children more able to transmit. Denmark, which largely kept schools open even during their first wave, struggled with small clusters of Alpha outbreaks in schools. Delta caused more problems in the UK in June 2021: because children were the only ones in the population left unvaccinated, it was within this group that the virus could spread.

But, overall, it's fair to say that scientific consensus – throughout the pandemic – was that children were not key players in transmission or driving the spread of COVID-19. Yet they came to be easy targets for blame and experienced the fallout and harm of restrictions more than any other group. A Brown University study on 672 children born in the US state of Rhode Island found that children born during COVID-19 had significantly reduced IQs and verbal and motor skills compared with children born in the decade pre-COVID. It is widely recognized that the early years (especially the first hundred days) are crucial for child development; unfortunately the children born during COVID-19 had more stressful home environments (with parents juggling work, childcare and financial strain) with limited access to playgrounds, nurseries and other adults. Those infants from poorer backgrounds had more of a cognitive loss over that period. Will these children recover from this early life stress? It is unclear.

## *Getting Schools Back Open*

With such divergent and strong opinions on the role of children in transmission as well as uncertainty about the short- and long-term consequences for their health, debates about the reopening of schools were fraught. While paediatricians and child health experts continued to raise concerns about the harm to children from being out of the educational system, especially those from vulnerable backgrounds, teachers resisted the push to go back into work and reacted badly to the language of 'martyrdom' that was used by leaders such as Trump to categorize them as essential workers. Schools became a political football, and even scientists took the extreme positions of schools either being completely safe or not safe at all. The role of children was not fully understood for months.

From March 2020 our research team at Edinburgh University was immediately focused on the issue of children and schools, with Dr Adriel Chen and Dr Ines Hassan gathering international evidence and assembling the latest data. Working with the Royal Society, Ines wrote an insightful report on schools and children that highlighted the harm from schools being closed: increased educational inequalities, a rise in food insecurity and the loss of future income. The report recommended key safety measures such as ventilation of classrooms, children to be kept in small bubbles and, in some situations when warranted, masks as well. Their input would help me to provide advice to the Scottish government while I was serving on the Advisory Subgroup on Education and Children's Issues.

During the first wave of COVID-19 in the first quarter of 2020, many countries closed all schools, while some closed them only in specific places or for specific year groups. In total UNICEF estimates at least 1.5 billion schoolchildren were affected by closures, which prompted their Executive Director, Henrietta Fore, to note, 'The sheer number of children whose education was completely disrupted for months on end is a global education emergency. The repercussions could be felt in economies and societies for decades to come.'

In August 2020 WHO conducted a survey of 39 countries in Sub-Saharan Africa and found that schools were fully open in 6 countries, partially open in 19 countries and closed in 14. The disruption was not only about lost learning but also about children missing school meals, increased stress, increased exposure to violence and exploitation, increased childhood pregnancies (the pregnancy rate doubled during the Ebola outbreak in Sierra Leone) and overall challenges in mental development caused by not enough social interaction. Dr Matshidiso Moeti, from WHO, warned: 'We must not be blindsided by our efforts to contain COVID-19 and end up with a lost generation.'

In January 2021 UNESCO reported that, globally, schools had been closed an average of 14 weeks since the start of the pandemic, and an average of 22 weeks when partial closures were taken into account. The duration of closure varied greatly by region: Latin America and Caribbean schools were fully closed for 20 weeks on average, European schools 10 weeks and only 4 weeks in Oceania.

School closures have far-reaching and detrimental effects. Many children, especially in poorer countries, will never return to formal schooling again and instead must take up jobs to help support their families. Even before COVID-19 more than 50 per cent of children could not read by the age of ten. In Miami (where I grew up), state schools opened with 10,000 fewer students than in 2019. Despite sending out teams of social workers to locate students, by March 2021 a thousand students could still not be found.

While remote learning was put in place, especially in the US, this was for the privileged rich and excluded the poor. I say 'privileged' but middle-class parents struggled in their own way too, juggling home working and home schooling their children. A colleague said to me that his most challenging day was 'being on a call with ministers while my wife was too unwell to get out of bed. Back in April 2020. My then five-year-old daughter opened the door in a tutu and cat mask. Quite funny. Took the mask off to reveal full makeup. Bit funny, bit disturbed. Then told me she'd burnt her hand with matches in the kitchen. Which was awful. And once I'd checked her hand and run down to check there wasn't a fire, I still had to wave her away and get back on the call. That was soul-wrenching.'

Returning to the challenge of online learning, at least 463 million students worldwide were unable to access remote learning, because they didn't have either computers or the internet. In the US about two thirds of state-school students were in home learning. But this was inaccessible to the most disadvantaged: 14 per cent of children aged 3–18 don't have the internet at home.

Clayton Burch, West Virginia's Superintendent of schools, noted, 'We have a lot of families and teachers who want that idea of virtual and remote learning to work, but connectivity is so poor, it just hasn't. I don't think we've made a dent in high-quality, equitable access in everyone's homes.' In Los Angeles Jaime Lozano had been trying to teach online classes but nearly all his students were from poor backgrounds and couldn't afford the internet. Laura Stelitano,

Associate Policy Researcher at the RAND Corporation, observed that 'Kids without internet access are more likely to suffer and not even be in contact with their teachers.' And it's not only internet access that is denied to them but also access to computers, tablets and laptops.

Even if internet and equipment issues can be obtained, students may also have to do chores within the household, take care of younger siblings and figure out how to set up the technology, often in households where parents are working outside of the home and therefore absent.

In-person schooling is also about providing children with heated spaces during the day, food, books, outdoor play areas and access to adults trained in teaching and in interacting with children. Even in Europe, a wealthy region of the world, 10.2 per cent of children live in homes that cannot be heated well, 7.2 per cent have no access to outdoor play areas, 5 per cent have no space in their home to do homework and 5 per cent have no access to books. In New York 10 per cent of students were homeless or experienced housing instability during their remote learning.

The pandemic sadly resulted in homelessness levels in New York City being as high as during the Great Depression. Despite banning evictions, tens of thousands lost jobs because of lockdown and restrictions, so they could not afford to pay their rent. Many children also have their only hot daily meal at school: 6.6 per cent of households in the EU and 5.5 per cent of households in the UK cannot afford a proper daily meal with a meat, fish or protein source. This number is 14 per cent in the US.

The US already had severe educational inequalities that the pandemic increased. A June 2020 analysis predicted that the average student would fall seven months behind because of COVID-19, with Hispanic students falling nine months behind and Black students ten months behind. Kayla Patrick of the Education Trust explained, 'We already knew that Black and brown students weren't getting the support that they need even before the pandemic. And then the pandemic made all of that worse.'

And there are clear gendered effects too, with school closures putting girls at risk. Globally, before the pandemic hit, the number of girls dropping out of primary school had almost been cut in half. In 2015 women and girls more than fifteen years old spent on average seven years in school, compared with five years in 1990. However, as a result of the school closures, these gains were reversed, with more girls dropping out of school – making them vulnerable to domestic violence, sexual abuse, child marriage and teen pregnancy, and pushing them into the labour force. UNESCO projected that 11 million girls may never return to school after the disruption from COVID-19.

In addition to those challenges, school closures also lead to a loss of play and social interaction, an increase in physical inactivity, delayed accessing of paediatric care and more mental health issues. Adolescents felt like the forgotten group: teenagers on the brink of adulthood as well as young adults who faced a year of no jobs, no university in-person learning, no travel and no real interaction or parties with their friends. Loneliness was already seen as a crisis among this age group pre-COVID-19, given its negative effects on health.

Considering the harm associated with school closures, why were they even closed in the first place? As noted above, I worked closely, offering scientific advice, with the Scottish government from April 2020 onwards, and the concerns focused on four key questions.

The first was 'What is the risk to children if they get the virus at school?' At the start there was very little data (as noted in the previous section) about the health impact on children themselves, so the cautious approach would be to close schools, especially given the early reports of PIMS in children. But by June 2020 the risk to children themselves seemed low, if community prevalence was low; there was less of a risk from COVID-19 than from many other infectious respiratory diseases that children face in school. In addition there were limited instances of child-to-child transmission, although this changed with the Alpha and Delta variants. So here there was largely a green light to move ahead, especially from the paediatric community.

On the 16th of June 2020 I tweeted, 'If COVID-19 numbers can be brought low enough in Scotland by 11 August (under 20 confirmed cases), & with appropriate "test and protect" policies, my personal view is that schools should reopen as normally as possible (kids back full time & able to play/interact together).' The Scottish media immediately picked up on this as criticizing the Scottish government's original plan for blended in-person and online learning. First Minister Nicola Sturgeon then tweeted, 'Right now (like other UK nations), we must plan for a school model based on physical distancing. But as @devisridhar says, *if* we can suppress virus sufficiently & have other measures in place, nearer normality may be possible. It's why we must stick with plan to suppress.'

I personally was baffled by certain journalists' take on my tweet, attempting to sow division between advisers and government; in fact, Sturgeon and I spoke regularly by phone about key issues and were generally aligned on the need to suppress and get cases as low as possible through the summer. She never pressured me on any issue, and instead often listened carefully to my independent advice as she reflected on what to do next. I supported Sturgeon's tweet, noting that we were aligned and we both wanted to be cautious on easing lockdown, and make any decisions on schools based on data.

I had hoped that this would be the end of it, but my response triggered Ruth Davidson, the former Scottish Conservative leader, to say, 'Guess someone got the hairdryer treatment.' Being American, I had no idea what the hairdryer treatment even meant. For all non-British readers, this is a term used in football (i.e., soccer) to describe an angry verbal reprimand by a manager to a player or to the team. It's a metaphor for the loud blowing of air from a hairdryer. That was just the start of continual attacks by Scottish Conservative politicians on almost anything I said about COVID-19. A year later, in July 2021, Ruth and I sat on a *Good Morning Britain* panel where we agreed on the need for caution around reopening nightclubs. Which became awkward when we ran into each other at 2 a.m. at the Polo Lounge in Glasgow a few months later. (I'm joking. That never happened.)

Returning to schools, the second question for the advisory group was 'What is the risk to their families if the children bring it home to parents and grandparents?' This was a larger worry, because younger children are often in close contact with their household members and could in some instances infect them. However, the evidence on this was far from straightforward. A study in Scotland found that households with primary school age children were less likely to test positive than those without children, leading to the hypothesis that these parents and households had probably picked up colds (common coronaviruses) recently and built up some cross-immunity for SARS-CoV-2. Secondary school age kids were a different picture, as those aged twelve and older were biologically more similar to adults than the younger groups. The solution seemed to be to ensure mitigations were in place in school to prevent transmission, and also to keep community prevalence low.

The third concern was 'What is the occupational risk to teachers of being close to children?' This was the main obstacle to reopening schools, because unions were concerned about the safety of their teachers, and quite rightly so. And it unfortunately became politicized in both the US and the UK. Instead of engaging with teachers on what measures would make them feel safe in classrooms, leadership in both countries tried to force schools open with pleas to 'martyrdom' (sacrifice for your country) and guilt (kids are suffering because of your selfishness). This was unfair on teachers, who in general chose a teaching career to help children and support them, and not for monetary or status reasons.

For example, former President Trump pushed hard for schools to reopen ahead of his November 2020 re-election race, asserting, 'I think schools have to open. We want to get our economy going . . . I think it's a very important thing for the economy to get the schools going.' But this led to major pushback from teachers. Ashley Newman, a 29-year-old primary school teacher, quit her job rather than gamble with her family and her health. She said, 'Our options are either deal with it and risk your health and your family's health, or you can leave. If the teachers aren't safe, then the students aren't safe. And then the community isn't safe. It's a ripple effect.'

Trump even went so far as to declare teachers 'critical infrastructure workers' (alongside nurses, police officers and meat packers), meaning that they wouldn't need to quarantine even after being exposed to COVID-19. This was to avoid teachers having to isolate and the subsequent school closures that would follow. But teachers' unions pushed back, saying that this was 'an excuse to force educators into unsafe schools' and arguing 'If the President really saw us as essential, he'd act like it. Teachers are and always have been essential workers – but not essential enough, it seems, for the Trump administration to commit the resources necessary to keep them safe in the classroom.'

The fourth concern was 'What are the implications of schools re-opening for community transmission?' Here, it seemed, in-school transmission did not play much of a role. It was more about the activities and mixing that schools triggered: playdates among families, more use of public transport, more adults back in offices and more of a social life for parents. In short, parents gone wild when kids were back in school.

We also discussed key mitigations, but it was always a benefit–cost, trade-off decision. We could outline the options, but ministers had to decide which one to choose. First, masks were seen as an easy way to try to stop transmission, but in younger children they also carry harm in terms of child speech and social and emotional development. We assessed that, in the light of low prevalence and low child-to-child transmission risk, masks would not be needed in primary schools but would be needed in secondary schools if prevalence increased. Even in teenagers, masks can result in skin issues and acne, so are not completely without issue, because they would have to be worn all day. It felt uncomfortable to ask all children to wear masks all day, although it became necessary in certain year groups. There's a difference between a child wearing a mask for a short length of time – in a shop, or on public transport, or in medical settings – and a child wearing one for a full school day.

Second, bubbles were introduced to assist test and trace in knowing which children would need to isolate. This was straightforward in primary schools, which typically have one teacher and a class of thirty or so pupils, but much more complicated in secondary schools, where students interact with numerous teachers and move around the building to take different subjects. There were both fairness and practical issues. On fairness, there were multiple instances of families coming back from holidays abroad and not isolating as required, thus causing an entire school bubble to isolate when the children returned to their classrooms. You can imagine how popular those families would be because of that selfish decision. Practically, one secondary school teacher testing positive could cause up to 200 students to isolate, with major educational disruption as a result.

Third, distancing was also far from straightforward. Two metres made full-time return of pupils impossible, because schools, especially in deprived areas, just didn't have the space. The result would have been those in the poorest catchments having in-person learning only once or twice a week, while those in posh private schools had a normal return. All these students would then have to sit the same entrance exams to university. This approach would have increased existing educational inequalities and hurt those children most in need of school.

Fourth, so-called 'blended' learning – a bit of in-person tuition, a bit of online tuition – itself carried the harm as noted above of unequal access to devices, the internet and a supportive household environment. It was a very poor alternative to in-person learning. And, finally, ventilation was essential to avoid transmission, but difficult to achieve in old school buildings, where windows did not open, or, when windows could open, classrooms became freezing cold in winter. The push to outdoor learning, which sounded lovely in summer, resulted in children being forced outside in rain or snow, and, again, some of the poorest children suffering through a lack of warm-enough clothing, waterproof jackets or appropriate footwear. Local councils started asking for donations for warm coats for those children who found it difficult to learn because they were freezing during the day from open windows and long outdoor breaks.

The point I am trying to make is that there were no easy answers on schools. I gave an interview on the 18th of March 2021 to Professor Eric Topol of Stanford University:

> The schools have kept me up at night because there are so many conflicting data points, and there are so many stakeholders, and they're all not wrong. Sitting in meetings where unions are telling you how

> teachers are sobbing because they don't want to go into classrooms, because they're scared they're going to be infected. You are wondering as a teacher if you're going to get infected and die because we gave the wrong advice.
>
> At the same time we get every week a report from the child welfare officer. You hear about the rises in child abuse, and the children who are getting beaten up at home because they're not able to get out, and they have parents who have lost their jobs, and they're angry, stuck in tiny flats, and they have no money. You're thinking, well, kids need to get back to school. No one is wrong; everyone is right in what they're saying to you. How do you give the best scientific advice? What keeps me up, the one thing that does stress me is, am I giving the best scientific advice? Where are the holes in my argument?

My take quite simply, after much reflection and analysis, is that, while schools were not 'super spreading' locations like bars or clubs, nor did they operate in a bubble. They were part of a community and mirrored what was happening in terms of cases. Quite simply, the more cases in the community, the more likely it was that cases would come into a school and students would have to isolate, or that the school would have to shut completely. The best way to keep schools open was to keep cases as low as possible in the community.

In Scotland 76 per cent of schools had absolutely no cases from August until mid-October 2020 because of the low prevalence achieved over the summer. The schools with cases often just had one or two as a result of the rapid outbreak response. Eventually schools had to shut again from January to mid-February 2021 as the Alpha variant took off and Christmas mixing occurred, all of which led to the winter wave. Sturgeon made it clear that schools would be the first to open and last to close in Scotland, a phrase that Johnson would repeat several weeks later with regard to England; they did so in February 2021.

This contrasted with New York City. On the 18th of November 2020 Mayor Bill de Blasio announced schools would close, while indoor dining, gyms and other businesses were allowed to remain open. Robin Lake, Director of the Center on Reinventing Public

Education, said, 'Academically and economically, it's a little bit crazy to prioritize bars over schools.' Many parts of the US embrace this model, with large parts of the economy open, including riskier indoor settings like bars and restaurants, while schools remained online.

During the pandemic the schools debate became unnecessarily personal and ego-driven for many scientists, who seemed intent on proving their own hypothesis instead of looking at what would be best for children, teachers and families. In short, when it became clear that children were not major players in COVID-19, restrictions should have been eased on them to allow them to have as much freedom as possible.

This is what a 'child first' lens means, and it is one that was politically popular too. In Scotland this meant letting children under twelve play in groups outside from July 2020 onwards, even during the various lockdown measures. The benefit of children playing and interacting outweighed the risk of child-to-child transmission outdoors. And this meant that children could attend outdoor tennis, football, and other camps and activities even as adults had to restrict their mixing. Adults were adults and modified their behaviour, while kids were allowed to be kids. Unfortunately, this meant that teenagers felt left behind, as noted above, and had to face their own combination of challenging and sometimes potentially harmful circumstances.

The larger lesson moving forward is that pandemic planning must include how to keep schools safely operating for in-person learning and incorporate the use of outdoor spaces, face coverings, testing, bubbles, distancing and other measures. Home or remote schooling just doesn't work and shouldn't be seen as an acceptable outcome for children. Now is the time to start making those plans. Just as hospitals, supermarkets and other essential institutions stayed open even during the strict lockdown measures, so should schools. This could involve looking at how to repurpose public spaces like stadiums, libraries and conference centres for emergency situations, how to ensure there are enough teachers and classroom assistants for this transition, and how to allocate appropriate financial resources to facilitate this response and ongoing in-person learning for children.

## *Rethinking Good Pandemic Preparedness*

The US response showed that leadership, good governance, proper planning and social cohesion are key markers of a strong COVID-19 response and not, as previously thought, country wealth, technical ability or capacity. The US's weak position was caused by massive income inequalities, deep-seated race issues and no universal health insurance scheme, as well as no protection for workers who were ill or needed to isolate. Add to that a president who ignored scientific experts and was concerned primarily about re-election and you have a recipe for a haphazard and devastating response. This means looking at pandemic preparedness in a broader light than just core capacities, as WHO's International Health Regulations do. The role of political leaders in crises needs to be taken into account, as do the social and economic policies that impact directly on readiness for health emergencies.

In the case of the US its major weakness was not having universal free health care or a strong social benefit system to enable people to take time off work and isolate. In the first half of 2020, 30 million people living in the US did not have health insurance. These people are disproportionately likely to be from Black or Hispanic backgrounds, younger in age, and on low incomes. Not having health insurance meant that going into hospital had to be paid for directly, and the costs were extortionate. Hospitalization with COVID-19 cost between $42,486 and $74,310 for those without insurance in the US, according to estimates in April 2020. This fear of going into hospital, or seeing a doctor, meant that many people tried to manage their infections at home.

Like the government of the UK, the Biden Administration would turn its focus to vaccinations as the route to containment and recovery. As will be discussed in Chapter 9, the FDA quickly approved Pfizer/BioNTech, Moderna and Johnson & Johnson (J&J) vaccines, and the CDC pushed a large public campaign for high take-up. Incentives were created for getting vaccinated such as free beer, free metro or subway passes, college tuition lotteries and, for a short spell of time (before it was reversed by the onset of the Delta variant), not having to wear a mask or follow distancing guidance. As was the case in the UK, vaccines ultimately paved the way 'out'.

While many countries – like the US and UK – suppressed the virus in the short term without a longer-term plan, advisers in both Sweden and New Zealand did not think a vaccine was anywhere on the horizon, and so charted a different course. In Sweden's case they decided to let the virus sweep through the population. In New Zealand they decided they'd stamp out each and every case. In the next few chapters we'll take a closer look at the varied responses of different countries from around the world. It's something governments should have been doing throughout the pandemic.

# 7. Elimination versus 'Letting Go'

As the pandemic progressed, countries started to diverge on how to best address the virus. There were only three strategic choices for responding to COVID-19 in the absence of a vaccine or effective therapeutic. The first was mitigation, which meant the pandemic spread through the population until 'herd immunity' was reached, as we saw in Britain. The second was suppression, which aimed to keep case numbers low with prolonged control measures such as lockdown cycles. And the third was elimination, which was to stop community transmission within borders and then put in place safeguards so the virus was not reimported.

Most countries had a suppression strategy that fell somewhere in between these choices, but Sweden made the deliberate decision to pursue natural herd immunity without a vaccine. By contrast, New Zealand's Prime Minister, Jacinda Ardern, pushed for full elimination of the virus from the islands and catching imported cases with strict border controls.

In this chapter I tell the stories of New Zealand, Australia, the Netherlands and Sweden. Why did advisers in these countries recommend such divergent approaches? How did it play out in 2020 and into 2021? Across the world the initial focus was put on mortality and preventing deaths, yet in late spring 2020 there was growing recognition of another tragic consequence of COVID-19. As we will see, it turns out that not dying from it wasn't necessarily the end of the story for swathes of COVID-19 survivors around the world. So-called 'long COVID' is a nasty syndrome. But, first, to New Zealand.

## *The Paradise of New Zealand*

New Zealand initially followed their 2017 pandemic planning, which was similar to the planning that the UK and European countries were using. This was developed for an influenza pathogen, widely considered to be the most likely cause of a large-scale health emergency.

The original strategy, underpinned by the need to protect health and the economy, aimed to slow entry and initial seeding, prevent initial spread and to 'flatten the curve', i.e., to bring the peak of hospitalizations within health care capacity limits. Throughout February 2020 New Zealand was closely aligned with other Anglophile countries in seeing this virus as one that needed to flow through the population while ensuring enough health care capacity for everyone who needed it. Pandemic influenza, the nearest comparator in planning, was, at that point, seen as unstoppable, as discussed in Chapter 5; therefore, widespread community transmission was anticipated.

Most countries in Europe and North America started with mitigation but pivoted quickly to suppression once their health services became overwhelmed with COVID-19 patients. New Zealand advisers were watching Europe carefully, in particular Lombardy. After the release of the WHO Mission Report to China – which outlined how China had almost reached elimination of the virus and which provided the first clear epidemiological description of the disease in terms of who it impacted and the associated symptoms – one of New Zealand's key health advisers, Professor Michael Baker, quickly pivoted to a SARS elimination plan. They could already see the success of elimination strategies in Taiwan, Hong Kong and Singapore. Baker said, 'If China can protect a population of that scale, surely New Zealand can protect five million people.'

Baker found a receptive audience in the empathetic and public health-orientated Jacinda Ardern, who was equally unhappy with both mitigation and suppression. She reflected in a TV interview, 'I remember my chief science adviser bringing me a graph that showed me what flattening the curve would look like for New Zealand. And where our hospital and health capacity was. And the curve wasn't sitting under that line. So we knew that flattening the curve wasn't sufficient for us.'

When scientists in other countries such as the UK and the Netherlands heard that New Zealand was attempting this approach, they remained sceptical as to whether it would be possible to eliminate a virus that transmitted much like the common cold or flu. Baker pushed for a move away from theoretical modelling and towards observational data: 'We had to move away from the normal levels of scientific certainty and say, well, on balance of evidence, we know this has worked in China.' Ardern agreed and noted that, even if they couldn't achieve full elimination, an elimination approach would still save lives. Others, like the Faroe Islands, Iceland, Greenland and Fiji, decided they too would adopt this strategy, especially given their geographical situation as islands able to manage borders well.

Watching from afar as this strategic choice was being made, I tweeted on the 3rd of April 2020: 'Sounds like a movie -> New Zealand might be the only country able to eradicate SARS-CoV-2. With its borders closed to overseas passengers, people on the island can go back to normal society on the isolated island and wait for a vaccine, or other solution, for COVID.' I got pushback from many who said New Zealand would never be able to achieve this fantasy.

To move back to the start of the epidemic in New Zealand: on the 28th of February 2020 the first COVID-19 case was reported, and soon after, on the 14th of March, all international arrivals into the country needed to self-isolate for fourteen days, unless arriving from the Pacific, and cruise ships were banned. On the 19th of March borders were closed to all but New Zealand citizens and long-term residents.

On the 22nd of March an alert system with four levels was put in place based on the COVID-19 situation, and on the 25th of March a State of National Emergency was announced that equated with a strict stay-at-home lockdown. This meant gatherings were cancelled, public venues closed, businesses closed except for essential services, educational facilities closed, travel severely limited and only safe recreational activity like walking or running in a local area allowed. It also meant no social interactions outside one's bubble, defined as one's household and very close contacts. During this time, when people's contacts were reduced, the government increased testing and tracing to aggressively go after the virus and find all cases.

In addition, the government launched a massive communications campaign to explain lockdown measures to residents and to recruit them into a collective effort. No 'war on COVID' was declared; instead the government's message was about the country coming together, with the campaign slogan being 'Unite against COVID-19'. A website acting as the single source of messaging and information was launched and reached 700 million views in the first three months. Sarah Robson, a journalist at Radio New Zealand, explained, 'Because [Ashley Bloomfield, Director-General of Health] had clearly communicated the trajectory we were on in terms of the increase in the number of cases, when Jacinda Ardern said we were going into lockdown, people understood why.'

New Zealand also implemented incredibly strict border measures. From the 9th of April managed isolation in hotels was introduced, with those travelling to New Zealand having to book a place in a hotel before they could book their flight. They had to stay in this facility for 336 hours (at least fourteen days) and undergo regular health assessments and COVID-19 testing.

By June 2020 Jacinda Ardern was able to celebrate the fact that New Zealand had achieved elimination and would keep in place strict border controls. Other island nations soon followed. The country reached 102 days without cases between early June and August 2020, until there was an unexplained outbreak in Auckland. This outbreak led to an eighteen-day Level 3 lockdown in the Auckland region until all cases were found. Auckland would face several short, sharp lock-downs to catch cases as they emerged. Ardern kept the public on-side in briefings: 'As a team we have also been here before. We know that if we have a plan and stick to it, we can work our way through difficult and unknown situations.'

Ardern's leadership was widely praised, with 78 per cent of New Zealanders saying that they had increased trust in government as a result of its strong pandemic management. Subsequently she won a second term in office after national elections were held in October 2020. Her party garnered 49.1 per cent of the vote, taking 64 seats in the 120-seat assembly and an outright parliamentary majority.

But their approach was not without its challenges. First, the continual flare-ups in Auckland led critics to point out that the strategy hadn't worked. Trump for example revelled in the struggle by New Zealand authorities to squash cases: 'The places they were holding up, they're having a big surge. Do you see what's happening in New Zealand? "They beat it, they beat it," it was like front-page news because they wanted to show me something. The problem is "big surge in New Zealand", it's terrible, we don't want that.'

And, just like in other parts of the world, New Zealand health authorities faced people refusing to cooperate with testing and isolation, in this instance in an evangelical church cluster, because these people did not believe in the existence of the virus. Ardern commented on the challenge to keep going: 'We thought we were through the worst of it. And so it was a real psychological blow for people. And I felt that too. So it was very, very tough.'

Second, like in other parts of the world, strict lockdown measures took a psychological toll: one study showed that about 30 per cent of participants in a health and wellbeing study experienced moderate to severe psychological stress and 16 per cent reported moderate to high levels of anxiety. However, 62 per cent of participants also reported enjoyable aspects, including spending more time with family and living in a quieter, less polluted environment. Professor Susanna Every-Palmer, Head of the Department of Psychological Medicine at Otago University, said, 'New Zealand's lockdown successfully eliminated COVID-19 from the community, but our results show this achievement brought a significant psychological toll.'

The third major issue was the closure of borders, which both decimated the tourism industries and harmed families that found themselves split across the world. In 2019 New Zealand had more than a million international visitors; since March 2020 until the time of this book's writing the number has been reduced to nearly zero. Sectors reliant on those visitors collapsed. For example, Shaun Kelly, owner of a hostel in Queenstown, said he's only operating at 25 per cent occupancy and 'In terms of the business environment, to be honest, it's awful . . . Any evidence to say that we're doing okay is wrong.' Families have also had to be split apart for over a year, and Charlotte te Riet Scholten-Phillips, founder of the think tank The Fair Initiative, has said, 'No one's suggesting we just throw the border open and just let everybody in – but people want to know, will it be a month, two months, six months, before I see my family again?'

The key to New Zealand's fully reopening its borders was the complete vaccination of its population. In the meantime, travel bubbles with other countries that had achieved elimination could be set up. As Bloomfield said in a briefing, 'There's no doubt that having as much of the population vaccinated as possible is key to us being able to open the border', and Ardern agreed: 'I have said that 2021 is the year of the vaccine.'

New Zealand managed to reach private agreements for the following vaccines: Pfizer/BioNTech for 10 million doses; J&J/Janssen for 5 million doses; Oxford/AstraZeneca for 7.6 million doses; and Novavax for 10.72 million doses (we will return to these in Chapter 9). On the 17th of December 2020 the New Zealand government announced it would provide free vaccines to neighbouring Pacific Island nations, including Tokelau, Cook Islands, Niue, Samoa, Tonga and Tuvalu. Ardern also announced 1.6 million doses of COVID-19 vaccines would be donated to the COVAX facility (a multi-country global mechanism for sharing vaccines), making it the first country to pledge doses and not just money. COVAX will be discussed in detail in Chapter 10.

When criticized about New Zealand's slow rollout of its vaccine, Ardern claimed it was more important for countries with high infection rates and especially developing countries to have the limited number of doses available, particularly as New Zealanders remained relatively unharmed: 'It's a full-year programme we have only just begun. We're not in a race to be first, but to ensure safe and timely access to vaccines for all New Zealanders.'

## *Australia Attempts Elimination*

Australia's response in 2020 and the first half of 2021 ended up looking a lot like New Zealand's 'elimination approach', although the path to get there was more fraught with difficulties and less direct.

Like New Zealand, Australia started with pandemic flu planning. On the 18th of February 2020 the Australian Health Sector Emergency Response Plan for Novel Coronavirus was published and updated on the 23rd of April 2020. This plan was based on the Australian Health Management Plan for Pandemic Influenza (2014; updated August 2019). In addition to developing the health sector emergency plan, the Australian government also invested in a test/trace/isolate system, rolled out a public health campaign and created an economic safety net.

The health sector emergency plan had two stages: action and stand down. Action included gathering and providing information about disease spread and preparing health services for patients, while stand down included a return from the emergency response to normal business.

On the 27th of February 2020 Australia's government declared that COVID-19 would become a pandemic, but the Prime Minister, Scott Morrison, told people not to worry, as Australia remained in the containment phase. At this time a travel ban to China had been in place for four weeks. Morrison said, 'You can do all of these things [go to the football and cricket and play with friends down the street] because Australia has acted quickly, Australia has got ahead of this at this point in time. But to stay ahead of it we need to now elevate our response to the next phase.'

On the 16th of March 2020 large gatherings of over 500 people were cancelled, and all international arrivals were required to self-isolate for fourteen days. On the 18th of March, strong advice not to travel was introduced, and indoor gatherings of more than a hundred people were banned. On the 20th of March 2020, Australia's borders closed except for non-residents. On the 23rd of March 2020 closure of all non-essential businesses was mandated, with a stay-at-home lockdown on the 24th of March 2020. On the 28th of March 2020 hotel quarantine measures were implemented. These measures were effective at keeping case numbers low, and on the 8th of May 2020 lockdown easing was announced.

After the first lockdown and reopening, there were no more national lockdowns in Australia; rather, individual states starting their own plans towards elimination, border closures and lockdowns. For example, Sydney's Northern Beaches was in lockdown between the 19th and 24th of December 2020. On the 8th of January 2021 Brisbane went into a three-day lockdown after a quarantine hotel breach. Another three-day lockdown went into effect on the 29th of March 2021 to curb community transmission. With these kinds of short, sharp lockdowns, enforcement was essential, but some made accusations that the police were too heavy-handed and discriminatory in their use of powers to stop people and judge their reasons for being outside. In New South Wales a man was fined when he stopped to buy a kebab on his way home from a run.

## *The State of Victoria*

The state of Victoria had a particularly hard time reaching elimination after a massive flare-up in August 2020. Case numbers in Victoria were at a similar level to those in Britain at the time; but in Britain these were seen as a 'success' while in Victoria they were a 'failure'. And Britain opened up with these case numbers, while Victoria went into a harsher lockdown to get to zero. In addition to the initial national lockdown measures, a night-time curfew between 8 p.m. and 5 a.m. was introduced, as well as reduced workforces in factories and construction sites. A payment of US$1,070 was introduced for those who had run out of sick leave and who could not access other benefits if they were instructed to isolate for fourteen days.

The first resurgence of cases came from a failure in hotel quarantine, with staff at hotels in which travellers quarantined bringing the virus into the community. In Victoria staff of the quarantine hotels were hired through private security firms, unlike neighbouring state New South Wales, where the police force was used to secure hotels.

One of the security guards at the five-star Stamford Plaza Hotel, Shayla Shakshi, said the virus was treated casually by the guards, who were poorly trained. She herself was recruited from WhatsApp. She said, 'I just knew something would happen because it's just like the guards were [playfully] hitting each other. They were hugging each other. They were touching each other. They weren't actually serious about how serious COVID-19 is. They were taking it as a joke. Like, "It's just some virus that anyone can get. We're not going to get it."'

Claims were even made that security guards, tasked with ensuring people stayed in their rooms and did not mix, were having sex with infected guests quarantining at the hotel. These guards then went home to their lives in the communities and continued to spread the virus. Genomic sequencing, which can be used to trace chains of transmission, revealed that cases in the community were directly linked to security guards working at quarantine hotels.

Community spread also continued because people were not self-isolating when they were waiting for their test results or they waited too long after developing symptoms to get tested. While there were rare instances of people claiming lockdown rules breached their human rights and refusing to adhere to the rules, in most cases, people broke rules because they said they could not afford to stay off work even with the state payment. Victoria's biggest problems emerged in aged care homes, meat factories, schools and public health estates, in effect among those who couldn't work from home. Daniel Andrews, Premier of Victoria, kept pushing people to get tested. 'If you've got symptoms, the only thing you can do is to get tested. You just can't go to work. Because all you'll be doing is spreading the virus.'

Testing was essential to Victoria's elimination strategy. In November 2020 door-to-door asymptomatic testing was conducted in two of Melbourne's hardest-hit areas: Hume and Wyndham. In twenty-four hours 10,000 swabs were processed. The asymptomatic testing was done on top of the usual symptomatic testing throughout Melbourne.

Testing and lockdown measures meant that cases continued to fall, and lockdown could be lifted on the 8th of November 2020. But, again, on the 12th of February 2021, a cluster emerged and a five-day lockdown was announced, during which mass testing would be conducted to find all cases. This was successful and lockdown was lifted on the 17th of February.

The lockdown occurred right in the middle of the Australian Open, which had been going ahead with half the crowd of the previous year and extensive preparations. As Craig Tiley, CEO of Tennis Australia, said, 'We had to bring in 1,000 people from over a hundred countries around the world on seventeen charter flights from different cities into Melbourne, and quarantine that number of people for fourteen days and then every single day, the athletes, enabling them to get outside of the room for five hours.' The Melbourne tennis site was divided into three sections to enable the authorities to conduct tracing, and masks were required indoors.

Several of the tennis players complained about the quarantine and asked for VIP exemptions. Andrews was unapologetic. 'That was the condition on which they came. So there's no special treatment here . . . because the virus doesn't treat you specially so neither do we.' The complaints of the players fell on deaf ears: Victorians had suffered a long, harsh lockdown to reach the normality they enjoyed with elimination, and they did not want it ruined by incoming players. And the officials were proved right, as several people linked to the Open did test positive while in quarantine.

In a bright spot for sports enthusiasts during COVID-19, tennis fans across the world could watch Naomi Osaka, Rafael Nadal and Serena Williams play in front of crowds, even if it felt like watching life on another planet, given the harsh winter lockdowns at the time across Britain, Europe and parts of the US. Former No. 1 US tennis player Andy Roddick said, 'It makes my heart full seeing fans in the stands. Well done, Australia!' Overall the tournament was a massive success for international elite sports, with Osaka winning the women's title and Novak Djokovic taking the men's.

And it was only possible to hold this kind of event at that point of the pandemic (pre-vaccine rollout) because of the elimination approach that had been taken by the Australian government. Professor Brendan Crabb, Director and CEO of the Burnet Institute, a medical research group, said that, while he wouldn't have originally supported mass events, it was possible because they had achieved zero COVID-19 infections. As Crabb said, 'This is huge, I don't mean for Australia, I mean for the philosophy of COVID zero. This is a very sound way to live, it's sound for health, it's sound for your economy and it's also sound to limit the number of mutant viruses [developing]. So you've got this showcase event demonstrating to the world how valuable COVID zero is.'

Australia faced many of the same challenges as New Zealand in the end. How to manage new, more infectious variants like Delta? How to open borders safely? How to acquire vaccines and vaccinate enough of its population? How to cope with families split apart, not only internationally but even within the country because of intra-country border closures between states? But it profited in many ways: good domestic economic performance, not much pressure on its health services and very little loss of life. Elimination is a short- to medium-term approach that seemed a fruitful one for those countries that could manage it before vaccinations were available.

## *Sweden's Gamble for* Flockimmunitet

By stark contrast, Swedish public health authorities led by Dr Anders Tegnell and Professor Johan Giesecke decided that the only sustainable way to deal with this kind of respiratory pathogen would be to let it flow through the population, and avoid the economic and social costs of lockdown. In a May 2020 article in the *Lancet*, Giesecke, a Swedish physician and Professor Emeritus at Karolinska Institutet, wrote that the virus was spreading rapidly and that there was little that could be done to curb its spread. He expected that lockdown would decrease severe cases for a while, but that cases would reappear once restrictions lifted. He expected that the number of deaths due to COVID-19 would be similar in different countries after a year, regardless of the restrictions that had been put in place. According to Giesecke, the most important task was not to stop COVID-19 spreading but to focus on optimal care for its victims.

The ideas set out by Giesecke seemed to form the basis of the Swedish strategy. The Public Health Agency, and state epidemiologist Tegnell, focused on flattening the curve to ensure health systems were not overwhelmed. Rather than legally enforced shutdowns, Swedish Prime Minister Stefan Löfven said Sweden's approach would be based on trust and 'common sense'. Leaked emails from Tegnell indicated that herd immunity (*flockimmunitet* in Swedish) was discussed as a desirable outcome; Tegnell estimated that, by the end of May 2020, 40 per cent of Stockholm's population would have antibodies and immunity against COVID-19, which would blunt the severity of future waves. Tegnell said in an interview with the *FT* in May 2020, 'In the autumn, there will be a second wave. Sweden will have a high level of immunity and the number of cases will probably be quite low.'

In practice this meant that Sweden tried to stay 'open', while the rest of Europe and its Scandinavian neighbours shut down. Sweden's policies provoked horror among the governments of Norway, Denmark and Finland, which had all effectively contained the outbreak through early lockdowns and aggressive testing and tracing.

In March 2020 crowds were first limited to 500 people and later to 50 people. Bars, restaurants, shops, day-care facilities and schools for those under sixteen stayed open. Schools, colleges and universities for those aged over seventeen started distance learning. Unlike in most countries, quarantine was not implemented for those returning from travel abroad; nor did the family members of those who had tested positive for COVID-19 need to isolate. Face coverings were also not recommended, not even for visitors to elderly care homes (they were eventually recommended in November 2020, almost nine months after Germany and the Czech Republic). Similar to the UK's early advice, those aged over seventy or with underlying conditions were asked not to mix outside their households and not to go to shops, including supermarkets and pharmacies.

Sweden's approach was defended by Lena Hallengren, Minister for Health and Social Affairs, as a balanced one of not overreacting to COVID-19's threat. She said, 'We should put lives and health first, protect the health care system as much as we can, and make sure that they have the resources they need. But we have also stressed the importance of securing society's other important functions. We believe that once this pandemic is over, society should be able to continue to function.'

How unique was Sweden's approach? Denmark, Norway and Finland all took a precautionary approach to containment, using early intervention and legal measures. For example, in March 2020 Denmark,

Norway and Finland closed workplaces, schools, hair salons, bars and restaurants. Sweden, by contrast, tried to have a more hands-off approach to the spread and just offered recommendations to the public. Johan Erik Lallerstedt, a filmmaker in Stockholm, said in April 2020, 'In Sweden, the approach has always been to make suggestions and let the public decide. Whatever guidelines the government may give are just that: suggestions. Nothing is enforced.'

Meanwhile Sweden (in line with not being concerned about community spread) did not build its testing and contact tracing capacity. It consistently had among the lowest testing rates in Europe. By the end of May 2020 Sweden tested only 20 per cent of the number of people per capita, compared with its neighbour Denmark. Sweden did not try to stop imported infections until the 7th of February 2021, when foreign nationals needed to present a negative COVID-19 test upon arrival into Sweden. This was nearly a year after East Asian countries introduced travel screening.

What Sweden attempted was to 'shield' the vulnerable in care homes and in society. On the 31st of March 2020 visits to care homes were banned. However, infection rates in care homes soared, with half of all COVID-19 deaths occurring there. Several reasons have been suggested for this: first, initially there was no recommendation for staff members in care homes to wear PPE and, when it did become available, it was limited in supply and late. Second, asymptomatic transmission was not adequately recognized by the Public Health Agency.

Third, given the unstable nature of their employment, some care home staff still went to work despite feeling unwell. One care home worker said, 'Where I'm working we don't have face masks at all, and we are working with the most vulnerable of all. We don't have hand sanitizer, just soap. That's it. Everybody's concerned about it. We are all worried'; and another noted, 'Staff often work for fourteen hours with substandard protection and continue working despite exhibiting symptoms.'

In addition, triage criteria for hospitals meant that elderly patients were not automatically taken to hospital, and regional health authorities discouraged staff from sending care home residents to the hospital for treatment. Latifa Löfvenberg, a nurse who worked at care homes during the first wave and a local politician with the Sweden Democrats, said, 'They told us that we shouldn't send anyone to the hospital, even if they may be sixty-five and have many years to live. We were told not to send them in. Some can have a lot of years left to live with loved ones, but they don't have the chance . . . because they never make it to the hospital. They suffocate to death. And it's a lot of panic and it's very hard to just stand by and watch.'

The shielding guidance in the community was for those aged over seventy and in at-risk groups. They were encouraged to avoid mixing and remove themselves from society for the foreseeable future, with suggestions even made that those who were vulnerable should stay at home indefinitely until either a vaccine or antiviral therapy was found or natural 'herd immunity' reached. Completely neglected in public debates were the serious ethical and moral questions around building a society in which the healthy and young are left to circulate, and the elderly, the disabled and the vulnerable are hidden away.

When reflecting on the mental and physical isolation felt by those isolating, Johan Carlson, the Director-General of the Public Health Agency, acknowledged, 'It is not reasonable that risk groups should have to bear such a heavy burden for society in the long run, especially when we can see that the mental and physical consequences are significant for those who have been isolated.' During the shielding period an estimated 1.5 million self-isolated (out of a population of 10.2 million), which was considered the largest factor in case numbers coming down in summer 2020.

## *Sweden Pivots to Containment*

In November 2020 Sweden had a second winter wave that hit harder than health officials and politicians expected, leading to a high number of cases and deaths. ICU facilities were full and hospitals on the brink of collapse. In the month leading up to the 8th of January 2021, more than 2,000 deaths were reported in four weeks, as compared with 465 deaths in Norway (which has only half the population of Sweden) during the entire pandemic.

King Carl XVI Gustaf, as well as the Swedish Prime Minister, publicly acknowledged that the approach taken had failed to protect lives. The King said, 'I think we have failed. We have a large number who have died and that is terrible. It is something we all suffer with.'

Changes made to the strategy included cutting gathering sizes from 50 to 8 and a face mask policy on public transport; and, in early January 2021, an emergency law was proposed to lock down parts of society and close schools for children over thirteen. Sweden did not escape the economic impact of its earlier relaxed approach: it suffered economic costs that were similar to those of its Scandinavian neighbours in 2020, when members of the public started to make their own decisions on not circulating, consuming and protecting themselves. It was the virus that was killing the economy, not restrictions alone, which is a theme I will return to at the end of the book.

Internationally Sweden's approach was depicted as both visionary and catastrophic. At first, former US President Trump pointed to Sweden's 'living with COVID' to stop lockdowns; he then used the example of Sweden to defend US lockdowns. He tweeted on the 30th of April 2020: 'Despite reports to the contrary, Sweden is paying heavily for its decision not to lockdown. As of today 2,462 people have died there, a much higher number than the neighbouring countries of Norway (207), Finland (206) or Denmark (443). The United States made the correct decision!'

Trump even criticized the 'herd immunity' approach taken. 'Now they talk about Sweden, but Sweden is suffering very greatly. You know that right? Sweden did that, the herd. They call it the herd. Sweden is suffering very, very badly, it's a way of doing it.' *Time* magazine captured the situation best with its article 'The Swedish COVID-19 response is a disaster. It shouldn't be a model for the rest of the world'.

But anti-lockdown journalists and groups around the world found solace in this different approach to COVID-19, and the idea of 'Sweden' developed a cult-like following among this community. The *FT* reported 'Architect of Sweden's no-lockdown strategy insists it will pay off'. The *Economist*, for example, covered how normal life was in Sweden compared with other countries: 'While Sweden's fellow

Scandinavians and nearly all other Europeans are spending most of their time holed up at home under orders from their governments, Swedes last weekend still enjoyed the springtime sun sitting in cafés and munching pickled herrings in restaurants. Sweden's borders are open (to EEA nationals), as are cinemas, gyms, pubs and schools for those under sixteen.'

Did this lax approach work in Sweden? The debate is polarized and depends upon how one values many things: human life; economic performance; and the basic freedom in a democracy to open a business, live and make decisions as one wants, and travel freely. But Giesecke's view that every country would suffer the same death toll over a year would not be correct, especially as vaccines became available much sooner than anticipated. Swedes paid a heavy price in that lives were lost unnecessarily. And, as the year progressed, Sweden went the same way as its Scandinavian neighbours – into suppression – with the *FT* carrying the article 'Sweden's distinctive COVID strategy nears an end as lockdown proposed'.

## *The Netherlands' 'Intelligent Lockdown'*

The Netherlands' response started similarly to Sweden's but ended looking like the rest of Europe's. Dr Jaap van Dissel, possibly the country's most prominent public health official and Director of the Dutch National Institute for Public Health and the Environment (RIVM), initially downplayed the risk of COVID-19. On the 18th of January 2020 an RIVM spokesperson said the chances of the virus showing up in Europe were 'very small', adding that there were no direct flights from Wuhan to the Netherlands.

On the 27th of January 2020 COVID-19 was classified as a major public health threat, which gave the government additional powers to intervene. An Outbreak Management Team, consisting of medical experts to advise on the strategy to take, was established, and the Prime Minister, Mark Rutte, said that this advice would be implemented and not just considered.

On the 27th of February 2020 the first COVID-19 case was

confirmed in the Netherlands: it was a patient recently returned from Lombardy. Soon after, cases in two Southern Netherlands provinces (North Brabant and Limburg) surged, and ICUs became overwhelmed there, necessitating the transfer of patients to hospitals in other parts of the country as well as to Germany.

Like the UK, the Netherlands started with containment but moved to mitigation on the 16th of March 2020, with the goals of building up herd immunity and preventing ICUs from running out of beds. The first lockdown was called an 'intelligent lockdown' by Rutte. He said, 'We can slow down the spread of the virus while at the same time building group immunity in a controlled way.' Van Dissel noted, 'We want to spread the virus to people who are not really affected by it . . . The disadvantage of such a total lockdown is that the virus will form a kind of peat fire, that there is insufficient build-up of immunity and that you will therefore always remain sensitive to reintroduction of the virus in the future. If everything is opened up again, foci of infection can re-emerge elsewhere.' In response to a public backlash against the idea of 'herd immunity', van Dissel claimed that this was not an aim of the strategy but a consequence of living with the virus.

As infections rose quickly, on the 12th of March 2020 the government introduced restrictions, including cancelling events with more than a hundred people and closing museums, concert venues and sports clubs. It asked everyone with symptoms to stay home and encouraged those in vulnerable groups and the elderly to avoid public transport and mixing. Primary and secondary schools stayed open, while universities and higher education switched to remote learning. On the 15th of March 2020 more measures were introduced: schools, hospitality venues, gyms and sex clubs were all closed. Schools and childcare were available only to the children of key workers. On the 18th of March 2020 Dutch borders closed to all non-essential travellers from outside Europe, and a day later visits to nursing homes were banned. On the 23rd of March 2020 even more businesses, such as hair salons, were asked to shut, and the maximum group size was lowered to three.

Even 'flattening the curve' required extensive restrictions. Van

Dissel acknowledged, 'There is very little difference in the approaches between European countries. Some countries say they are in lockdown, but if you look at the measures they're taking, you can say they are not.' This was one of the problems of using the word 'lockdown' in a binary sense, like an on–off switch, rather than in the sense of a gradient, like a dial that could be adjusted to different degrees of mixing and activity. One generic term could not capture exactly what countries were doing nationally.

The Netherlands also faced challenges in its response. Here are just a few. First, the Minister for Medical Care, Bruno Bruins, responsible for tackling COVID-19, collapsed in Parliament on the 18th of March 2020 during a debate about the government's handling of the crisis. The next day he resigned, saying he was exhausted, needed rest and was no longer physically up to the task. Bruins said, 'And I have concluded that my body can no longer handle this due to exhaustion.' Rutte, accepting his resignation, said: 'I want to say here that we have seen this man working very, very, very hard since he took office. He's been in charge of the whole process of getting it going. I want to thank him incredibly for that. I think it's incredibly sad.'

Second, curfews and riots erupted over the restrictions. Koen Simmers, a Dutch Police Union official, observed that 'We haven't seen so much violence in forty years.' On the 23rd of January 2021 a night-time curfew was introduced to help manage the second winter wave. This led to three nights of serious violence, with rioters setting fire to cars, smashing windows, throwing rocks and fireworks at the police, and looting supermarkets and other shops. In the fishing town of Urk a coronavirus test centre was set on fire, while in Enschede rocks were thrown to break the windows of a hospital. Police used teargas and water cannons to disperse the crowds. Rutte declared that it was 'unacceptable. This has nothing to do with protesting. It is criminal violence and that's how we'll treat it.'

Viruswaarheid (translation 'Virus Truth'), which helped to organize the protests against restrictions and promotes conspiracy theories about the pandemic, launched a court case against the Dutch government's curfew. A lower court judge ruled that the curfew had to be lifted immediately because emergency powers had been wrongly used to invoke it. However, later that day, an appeal court ruled that it should be maintained until a full court hearing on the legality of the issue later that week. During the court case the government won the ruling: limiting constitutional freedoms is justified to tackle COVID-19, it was said. The appeal court judge noted, 'In this case, the state's interests weigh more than that of Virus Truth's.'

Finally, the Dutch government was one of the first to start getting people back into nightclubs, theatres, football stadiums and pubs. These natural experiments came into being through a collaboration between government, industry and scientists called Fieldlab. For example, a dance party was organized where participants were divided into bubbles of 250 people, and one bubble of 50. Each bubble had to adhere to different rules, and their movements were tracked with the help of a tag. Everyone attending the events had to have had a negative COVID-19 test forty-eight hours before the event, and all participants were asked to take another test five days after the event. The initial studies indicated that, with pre-event testing and good ventilation, risks in these types of venues could be brought down to an extremely low level.

Was it ethical to use humans as guinea pigs in this kind of natural experiment? A panel at the Radboud University Medical Centre ruled that Fieldlab did not meet the legal definition of medical research and that ethical approval from a medical ethics committee was therefore not needed. This was challenged by a group of 350 researchers who wrote an open letter complaining that the Fieldlab study had a lack of peer review, an opaque set-up and ethical failings. They mentioned that guidelines for research in social and behavioural sciences should have been followed, and participants should have been able to give informed consent. Professor Denny Borsboom, of the University of Amsterdam, said, 'Not a single behavioural scientist is involved. If they were, this would have never happened.' Similar experiments started to take place in the UK, with live concerts and even the tennis tournament Wimbledon being part of larger research on infectious disease spread at mass events.

## *Long COVID*

When COVID-19 first emerged, the concern was largely about mortality: how many people would die from the virus, either with or without medical care. In those countries that allowed the virus to spread, such as the Netherlands and Sweden as well as the US and the UK, a post-viral condition became increasingly recognized as younger individuals struggled to recover from COVID-19.

One of the first accounts about long-term COVID-19 symptoms and slow recovery was published in April 2020 in the *New York Times* by Fiona Lowenstein. A month later, in May 2020, Professor Paul Gardner, from the Liverpool School of Tropical Medicine, wrote a ground-breaking blog for the *British Medical Journal* about his experience of having had a variety of COVID-19 symptoms for seven weeks after falling ill. He said, 'The symptoms changed, it was like an advent calendar, every day there was a surprise, something new.' Pulitzer Prize-winning journalist Ed Yong interviewed nine people whom he called 'long-haulers' in an effort to bring attention to this issue; his article received 1 million views.

The term 'long COVID' itself was first used by a long COVID sufferer, Elisa Perego, on the 20th of May 2020 on Twitter as a hashtag and was taken on by patient groups worldwide. It was first used in print media in the UK in a Sky News article, and by the summer of 2020 had become the way the media described patients with long-term COVID-19 symptoms. Scientists were concerned but didn't have solid data on how common this condition was in those who had had COVID-19, nor did they have a good scientific understanding of why people were suffering. Professor Danny Altmann, of Imperial College London, said, 'We have thousands of people reminding us that this might not just haunt us for this summer, but it might haunt people for the rest of their lives.'

No widely accepted data was available (at the time of writing of this book) on the prevalence of the condition, or whether it is one condition or multiple syndromes rolled together. Dr Hans Kluge, WHO Regional Director for Europe, put estimates of prevalence on the high end: 'The burden is real and it is significant: about 1 in 10 COVID-19 sufferers remain unwell after twelve weeks and many for much longer.' A study in the UK found that 1 in 5 people who had tested positive for COVID-19 had symptoms for five weeks or longer, with 1 in 10 experiencing symptoms for more than twelve weeks. In an Italian study of those who had been hospitalized for COVID-10, only 12.6 per cent of patients were symptom free sixty days after feeling unwell.

What does long COVID-19 involve? A common clinical case definition seems important, but at the time of writing it is a mix of different syndromes: post-intensive care syndrome, post-viral fatigue syndrome and long-term COVID-19 syndrome. The UK government's National Institute for Health and Care Excellence (NICE) guidelines state that people who have symptoms for fewer than twelve weeks have acute COVID-19, and those who have symptoms for longer than twelve weeks have long COVID. The symptom list is long, with over fifty identified in relation to long COVID.

Dr David Petrino, a rehabilitation specialist at Mount Sinai Hospital in New York, said, 'It's like every day, you reach your hand into a bucket of symptoms, throw some on the table, and say, "This is you for today."' It seems to be a mix of fatigue, shortness of breath, chest pain and tightness, joint pain, loss of appetite, recurrent fevers, chronic coughs, headaches and hair loss, as well as problems with organs such as the heart, pancreas, brain, liver, lungs and kidneys. One long COVID patient said, 'It's almost like there's inflammation in my body that's bouncing around and it can't quite get rid of it, so it just pops up and then it goes away and pops up and goes away.'

Dr John Nkengasong, Director of the Africa CDC, noted, 'This is a new virus. I'm a biologist of thirty-two years; I worked on HIV/AIDS for twenty-nine years. You thought you knew it, but you never know viruses, how they will impact.' Comparisons have been made between long COVID and ME, or chronic fatigue syndrome.

Because of the mixed symptoms, and given the focus on living versus dying, many long COVID sufferers have felt misunderstood and stigmatized. Sceptics of long COVID have portrayed it as a mental health condition or even called it overreacting and attention-seeking (of course, even if it were 'only' a mental health condition, the widespread suffering would be equally abhorrent). One long COVID patient described it best: 'I don't think people are aware of the middle ground, where it knocks you off your feet for weeks and you neither die nor have a mild case.'

Online support groups formed to help those suffering to find each other and feel understood. Stories were shared about the lack of appropriate medical investigation and the lack of sympathy and support from friends, family and doctors. One patient said, 'I've had messages saying this is all in your head or it's anxiety.' This was compounded by many individuals not even having had confirmation that they'd had COVID-19, given the lack of testing in countries like Sweden, the Netherlands, the UK and the US at the start of the pandemic. A survey of patients with ongoing COVID-19 symptoms revealed that 47.8 per cent had not been tested when they were first unwell. Of those who had been tested and received results, 46 per cent tested positive and 54 per cent tested negative.

The actual mechanism for explaining the wide variety of symptoms is still being investigated by scientists, specifically immunologists. Some think that the immune system goes into overdrive, causing an ongoing reaction. Others think that the virus lingers in the body, causing continual flare-ups. And Dr Petrino, mentioned above, thinks that the symptoms experienced resemble dysautonomia, an umbrella term for disorders that disturb the autonomic nervous system, which could explain why many patients struggle for breath when their oxygen levels are normal, or have unsteady heartbeats when they aren't feeling anxious.

The personal stories are heart-wrenching. Take Lere Fisher, forty-six, who woke up on the 20th of March 2020 feeling like he'd been 'run over by a steam train'. He had fatigue, headache, chest pain and a sore throat for two weeks. When he felt slightly better, he tried to go for a walk, but his 'lungs' were on fire and he struggled to breathe. He had to call NHS 111 – the UK's non-emergency number – and have paramedics take him home. He then developed brain fog, loss of taste and smell, and bad stomach pains. The paramedics identified these as typical COVID-19 symptoms, but the test came back negative. Six months later, in October 2020, Fisher was still not able to work and spent most of his time in bed. The unpredictability of how he will feel makes it hard for him to plan his hours, let alone the days ahead.

Some have argued that the true toll of this pandemic won't be the lives lost, but those who struggle to live each day while debilitated. These are largely younger individuals who might need to receive sick pay and benefits, and rely on physio services and health care for the coming months and years. The costs associated are not just in human suffering but also to the economy and to health care services.

Yet, because of the difficulty in getting concrete data on this condition – there is no common clinical diagnosis or definition, and a lack of testing linked to patient follow-up – it has been hard to get a clear scientific understanding of how best to manage it. None of the modelling at the start of the pandemic considered morbidity, only mortality. This is a lesson for future pandemics and a clear warning about letting a new virus spread in a population without fully understanding the longer-term effects on the body.

It is likely that Sweden and the Netherlands will have hundreds of thousands of people coming forward for support in the future; Australia and New Zealand, on the other hand, avoided this wave of serious illness. The good news is that the data emerging from Yale University indicates that vaccines seem to be helping those with long COVID to clear their symptoms, and research is ongoing to explain why this is the case. One possible explanation is that the vaccines reset the immune system so it stops overreacting, while another is that the vaccine helps the body finally fight off the virus.

## *Buying Time for a Solution*

New Zealand's and Sweden's contrasting approaches were both influenced by their scepticism that a vaccine would be available soon. There were no vaccines for any other coronaviruses, and it usually takes years, if not decades, to develop an effective one. New Zealand decided that elimination was thus the safest path, given the considerable uncertainty over what 'living with COVID' would mean, while Sweden decided that developing natural immunity was the most sustainable path. Not all countries could choose the 'elimination' path: New Zealand and Australia benefited from their geographical ability (being islands and remote) to bubble themselves off from the world. As I tweeted on the 31st of October 2020: 'Oh – to be an island in 2020'; and again on the 6th February 2021: 'To be an island in 2020 or 2021 is probably the greatest geographical gift you could have.'

While governments were taking decisions on strategy, vaccine studies were racing ahead. Hundreds of vaccine candidates started being developed in countries like the US, UK, Germany, South Korea, China and even in low-income settings such as Nigeria. Watching the vaccine race under way and the potential game-changer this would be for the world, I started to advocate for a similar 'elimination' approach to New Zealand and wrote my *Guardian* column on the 22nd of April 2020 on why 'Crunching the coronavirus curve is better than flattening it'.

I presented several different future scenarios in the article, including the best one: a safe and effective vaccine that would be available in the next eighteen months. The countries that managed to buy time for this solution with the minimal loss of life and with the fewest economic restrictions would have done a good job at managing the pandemic. I pointed at Australia, New Zealand and numerous Asian countries that had their outbreaks under control by 'crunching' the curve, and were holding and waiting for a scientific solution.

While the race for the vaccine was ongoing, as we will explore in Chapter 9, middle-income countries in the global south struggled to manage their waves of infection, especially as SARS-CoV-2 mutated into new versions, called variants, whose spread was more difficult to stop. We will now look at how three countries with large populations, land borders and limited resources tried to respond.

# 8. Variants and the Global South

Much of the last few chapters focused on the Western world, largely because what unfolded was unpredictable and preventable (unlike in poorer countries, where it was predictable and largely unpreventable) but also because of my role in advising high-income countries.

COVID-19 was far from straightforward yet it was still an easier challenge in small and rich countries. What about the populous countries with huge inequalities and large pockets of poverty? Remember that most poor people live in middle-income countries: they are home to 75 per cent of the world's population and 62 per cent of the world's poor. May 2020 saw a clear shifting of the epicentre of the pandemic from Europe to middle-income countries with large populations. Countries such as India and South Africa immediately went into lockdown, copying the efforts of China and Europe to stop spread and keep health services running.

In this chapter I turn to the challenges of three large countries that are also regionally important: Brazil, India and South Africa – not only because of what the pandemic revealed about their leadership but also because these countries saw the emergence of dangerous variants. President Jair Bolsonaro in Brazil was a particularly horrible leader, given his neglect of science, refusal to cooperate with WHO and continual denial of COVID-19 being a serious disease. Similar to President Trump in the US and Prime Minister Johnson in the UK, certain populist and right-leaning leadership styles had a massive influence on the public health response adopted. And, with lockdowns becoming the crude response mechanism to COVID-19, one question was continually raised: was the cure (lockdown) worse than the disease? We will explore all these issues in this chapter, starting with variants.

## *The Rise of Variants*

The year 2021 was one not only of vaccines, as will be discussed in the next chapter, but of variants. On the 20th of January 2020 the *New York Times* chose something I said in an interview as their quotation of the day: '2021 is going to be a cat-and-mouse game to see if we can vaccinate people quickly enough to stay ahead of the variants.'

The word 'variants' was not widely used outside of the scientific community before COVID-19, but it became a part of daily conversation in 2021, because of the serious problems variants presented to governments – and people – as the pandemic progressed. Variants were used to describe slight variations in SARS-CoV-2, often specific mutations, that made COVID-19 different.

Originally variants were identified by the place of their origin, giving rise to names such as 'the Kent variant' for Alpha and 'the Indian variant' for Delta, but this led to concerns that xenophobia would be directed towards people from those places. The US, for example, had already seen violence towards Asian communities after former President Trump called COVID-19 the 'China virus'. The scientific community preferred using numbers such as B.117 and P.1, but this was confusing to the media and general public. In the end WHO decided to use the Greek alphabet to name 'Variants of Concern' and to have one international naming system. This led many people to revisit the Greek alphabet in anticipation.

Why did these arise? Every time a virus infects a new person, it replicates, and with this replication comes the possibility of errors being introduced into the virus's genetic code. These result in mutations, or changes to the virus, which are sometimes harmless and escape notice (variants that die out), but which at other times have an evolutionary advantage, such as making the virus more transmissible or able to infect those who have already had COVID-19. In short they allow a particular variant to out-compete others and spread widely.

While there are hundreds of variants of SARS-CoV-2, WHO started using the term 'Variant of Concern' to describe those specific variants that would require governments to track them and take notice. Scientists were concerned about three aspects of a new variant. First, was it more transmissible? Alpha and Delta were variants that increased transmissibility among people, leading to faster spread through populations. $R_0$ increased. Second, was the new variant leading to worse health outcomes or affecting younger age groups? This was more difficult to detect, as it required sequencing data on each person hospitalized with COVID-19 to identify the variant that infected them; this then had to be linked to data on age, co-morbidities, gender and disease severity; and then a rapid analysis was needed to determine the variant's impact. Data from Scotland on the Alpha and Delta variants showed increased hospitalization associated with them.

And third, and most worrying, might a variant evade the protective immune response that vaccines (or infection) had provided for populations? It is unlikely that a variant would render a vaccine completely ineffective, but it could have enough immune-evasion to lower the effectiveness of vaccines, which would result in more severe 'breakthrough' infections – that is people getting ill with COVID-19 even after being fully vaccinated. For example, the Delta variant caused more breakthrough infections than the original SARS-CoV-2. Another example is the Beta variant, which was identified in South Africa: it showed significant escape from the AstraZeneca vaccine, leading to South Africa abandoning its AZ rollout and shifting to J&J.

Variants basically meant the pandemic kept changing into a more and more difficult situation. Countries had to use harder suppression mechanisms (stay-at-home orders), vaccinate to a higher threshold to stop continual transmission (moving up from 60 per cent to 98 per cent), and continuously evolve guidance on schools and mixing. As science raced to find solutions and governments got more adept at their public health responses, the virus moved the finish line for the pandemic further away – and increased the hurdles that had to be jumped in the process.

However, public messaging around variants is tricky. When I first mentioned Alpha in December 2020 as the UK headed into the second wave, you could sense the frustration from journalists that the 'end' was further ahead than expected. When Beta started to spread, I warned on Twitter that the UK's main vaccine, AstraZeneca, would struggle to manage it because of its reduced effectiveness and that we needed to screen incoming travellers for this variant. My concerns over Beta and AstraZeneca triggered a backlash, including being called an anti-vaxxer. This was despite clear messaging in my thread that getting vaccinated was important, but also important was stopping the spread of a variant that our vaccines would be less effective against. Fortunately Delta spread more rapidly than Beta, and fully vaccinated individuals were largely protected from severe disease with Delta.

## *The Tragedy of Brazil*

In late 2020 a variant of concern emerged in Brazil. This country was consistently one of the world's hotspots for COVID-19 over the course of the pandemic, reporting one of the world's largest number of cases, hospitalizations and deaths, particularly among health workers, pregnant women and the indigenous population. In 2020 mortality rates among those hospitalized with COVID-19 in Brazil were high: 59 per cent among ICU patients, and 80 per cent among those needing mechanical ventilation support. This contrasts with Germany, where deaths were 47 per cent if in ICU and 57 per cent if ventilation was required.

Thousands of Brazilians also died at home without having access to medical care. On the 29th of March 2021, 17 of 27 federal states reached adult-ICU-bed occupation rates of 90 per cent. Thousands of people had to wait for an ICU bed, and 496 patients died while on the waiting list in Brazil's wealthiest state, São Paulo, alone. As a doctor in Porto Alegre said, 'I have a lot of colleagues who, at times, stop to cry. This isn't medicine we're used to performing routinely. This is medicine adapted for a war scenario.'

On paper Brazil was well equipped to handle the epidemic. It had universal health care, experience in managing disease outbreaks such as Zika virus and an established immunization programme. Dr Miguel Lago, Executive Director of the Institute for Health Policy Studies in Rio de Janeiro, concurred: 'We could have used the health system in a smart way to do contact tracing and to inform the population. But this did not happen.' The real problem was lack of leadership and not taking COVID-19 seriously. Even towards the end of 2020 the Ministry of Health had not developed a national plan to combat the pandemic, nor had any other federal government agency. States and municipalities that tried to respond on their own received inefficient assistance and support.

The federal government also ignored international recommendations for restrictions on mixing and mask use, and refused to establish a national mandate for isolation after testing positive. Even for those who wanted to protect themselves and wear masks, the price was prohibitive. The cost of a box of fifty masks rose from R$4.50 (£0.60) in January 2020 to R$140 (£56) by March. The federal government did not regulate the market in order to prevent price gouging and did not take a proactive role to negotiate with industry to meet the demand for masks.

At the federal level the main problem was President Bolsonaro, who was opposed to any measures to restrict the spread of COVID. He consistently repeated mantras such as 'just a little flu', 'only the elderly are at risk', 'the economy must come first' and 'social isolation is an extreme measure.' After he tested positive for COVID-19 himself – much like former US President Donald Trump – he tried to use this to show that COVID-19 was nothing to fear. He had a slight fever, muscle pain and tiredness, and said that using the drug hydroxychloroquine had cured him. He claimed, 'The majority of Brazilians contract this virus and don't notice a thing. You can't just talk about the consequences of the virus that you have to worry about. Life goes on. Brazil needs to produce. You need to get the economy in gear.'

Bolsonaro insisted that lockdown would be ineffective and terrible for the economy, describing it as 'an affront, inadmissible'. Almost all states decided that they would adopt restriction measures to curb transmission in any case, but they faced strong opposition from the federal government. Bolsonaro even filed a lawsuit with the Supreme Federal

Court against three governors who had temporarily introduced lockdowns, but the cases were dismissed by the court. Bolsonaro, like Trump, kept pointing to the need to keep the economy running and claimed that lockdown caused starvation, unemployment and social chaos.

Dr Daniela Ponce, of the Universidade Estadual Paulista Júlio de Mesquita Filho in São Paulo, Brazil, wrote in the leading science journal *Nature*:

> I am appalled that politics seems to have been prioritized over the pandemic in the past few months. While Brazil's mayors and state governors implemented measures to restrict the movement of people and combat the coronavirus, Bolsonaro appeared to focus on political battles. He had already lost two health ministers who were physicians: one was fired and the other resigned. In their place, he appointed Eduardo Pazuello, a general with no medical background, as interim Health Minister. In my opinion, Bolsonaro does not want to be held responsible for the worst economic crisis in the history of Brazil and sees prioritizing the economy as his best chance for re-election.

Some believe Bolsonaro was waiting on a silver-bullet solution. For example, his government spent millions producing and distributing hydroxychloroquine pills, which have shown no benefit in studies. Even so, he still endorsed it by posting a video on Facebook to show he was taking the drug after his positive test. He has also supported treatment with ivermectin to prevent hospitalizations, which had no evidence-base to support it.

Despite federal inaction, states responded. On the 23rd of March 2020 partial lockdown was implemented in São Paulo and then soon after in Rio, two of the most populated cities. Schools, universities, restaurants, beaches, shopping centres and all non-essential businesses were closed, public transport was limited, and mass events were cancelled. Supermarkets, banks, pharmacies, pet shops, construction, manufacturing and health services continued. This partial lockdown led to empty streets and public spaces as people stayed home. Government figures show the circulation of people in São Paulo and Rio decreased by 75–80 per cent.

However, given that almost no economic support was provided to businesses and individuals during lockdown measures, adherence to measures started to wane. As Dr Daniel Villela, of the Oswaldo Cruz Foundation, said, 'At the beginning of the pandemic, there was a lot of effort to mitigate the transmission of the virus and to reduce the mobility of the population. In bits of the country we observed flattening of the curve. But with this second wave, people are becoming fatigued with restrictions.' The combination of Bolsonaro playing down the severity of COVID-19 with a lack of federal support to states, plus a population needing to work and earn in the absence of economic packages, meant that the COVID-19 wave continued.

The tragedy of Brazil was that, again, this was a country that had the resources and capacity to prevent much of the illness and many of the deaths that occurred. It was largely preventable. Yet the leadership and political will were missing. As Dr Maria Helena da Silva Bastos, a doctor, said, 'As a Brazilian who values life, I do not know what scares me more: contracting COVID-19, or Bolsonaro's government trying to belittle the disease, forcing us out of the door to confront it, even when they know that we have not done enough as a country to make it safe for us to do so.' Dr Luiz Vicente Rizzo and Dr Nelson Wolosker agreed: 'As physicians and scientists, the COVID-19 pandemic has been disheartening: first and foremost for the lives lost to the disease, and second, because of the mistreatment of science and evidence-based medicine.'

In late November 2020 a new variant started spreading in Brazil, initially referred to as P.1, and later as Gamma by WHO. It was first reported publicly on the 10th of January 2021 by Japanese health officials after four travellers were detected with it at Haneda Airport in Tokyo. Two days later researchers published a preprint (an article that has yet to be peer reviewed) describing where this new variant came from: they pointed to Manaus, the capital of Amazonas in Brazil.

Manaus was (tragically) a good place to test the 'herd immunity' hypothesis, i.e., if COVID-19 were allowed to spread in an uncontrolled way, would it stop transmitting after a certain percentage of people had been infected, say, 60 or 70 per cent? A study out of Manaus by Brazilian researchers estimated that 76 per cent of people living there had antibodies, a good marker of recent infection. While there was initial good news in the summer of 2020 that case numbers and hospitalizations were falling, in November 2020 there was a surge of infections – including in people who had previously had COVID-19, indicating that reinfection was occurring.

Genetic sequencing to better understand this pattern was done, and a new variant called Gamma was identified. Between the 15th and the 23rd of December 2020 Gamma was identified in 42 per cent of sequenced samples. The sequencing research team went back to earlier samples from March to November 2020, but Gamma was not detected. What this showed was that a new variant had recently emerged and was responsible for the surge of cases in the winter of 2020.

Gamma was bad news indeed, and an early warning of the danger that new variants would bring to the world. A preprint posted on the 25th of February 2021 found that this variant was capable of producing a ten-fold higher viral load and that it was 2.2 times more transmissible.

But, even in the most depressing of circumstances, there were glimmers of humanity and hope. In the case of Brazil it was the response of communities in *favelas* (urban slums). The percentage of the urban population living in slums in Brazil was 16.3 per cent in 2019, equating to around 13 million people, often with more than three people per room, poor housing conditions, little access to clean water and very low income. As COVID-19 spread in these *favelas*, community organizations guaranteed the distribution of food and personal hygiene items, sanitation of alleys and the dissemination of information on the virus.

Women in *favelas* bore the brunt of COVID-19, while also being the backbone of the response. Pre-pandemic research on 800 women in Maré (a *favela* complex of sixteen communities, home to roughly 140,000 people in Rio) showed that many had limited education and restricted access to formal employment; 36 per cent had experienced gender-based violence and 76 per cent said it was a common occurrence. These women not only provided the main care for children and elderly individuals but also worked informal jobs as well.

Women organized themselves and started a campaign: 'Maré says

NO to coronavirus'. This included: the distribution of food baskets and personal hygiene/cleaning items to the poorest families and the most vulnerable; a catering programme to provide 500 meals a day for the sick who were unable to leave their homes, for homeless people and for those with substance abuse problems; and a mask-sewing programme, with the aim of providing two per resident.

Brazil suffered from a lack of federal leadership and strategy: much of the suffering was arguably avoidable. In the absence of national government leadership, individual states, cities and even communities did their best to take care of themselves. The challenge of containing COVID-19 became even more difficult with the spread of Gamma, and questions started to be raised as to whether 'herd immunity' could ever be reached with this virus.

## *India: Get COVID-19 or Go Hungry?*

India was also badly hit by the COVID-19 pandemic, managing to handle a first wave before succumbing to a devastating second, driven by a new variant: Delta. And various Indian states started to diverge in their policy responses. More advanced states like Kerala were able to test widely, track down cases and isolate those with the virus. Tragically, migrant workers from states like Bihar suffered, as they were sent home from cities like Mumbai and had to make the long trek home with no money and no transportation.

The first key aspect of the Indian response was offering a wide range of tests and easy access to those tests. In January 2020 India had only one COVID-19 testing lab (the Indian Council of Medical Research's National Institute of Virology in Pune), but, by the 20th of May 2020, 555 labs had been set up, rising to over 1,000 operational labs; 80 per cent of those were public and 20 per cent private. In addition, over 2 million N95 masks (high-grade medical masks) and 1.18 million PPE kits (gloves, aprons, goggles) were distributed across Indian states free of charge.

In mid-March 2020 President Narendra Modi ordered lockdown for twenty-one days, from the 25th of March to the 14th of April.

This entailed sweeping changes to society: all services, except pharmacies, hospitals, banks and grocery shops, were closed; offices switched to work from home; all non-essential private and public transport was suspended; all research institutions and training were suspended; and all social, political, sports, entertainment, academic, cultural and religious activities were prohibited. When the first lockdown period ended, some state governments decided to extend for longer, but in 2020 there were four nationwide lockdown phases: the 25th of March to the 14th of April (21 days); the 15th of April to the 3rd of May (19 days); the 4th of May to the 17th of May (14 days); and the 18th of May to the 31st of May (14 days).

On the 26th of March 2020 the government announced a relief package of US$22.6 billion to support the poor. The plan was designed to benefit migrant workers through cash transfers and initiatives for food security. However, a significant proportion of affected groups were unable to receive the support because of narrow eligibility criteria: only those registered with the federal food welfare scheme were able to secure benefits. This plan was expanded on the 12th of May 2020, with Modi announcing a package of US$266 billion to support the economy.

Lockdown had major economic consequences in India. In the first financial quarter of 2020/21, April to June, India's GDP growth rate was −23.9 per cent, the worst ever recorded in its history. Major sectors relied upon for growth were badly affected, such as manufacturing (−39.3 per cent), mining (−23.3 per cent), construction (−50 per cent), and the tourism and hotel industry (−47 per cent).

The closure of factories and workplaces resulted in millions of migrant workers losing income, which meant their families went hungry. With no jobs, thousands walked or cycled back to their native villages, but this resulted in some being arrested for violating lockdown restrictions, and hundreds dying from exhaustion and in road traffic accidents. Some states, like Uttar Pradesh, arranged free buses to transport migrants back to their villages. Lockdown deaths were reported from starvation, suicide, exhaustion, road and rail accidents, police brutality and denial of timely medical care; and most of those deaths were among migrants and labourers.

Dr Mike Ryan, of WHO, warned the Indian government that lockdowns alone would not eliminate coronavirus, adding that India must take additional measures to prevent a second and third wave of infections. Unfortunately his advice went unheeded: Modi prematurely declared success over coronavirus, saying that 'India was in the endgame of the pandemic', right before a deadly second wave started in the spring of 2021. A *Lancet* editorial commented: 'Hospitals are overwhelmed, and health workers are exhausted and becoming infected. Social media is full of desperate people (doctors and the public) seeking medical oxygen, hospital beds and other necessities . . . The impression from the government was that India had beaten COVID-19 after several months of low case counts, despite repeated warnings of the dangers of a second wave and the emergence of new strains.'

India did indeed see a new variant arrive, one that WHO would call Delta, and that would go on to become the dominant variant across the world in the summer of 2021. How was Delta detected? Following a surge of cases in the western states of India in January 2021, enhanced whole genome sequencing and analysis of spike protein mutations was undertaken to identify whether a new variant was responsible for this increase. Sequencing analysis revealed this to be a new lineage called B.1617, and had first been identified in October 2020 in Maharashtra, a state in India. This variant went on to displace Alpha and presented a clear warning to the world about the damage it could cause. Mathematical modelling indicated that the variant's growth advantage was most likely due to a combination of increased transmissibility and immune evasion. In an analysis of vaccine breakthrough – i.e., testing positive for COVID-19 even after being fully vaccinated – in over a hundred health care workers across three centres in India, Delta was responsible for greater transmission between health care workers, in comparison with Alpha.

While Delta substantially contributed to India's catastrophic second wave, the other factors in play included increased mixing at religious and political mass gatherings, fatigue with restrictions and mixed messages from government about the 'endgame' being in sight. At the time it was unclear whether Delta affected younger people more severely. Data from the Indian government showed that around 32 per cent of patients were aged below 30 in the second wave, compared to 31 per cent in the first wave. Hospitalization in those aged 20–39 increased from 23.7 per cent to 25.5 per cent, and in those aged 0–19 from 4.2 per cent to 5.8 per cent.

Across the world, Delta was a gamechanger, creating a whole new type of pandemic. It found its way into Zero COVID-19 countries like Vietnam, Taiwan and Thailand, causing large clusters of new infections, as well as accelerating the race between vaccine and virus in high-income countries. After its emergence in India, the UK was the first high-income country to be hit with Delta, which required the UK government to push towards a greater level of vaccination uptake to try to reach the increasingly mythical 'herd immunity' threshold, as well as to ensure people received both doses of their vaccine to get maximum protection against severe illness and death.

Even in the tragedy that India became, there were pockets of success. Led by the inspirational former Kerala Health Minister, Dr K. K. Shailaja, that state managed to keep deaths low. In May 2020 Kerala reported only 524 cases of COVID-19 and 4 deaths out of a population of 35 million and a GDP per capita of only £2,200. Compare this with the UK at the time, which had more than 40,000 deaths, and the US, with 82,000.

WHO has held up Kerala as an example to follow for other countries, pointing to its prompt response and innovative approaches, which were helped by previous infectious disease management experience and investments in emergency preparedness and outbreak response, after the 2019 Kerala floods and the 2019 Nipah outbreak.

The key components of Kerala's response were: early preparation with robust leadership and the government already declaring a health emergency in the state on the 30th of January 2020; early screening of all incoming passengers at all airports and sea ports; testing and intense contact tracing; rigorous isolation with regular monitoring; COVID-19 care centres established in all districts; psychosocial support and counselling provided to patients as well as frontline workers; and, finally, a successful mass media awareness campaign, 'Break the Chain', to promote distancing, face coverings and hygiene measures.

Dr Rathan Kelkar, a senior health official in Kerala, explained:

> Kerala's model of controlling epidemic has its roots in the strong health system we built over the years. Our Hon'ble Chief Minister and Health Minister led from the front and facilitated inter-sectoral coordination as well as community participation. We focused on the strategy of trace, test and contain with extensive screening and quarantine of all the incoming travellers . . .The battle is far from over; we would continue to remain vigilant and keep innovating to make sure that Kerala lives up to the high expectations of our people and the Country.

Much of the leadership credit for Kerala's response can be attributed to Health Minister Shailaja, also referred to as the 'Coronavirus Slayer' and the 'Rockstar Health Minister'. She has been widely lauded for showing the world that disease containment could be achieved in democracies and in relatively poor settings. She was the inspiration for the film *Virus*, which described how Nipah virus was managed in Kerala. Nipah is an extremely serious infectious disease with a case fatality rate of 40–75 per cent, and there is no treatment or vaccine. Given this recent experience, Kerala was on high alert, because those in authority understood that new infectious diseases must be treated seriously and with rapid action. Waiting and watching was just not an option.

On the 20th of January 2020 Shailaja was already on high alert to the dangers of COVID-19 and began to plan, which is what she attributes her success to. Compare this with the UK and the US at the time, when almost no one in government was concerned about COVID-19. On the 23rd of January she organized a rapid response team to prepare all of the medical officers in Kerala's fourteen districts to set up operations. On the 27th of January the first cases arrived via a plane from Wuhan, but screening was already in place and the temperatures of all those coming off the plane were taken. Three were identified as being ill and later tested positive in hospital for COVID-19.

This early action held off infections for only about a month: late February and March 2020 saw an increasing number of incoming passengers carrying the virus and going undetected, which set off clusters across the state. Kerala chased all these individuals using rigorous contact tracing and testing, with results back within two days and enforced isolation. It is a remarkable story that shows what leadership, strong health infrastructure and humility can achieve in the face of an infectious disease threat. I was invited by the Chief Minister of Kerala to attend an expert advisory group planning a post-vaccine strategy, and noted how much the world could learn from their response, including the speedy rollout of vaccines when they became available in 2021.

Sadly, the legacy of India will be in the humanitarian disaster that the Delta variant inflicted, challenging even the robust public health response in Kerala. The wave in early 2021 in India would result in hospitals running out of oxygen, relatives unable to find beds for their loved ones until someone in the hospital died, and pyres being set up to burn bodies in the street because the crematoriums were full.

It was painful to watch these scenes play out on TV and social media, and feel like there was little we could do to help those suffering. I partnered with the Disaster Emergency Committee Coronavirus Appeal in Scotland to support its fundraising efforts: it was aiming to raise funds for getting PPE to health care workers on COVID-19 wards, providing logistical support to hospitals and providing vulnerable families with soap, clean water and food.

An Indian professor, Nabila Sadiq, who was only slightly older than me at thirty-eight, died of COVID-19 after not being able to find a hospital bed quickly enough to stop her lungs being damaged. She had posted on Twitter days earlier, asking for help finding an ICU bed. She died a week after testing positive, and only ten days after her mother had died of COVID-19. It made me reflect on how having access to medical services, even for moderate COVID-19, was necessary in order to survive the disease. She would have lived had she been in Scotland, like me.

## *South Africa*

The third country from which a variant of concern (Beta) emerged was South Africa. Like Brazil and India, South Africa has strong technical expertise and a more advanced public health infrastructure than much of the subcontinent. On the 15th of March 2020 President Cyril Ramaphosa activated the Disaster Management Act (2002), making a Declaration of a National State of Disaster. In a statement that day he ordered immediate travel restrictions, including a ban on foreign nationals from high-risk countries and the cancellation of visas to visitors from those countries. Three days later schools were closed.

The National Coronavirus Command Council was established on the 17th of March 2020 to coordinate the response, but it was still hesitant about lockdown measures. But, on the 23rd of March 2020, a very strict lockdown, some say the toughest in the world, was announced: it began on the 27th and lasted for three weeks. The South African National Defence Force would support this, with President Ramaphosa noting that 'The nation-wide lockdown is necessary to fundamentally disrupt the chain of transmission across society.' He added, 'Those who have resources, those who are healthy, need to assist those who are in need and who are vulnerable.' People were ordered to stay home, except for health care workers, those in the security services and key workers.

The stay-at-home lockdown was strict. South Africans were not even allowed to take dogs for a walk, but instead had to walk them around their house or flat. Enforcement was taken seriously. In the first week of lockdown alone over 2,000 people were arrested for violations.

The sale of alcohol was banned, to reduce the number of alcohol-related health incidents putting pressure on hospitals and to stop people gathering. The country had high rates of violence and drinking and driving, resulting in intentional and unintentional injuries. This policy seemed to have worked in terms of hospital pressure, with trauma admissions falling by 60–70 per cent in April and May 2020. A similar ban on tobacco accompanied this.

However, there were unintended effects, such as illicit alcohol sales growing and people resorting to making alcohol at home. And there were other consequences too. Nokwanda Zenzile, a Johannesburg student, said, 'Many people, especially women, were complaining about their safety that men will harass us due to their frustration of alcohol being banned.'

On the 13th of April 2020 Dr Salim Abdool Karim, Chair of the Ministerial Advisory Committee on COVID-19, described an eight-step national plan to manage the virus, which involved: surveillance and lab capacity; primary prevention (including closing schools and closing borders to international travel); lockdown to restrict people mixing; active case finding and testing; focusing on hotspots; building hospital capacity; expanding burial capacity; and, finally, ongoing vigilance and preparing delivery networks for vaccines whenever they arrived.

Lockdown measures, while effective at taking pressure off hospitals, exacerbated food insecurity and poverty. A household survey from the first lockdown found that roughly 3 million people had lost their jobs, especially low-income workers, with women accounting for 2 million of these. By mid-July 2020 food shortages were widespread across the country, particularly in the 'forgotten province' of rural Eastern Cape, where children in Peddie were eating wild plants to survive. One mother said, 'My children will tell you the taste of every plant in this area, as for the last three months I have been feeding them these plants.' Hunger doubled in South Africa, with 1 in 8 people reporting frequent hunger, making it the worst food security crisis experienced there in twenty-eight years.

The President was aware of this. He said in his 23rd of March 2020 speech, 'The action we are taking now will have lasting economic costs. But we are convinced that the cost of not acting now would be far greater.' A month later his government announced a stimulus package to boost the economy and support those suffering from loss of income during lockdown, but this was seen as insufficient for the scale of the crisis. Even pre-COVID-19, South Africa was challenged by a recession, unemployment and poor growth figures. COVID-19 made these much worse. By December 2020, 42.7 per cent of small businesses had closed owing to lockdown measures.

In addition, concerns grew over the length and intensity of lockdowns and the associated infringement on civil liberties. Former Finance Minister Trevor Manuel questioned the erosion of democratic freedom: 'The tragedy of the behaviour of our security services in implementing the COVID-19 national lockdown regulations is that their conduct has so often gone against the letter and spirit of our Constitution.' The use of the military to enforce nightly curfews, the restrictions on outdoor exercise and bans on commerce were all heavily challenged.

But, even with the lockdowns and the public health infrastructure, South Africa still suffered a second devastating wave. Karim himself said, 'Government interventions have slowed the viral spread, the curve has been impacted and we have now gained time', but 'As soon as the opportunity arises for this virus to spread, we are likely to see the exponential curve again. I have to tell you that, as much as we have succeeded in stemming the flow of this virus in our communities and keeping community transmission at a reasonable low level – a success no one else has achieved – I have to tell you a difficult truth. Can South Africa escape the worst of this epidemic [and is] this exponential spread avoidable? The answer, sadly, is that's very unlikely . . . we cannot escape this epidemic.'

South Africa's challenges became even greater with the detection of the new variant B.1351, or Beta, which was reported to WHO on the 18th of December 2020. South Africa fortunately had strong research capacity for detecting variants after the establishment of the Network for Genomic Surveillance in May 2020. Scientists in the genomics team at the Kwazulu-Natal Research Innovation and Sequencing Platform made the discovery, realizing that it had been dominating samples collected over the past two months, which suggested that it had transmission advantage in spreading in comparison with other variants.

On the 29th of December 2020 the EU Centre for Disease Prevention and Control warned European countries that this variant might have increased transmissibility, but that, so far, there was no evidence of a higher severity of infection. More worrying, reports from South Africa indicated that it was able to evade antibodies, i.e., it could reinfect those who had already had the original SARS-CoV-2. There was also a small study indicating that the AstraZeneca vaccine was not sufficiently effective at stopping mild to moderate disease caused by the Beta variant. As mentioned earlier in this chapter, the vaccination rollout was delayed by the government's need to stop using

AstraZeneca and instead procure J&J doses, which showed higher effectiveness. I discuss vaccines in the next chapter.

A South African bioinformatician, Dr Houriiyah Tegally, said the variant probably arose 'from immunocompromised patients, whose immune system has a harder time suppressing infections. The virus replicates many more times in these patients and . . . that's how such a case of escape (to another human) can happen.' South Africa has a large number of immunocompromised patients: it has the highest number of people living with HIV in the world (7.5 million) and one of the highest number of people living with TB, with an estimated 360,000 falling ill each year with the disease.

Fortunately, Beta did not become dominant in other countries, as it was not as transmissible as Alpha or Delta, and it receded. A combination of Delta's transmissibility and Beta's immune evasion would have been devastating for the world, both increasing the race to vaccinate as well as undermining the effectiveness of the vaccines.

## *Is the Cure Worse than the Disease?*

In populations already suffering from economic insecurity, informal labour markets and daily hunger, lockdowns caused huge harm. As one woman in an Indian village said, she had a small chance of catching and dying from COVID-19, but a one hundred per cent certainty of dying from hunger if she was not able to earn her daily wage.

One of the themes to emerge in all countries was that lockdowns had major impacts, but the exact level of devastation depended on that country's income level. Higher-income countries could borrow large sums of money to set up economic relief schemes to support businesses and individuals in maintaining some income and improving their ability to survive the pandemic. The UK did well in this regard.

By contrast, in poorer countries, families were left to fend for themselves. Food security was at the heart of many problems. If breadwinners lost their jobs, or fell ill from COVID-19, the entire family was affected. Lockdowns exacerbated existing problems, which, in many cases, pushed people into poverty, homelessness and major debt. Jean Drèze, an economist in India, said that lockdown had been 'almost a death sentence' for the underprivileged of the country. 'The policies are made or influenced by a class of people who pay little attention to the consequences for the underprivileged.'

On the flip side, just letting the virus run uncontrolled was also a death sentence for the most underprivileged. It was an impossible situation for poor countries: all they could do was to build up testing and tracing systems, support those isolating, expand vaccination coverage, encourage face coverings and try to reduce mixing between households outside of necessary economic activity.

However, the debate over whether the lockdown 'medicine' was worse than COVID-19 itself also occurred in high-income countries like Finland, Australia, Norway, Denmark and Germany. These countries, because of aggressive test/trace/isolate policies and early lockdown, were able to avoid the large death tolls that their neighbours experienced. However, the result was that their populations turned on public health advisers and scientists for 'overreacting' to the crisis. This is indeed the paradox of public health: you intervene to avoid something happening, and, when it is avoided, you are criticized for overreacting. Norway's Prime Minister even apologized to her public for this 'overreaction', as economic concerns came to the forefront of the debate.

The struggle of many during lockdown, especially children and adolescents, as described in Chapter 6, was real and terrible. Yet, in the absence of a solution to COVID-19, either a vaccine or a treatment, or a robust test/trace/isolate system and border restrictions, how else could countries have protected their health services? It is easy to complain about a problem, less so to find answers.

The larger debate was about whether governments should prioritize economies, which affect tens of millions, or limit the direct harm of COVID-19 itself, which affects tens or hundreds of thousands. With time it became clear that this dichotomy was a false one.

This was reflected in a debate I did on the 10th of February 2021 on Channel 4 with a Tory MP who shouted about how he couldn't take lockdown any more in England and how angry he was about restrictions. My immediate response was to say that he was acting like a toddler having a meltdown because he wanted a unicorn, and no one could give him a unicorn. I continued that we all knew the harm of lockdowns and the harm of restrictions. We also knew that uncontrolled spread leads to health service collapse and large excess mortality. This is what forced governments into lockdowns. In fact, those countries that responded effectively and controlled the virus, like Taiwan, South Korea, Denmark and Norway, had faster economic recovery compared with countries like Britain, Spain and Sweden, which had allowed the virus to spread.

The way to escape the national lockdown/release cycle was through developing and rolling out effective vaccines or therapeutics to reduce the number of people being hospitalized. If you want to know about the real heroes of the pandemic, it was not those complaining about lockdown measures on TV, it was health workers, and supermarket staff, and delivery drivers, and all those who kept society running while exposing themselves to daily risk of COVID-19. It was also those who complied with restrictions, including most young people, in order to protect older and vulnerable members of their communities. And of course it was ultimately the scientists who had worked for years on immune research who provided the ultimate prize of a vaccine, as we will see in the next chapter.

# 9. The Race for a Vaccine

The clock started on the 5th of January 2020 for countries to prepare and hold off the onslaught of infections to come. Even for those countries with robust responses, like South Korea and New Zealand, managing the situation was a continually changing puzzle: as soon as the solution was found for one wave of infections, another wave would arise. It could be held off the first time, or the second, perhaps, but not forever. With the arrival of new variants like Alpha and Delta, the tidal wave became stronger and faster.

As the virus continued to spread across the world, scientists raced to find solutions, knowing that every minute, every hour and every day was crucial. Ultimately science was the only sustainable exit from this pandemic: either a treatment or, even better, a vaccine. Scientific teams had been laying the groundwork to move quickly to develop a vaccine for a novel pathogen, but they needed the sequence out of China to get started. The 10th of January 2020 was a landmark date, for it was when Zhang and Holmes, as discussed in Chapter 2, published the sequence online. The starting gun was fired, and the scientific race began; research teams from the US, UK, Germany, Australia, Russia and China started running.

The finish line was clear: a safe and effective vaccine that could train our bodies to recognize SARS-CoV-2 and develop protective immunity, thus making it possible to prevent hospitalizations and deaths. To reach this, research groups pursued different ways of creating a vaccine based on their previous efforts to develop vaccines for other pathogens. The technologies fell into three broad categories: novel mRNA technologies; viral vectors; and inactivated virus. Here we'll see how each works, and when (or if) the teams of virologists would get to the finish line in the race for a COVID-19 vaccine.

## *What is a Vaccine?*

Given all the misinformation around on vaccines, including that they're being used to microchip people (yes, this theory has been shared millions of times around the world), I would like to spend a moment getting to the heart of what a vaccine is, and how transformative they have been in saving lives over the past century.

Our immune system works by identifying a foreign agent, such as a virus or bacteria, inside the body and mounting a response to destroy it. It creates large proteins called antibodies. Antibodies track down foreign agents and mark them to be destroyed. Antibodies are specific to a pathogen and remain in the body after the infection has gone. This 'memory' of a disease means the body is continually primed for that infection and can quickly destroy it before someone gets sick and feels unwell.

Vaccination is the safest and most common way for us to gain immunity against a bacteria or virus that we might encounter in the future. While there are different techniques that can be used to create a vaccine, the basic idea is to have the body encounter a harmless form of the virus or bacteria (usually through an injection into the arm), which then trains the immune system to fight it through the creation of antibodies. These antibodies can quickly identify the real form of the virus or bacteria and destroy it before we become extremely ill.

Vaccines have been revolutionary at protecting all of us against several life-threatening diseases that used to be common, such as polio, diphtheria, whooping cough and smallpox, as well as tetanus and measles. Historically, vaccines have taken years, if not decades, to develop. We still don't have a vaccine against HIV or dengue. And a vaccine for malaria was approved only recently, in 2021.

Considerable uncertainty clouded whether a vaccine could be developed for COVID-19, whether enough doses could be manufactured for the world, and whether people would consent to get jabbed. Professor Michael Osterholm, Director of the Center for Infectious Disease Research and Policy at the University of Minnesota, said in an interview at the start of the pandemic: 'I think the goal of eighteen months is one that will be very, very difficult to achieve. But it may just be our moonshot.'

I was more optimistic. On the 14th of May 2020 I was on the BBC show *Question Time*, where lockdown measures were being hotly debated. One of the business voices on the panel said that the lockdown was pointless because there was no solution to COVID-19. I argued back that science had found solutions to most health challenges, and that the early results from vaccine efforts were looking promising. He responded that HIV still didn't have a vaccine. I pushed back that scientists had developed antiretrovirals that let people live decades longer while being HIV positive, indicating that some sort of solution would be coming, and therefore buying time made sense at that moment.

Early on WHO set the baseline for what would be considered an 'effective' COVID-19 vaccine: it had to be at least 50 per cent effective at stopping moderate to severe illness and death. A secondary goal was to stop vaccinated individuals being infectious and transmitting, which could help towards the build-up of population (or herd) immunity. Vaccinating to a certain herd immunity threshold is how we have managed to control and eliminate diseases such as measles, mumps and rubella, which are now described as 'vaccine-preventable' diseases. This means that they stop circulating because they cannot find susceptible hosts to jump to; and those who cannot be vaccinated (because they are immunocompromised, for example) are protected from infection by the rest of the herd's collective immunity.

## *Political Pressures*

These scientific efforts did not occur in a vacuum. Although scientific teams around the world relied on each other and collaborated in this race against the virus, national government leaders wanted to show that they would be the ones to win the race for a vaccine. Never one to be modest, former US President Donald Trump announced that the US was pursuing 'the most aggressive vaccine project in history.

There's never been a vaccine project anywhere in history like this.' French President Emmanuel Macron got dragged into this competition, citing the 'genius of Louis Pasteur' and hailing France as a 'great vaccine country'. But the ability of politicians to speed up the vaccine process was limited in the US, Britain and Europe, where there are strict national processes for how vaccines are researched, tested for safety and effectiveness, and approved for emergency use, as we'll see.

By contrast, Russia raced ahead with approving and using a vaccine in the summer of 2020. On the 11th of August 2020 Vladimir Putin, who was isolating at his residence in a forest outside Moscow, announced in a video conference that he had approved the use of Sputnik V – making Russia the first to have an approved COVID-19 vaccine. The name reflects the first satellite, launched by the Soviet Union in 1957, and was evocative of the superpower rivalry in space during the Cold War.

Several months later in 2020, Liu Jingzhen, Chairman of Sinopharm, boasted about China's scientific prowess: 'almost 1 million Chinese have been given an experimental COVID-19 vaccine developed by the state-owned Sinopharm under the government's emergency use scheme. Until now all our progress, from research to clinical trials to production and emergency use, we have been leading the world.'

Russia and China also knew that developing economies were desperate for vaccines, and they could actively work to fill this 'vaccine vacuum' by sending millions of their home-grown vaccines abroad. This was medical diplomacy, providing vaccines in exchange for political or economic favours.

The political pressure to be the 'first' had to be managed by scientists, who wanted to take the necessary steps and precautions to ensure research was conducted appropriately and transparently. An added difficulty for scientists was that much of their vaccine research efforts were funded by national authorities, which waited anxiously for news on progress each day.

The race to get a vaccine came into conflict with public concerns over safety, and whether the science was being conducted properly. In an in-depth podcast interview I was asked by the Channel 4 presenter Krishnan Guru-Murthy in late December 2020 why I was willing to take the just-approved Pfizer vaccine. I responded that there is a clear scientific process to ensure vaccines are safe and effective in the UK and went on to explain the regulatory process the Pfizer vaccine had been through.

First, lab and animal studies (preclinical) are conducted to see whether clinical trials are merited: is it even worth moving forward with a vaccine? Next, Phase 1 trials assess the safety of a vaccine in a small number of individuals (dozens) and how the body's immune system responds. Phase 2 involves testing on a larger group (hundreds) of people in order to identify optimal doses and timing. Phase 3 then tests the safety and efficacy of a vaccine in a large group of people (thousands), often in multiple locations, by comparing what happens in groups that receive the vaccine (intervention arm) versus groups that think they had the vaccine but just received a placebo instead. Even after a vaccine is approved for emergency use, data is continually collected to monitor for any adverse reactions, as well as for the effectiveness of the vaccine in real-world settings.

Usually the different trial phases are run sequentially, often with time between them for preparing protocols (i.e., how a study will be run) and funding applications for each step, then seeking the required national ethical and regulatory approvals. But to speed up development during a pandemic, the phases were sometimes run concurrently: 1 and 2 overlapped, as did 2 and 3. No steps were skipped but combining meant that the next stage of the trial could be started as soon as enough data was available from the previous phase and it had been reviewed by a national authority like the US independent Data and Safety Monitoring Board.

## *Pfizer/BioNTech*

One of the first vaccines to receive emergency approval, and then have widespread use, was the mRNA vaccine Pfizer/BioNTech. The Messenger RNA (mRNA) technology had not been used before for an approved vaccine. These vaccines work by giving instructions to our cells to make a 'spike protein' that resembles the one found on the surface of SARS-CoV-2. The platform works by injecting mRNA into immune cells, which then use these instructions to make a protein piece. After the protein piece is made, the cells break down the instructions and get rid of them. The cells then display this protein on their surfaces.

Our immune system recognizes that this protein doesn't belong there and begins to build an immune response and make antibodies, as it would do if one had really been exposed to the virus. At the end of the process our immune system has learnt how to protect against the spike protein on SARS-CoV-2 and can mount a strong immune response when faced with the real virus. It recognizes it and knows how to fight it, and our body avoids the severe symptoms of COVID-19.

Messenger RNA was first suggested in 1961 by two French scientists, Jacques Monod and François Jacob. However, the discovery of mRNA was not followed by much interest for more than forty years. Why did the mRNA targeted vaccine research result in such little progress for decades? Professor Drew Weissman, of the University of Pennsylvania, whose research was instrumental in the development of the Pfizer/BioNTech vaccine, has suggested, 'My guess is that people tried it and it just failed. It was too inflammatory, too difficult to work with, and they just gave up.' Since 1966 an mRNA vaccine for RSV (respiratory syncytial virus), which is an acute respiratory infection causing wheezing and hospitalizing millions of infants each year, has been a scientific struggle. A clinical trial of an early candidate failed by making the illness even worse in volunteers and caused the death of two babies.

In the 1980s the idea of mRNA vaccines was picked up by Professor Katalin Karikó, known as the 'mRNA hustler' by her colleagues. She had been researching RNA technologies in immunology and therapies at the Szeged Biological Research Centre in her native Hungary. She initially applied this technology to HIV, with American public health expert Fauci noting, 'She was, in a positive sense, kind of obsessed with the concept of Messenger RNA.'

In a series of articles beginning in the journal *Immunity* in 2005,

Karikó and Weissman published breakthrough findings, describing how chemically modifying mRNA tricked cells into thinking the molecules had been made inside the human body rather than in a lab. Karikó and Weissman were good collaborators, with her remarking, 'I am the enthusiastic, noisy one, and he's the quiet, thinking one. But we have different knowledge, and we educated each other. In the middle of the night, three o'clock, I'd send him something and he'd instantly respond. You felt that the other person was there, shoulder to shoulder.'

This mRNA technology would form the basis of the Pfizer/ BioNTech vaccine. Professor Uğur Şahin is the CEO of BioNTech (a German biotechnology company), and in early January 2020 he became convinced that SARS-CoV-2 would spread from China and become a global pandemic. He urged his researchers to begin designing an mRNA vaccine, which would target the spike protein on SARS-CoV-2. By the 25th of February 2020 he had led BioNTech in designing ten possible candidates. Şahin's partner is Dr Özlem Türeci, Chief Medical Officer of BioNTech, both Germans with migrant backgrounds.

When asked whether they could become role models for a generation of diverse Germans, Şahin said, 'I am not sure I really want that. I think we need a global vision that gives everyone an equal chance. Intelligence is equally distributed across all ethnicities, that's what all the studies show. As a society we have to ask ourselves how we can give everyone a chance to contribute to society. I am an accidental example of someone with a migration background. I could have equally been German or Spanish.'

In March 2020 BioNTech partnered with Pfizer, a major pharmaceutical company, to scale up the research, launching a Phase 1/2 clinical trial in May on two versions of the vaccine. The Head of Vaccine Research and Development at Pfizer, Kathrin Jansen, told Şahin, 'This [COVID] is a disaster, and it's getting worse. Happy to work with you.'

These vaccines had to be kept extremely cold in freezer storage – at −25°C to −15°C – easier said than done on a global scale. To distribute them around the world required creating a 'cold-chain' and containers that could be mobile enough to be deployed where needed. Eventually a suitcase-sized box was trialled that could hold around 5,000 doses; and, to see if it could work for moving vaccines, the company ferried containers on hundreds of trips around the world, taking in Dubai, Africa and even the doorsteps of Pfizer employees.

In May 2020 Phase 1 trials on volunteers showed that the first vaccine (partial spike) produced antibodies against SARS-CoV-2 and T cells (another part of the immune system that attacks foreign particles) that responded well to the virus. Excited by these preliminary positive findings, researchers emailed Pfizer on the 7th of June 2020 with the results of this early-stage trial. The reply: 'Really, really encouraging. When can I see the next data?'

On the 4th of July 2020 the CEO of Pfizer, Albert Bourla, was shown the successful cooling containers for his feedback; however, he felt they weren't appropriate for real-world delivery. He thought 5,000 doses was too large a batch for most places, and asked them to look at 25- or 125-dose packages instead. He remarked, 'What in the hell is CVS going to do with 5,000 doses?' Pharmacies run by the American CVS Health Corporation and doctors' offices would need to receive fewer batch deliveries, given their patient pool; otherwise vaccines would go unused and expire before getting injected into people's arms.

On the 23rd of July 2020 results from the second candidate (full spike) arrived, and, like the first, it generated a strong immune response. Even better, subjects reported fewer side effects like fevers and chills, so the next day Pfizer and BioNTech agreed to trial the second full-spike vaccine candidate. A senior Pfizer executive noted, 'I started to relax a little bit, which I haven't really done since the beginning of all of this. It is just a major, major, major decision to make.'

Observing these early positive trial results, the scientific community, including myself, began to realize that there was a good chance a vaccine could be rolled out within months. The challenge then was how to buy time until the end of 2020 – and prevent a devastating winter wave before the bulk of adults could be vaccinated. My proposed solution in Scotland, seeing the negligible case numbers, was to push for maximum suppression (keeping cases as low as possible) through restricting international travel and the reseeding of infection. As discussed earlier, this was difficult to implement both practically (geographically) and politically in a UK context.

On the 27th of July 2020 Pfizer and BioNTech announced the launch of Phase 2/3 trials but were pushed by US and European regulators to include more subjects: 30,000 individuals rather than 8,000. A few months later the trial was expanded to 44,000 participants. Even then, on the 27th of October 2020, the CEO of Pfizer felt that the volunteers in the trial had not been exposed to enough COVID-19 to determine if the vaccines worked. One of his executives noted the difficulty in trying to recruit participants in viral hotspots, because COVID-19 surges kept changing across the world before they could recruit participants. 'It is like a puzzle, and the puzzle is changing.'

On the 9th of November 2020 Pfizer and BioNTech released their preliminary analysis of the first ninety-four cases, with an astounding efficacy rate of 90 per cent. Elation filled my heart reading the trial data: the gamechanger had arrived. This data resulted in emergency authorization by the UK on the 2nd of December 2020, and by the US FDA on the 11th of December 2020. The FDA released its independent analysis of the clinical trials, noting an efficacy rate that was even higher, at 95 per cent, and no serious side effects – just a day or two of fatigue, fever and muscle aches.

My messaging in November and December 2020 changed towards asking people to hold off mixing and socializing over the holiday period, especially with elderly and vulnerable individuals, until they could be vaccinated. Which would happen within weeks given the UK's access to supply. I gave interviews on STV and wrote articles in the *Guardian* explaining that pushing Christmas back by a few months would be much safer, given we were moving into a post-vaccine era. Why risk mixing this Christmas, and lose future Christmases, when we were so close to the finish line?

These results were way better than any scientist could have hoped for. Şahin remarked, 'The vaccine hinders COVID-19 from gaining access to our cells. But even if the virus manages to find a way in, then the T cells bash it over the head and eliminate it. We have trained the immune system very well to perfect these two defensive moves. We now know that the virus can't defend itself against these mechanisms. We only have indirect clues so far [regarding the duration of immunity]. Studies of COVID-19 have shown that those with a strong immune response still have that response after six months. I could imagine we could be safe for at least a year.' This prediction was based on looking at the immunity a person gets from other circulating coronaviruses after having recovered, and knowledge that a vaccine would induce even longer-lasting protection.

Three months later, on the 31st of March 2021, Pfizer/BioNTech announced that the vaccine was also highly effective and safe in adolescents aged 12–15, which would become important for returning secondary schools in the autumn of 2021. Şahin noted that 'The initial results we have seen in the adolescent studies suggest that children are particularly well protected by vaccination . . . It is very important to enable them to get back to everyday school life and to meet friends and family while protecting them and their loved ones.'

How did the team manage to get a vaccine out so quickly? The running of trial phases concurrently has already been discussed. The combination of top science from BioNTech, which knew how to develop mRNA vaccines, with Pfizer's expertise in bringing products to the mass market (given their track record with other vaccines), together with quick action from regulatory authorities, helped to speed up the development process to ten months rather than years. Şahin said, 'There was practically no waiting time. Imagine you want to get from one end of London to the next and there are traffic jams everywhere. You would need half a day. For our project, the streets were empty.' The CEO of Pfizer remarked, 'I'm a true believer that people, they don't really know their limits. And, usually, they have the tendency to underestimate what they can produce.'

Speed of delivery and manufacturing huge quantities were prioritized at this point. On the 19th of February 2021 US President Joe Biden arrived to tour a plant and thank the workers: 'I want the American people to understand the extraordinary, extraordinary work that's been done to undertake the most difficult operational challenge this nation has ever faced. The whole process takes teamwork, precision, and "round-the-clock focus".'

Supply became the key challenge, with countries pushing for more and more doses. The CEO of Pfizer challenged his team to increase commercial production at least ten-fold, asking, 'Why can't we make more and why can't we make it sooner?'; the President of Pfizer Global Supply replied: 'What we're doing already is a miracle. You're asking for too much.'

The problems that Pfizer was having were a glimpse into the main bottleneck of vaccine supply, which we'll see more of later. Yet what was already clear was that one company alone couldn't vaccinate the world. The demand was enormous: doses would be needed for almost every person on the planet. Rather than competition, it was an all-hands-on-deck approach. Closely tracking alongside Pfizer in its efforts was the company Moderna, which produced the next vaccine to receive emergency approval.

## *Moderna*

The story of Moderna started in 2008, when Dr Jason McLellan, a virologist at the University of Texas, joined the Vaccine Research Center at the National Institutes of Health (NIH) in Bethesda, Maryland, as an early-career researcher. He was working with Dr Peter Kwong on creating an AIDS vaccine. HIV mutates rapidly, so the researchers tried several different ways to develop vaccines, but ultimately failed to create one that elicited an immune response. McLellan said, 'You didn't know whether it was because the virus was too good or the ideas were bad.'

On the second floor of the same building was Dr Barney Graham, who had been working for over twenty years on RSV (respiratory syncytial virus). Both RSV and SARS-CoV-2 feature genomes made of RNA. Although they are only distantly related in the viral family tree, they share this physical trait.

In what the pair now refer to as a happy accident, Graham and

McLellan ended up working close to each other when Kwong's fourth-floor lab became too crowded for McLellan, and he moved down to the second floor. Graham said, 'It didn't take long for him to come to me and say, "I'd like to work on something other than HIV."'

Past unsuccessful attempts to neutralize RSV with a vaccine focused on the virus's F protein, but this protein is always changing shape, 'like a Transformer toy', said Graham. It can look one way before the RSV virus infects and enters a cell, and another way after the virus multiplies and escapes. These two identities are known as the 'pre-fusion' and 'post-fusion' states, and all vaccine attempts until this point focused on the latter. Part of the problem with the focus on the 'post-fusion' state is that it produces potentially harmful antibodies, such as those that can cause an overreaction of the immune system and serious illness (referred to as immune enhancement).

The pre-fusion form is extremely unstable and can irreversibly and spontaneously snap to its other state, so Graham and McLellan suggested that perhaps an RSV vaccine might be more successful if they could lock in the pre-fusion state. At this point no one knew what the pre-fusion state looked like. McLellan used X-rays to capture an image of the pre-fusion protein for the first time, and with this image they could build a new molecule themselves, a process called bioengineering.

When they tested this new molecule, the immune system reacted well, with almost fifty times the neutralizing power against the RSV virus of anything they had tested before. This was a major breakthrough and won McClellan the runner-up prize in *Science* magazine's 2013 Breakthrough of the Year award. This work built on Karikó and Weissman's earlier research on mRNA.

The next virus they attempted to work on was MERS, trying to design pre-fusion structures of the MERS-CoV spike antigen. Graham's lab partnered with Moderna, a biotech company, to design an experimental mRNA vaccine for MERS, which was their second collaboration with Moderna after a separate project on Zika virus a year earlier.

On the 4th of January 2020 McLellan was at Utah's Park City Mountain Resort during a snowboarding holiday, when Graham called him with an urgent request. Graham briefed him about the new cluster of cases out of Wuhan, which looked like part of the betacoronavirus family, and asked 'Are you ready to get back in the saddle?' McLellan then messaged some of his team on WhatsApp: 'Barney [Graham] is going to try and get the coronavirus sequence out of Wuhan, China. He wants to rush a structure and vaccine. You game?' Years of their work enabled them to mimic precisely the spike protein on SARS-CoV-2, which – as we will see – would then form the basis of other vaccines, like Novavax and J&J. Even Pfizer reached out about their work.

Two days later, on the 6th of January 2020, Graham contacted Moderna's CEO, Stéphane Bancel, who was on holiday in France. Bancel urged Graham to push ahead with a vaccine, saying, 'If it's a SARS-like coronavirus, we know what to do. This would be a great time to run the drill for how quickly you can have a scalable vaccine.' Bancel recounted, 'I remember pulling out my iPad and asking, "Where is Wuhan?" And then I went on Google Flights and I saw there [had been] direct flights to all the capitals in Asia, to the West Coast cities in the US, to all the capitals in Europe. Everything fell into place and I thought, "Shit, this is already everywhere – this is going to be a pandemic like 1918."'

But they needed the sequence out of China, and the Chinese government wasn't sharing much information. Fauci knew they could accelerate quickly with a vaccine but that they needed that sequence. 'I think the perspective that I had was seeing the link from the years of fundamental basic and clinical research [vaccine development research in labs] that got us to that point, in the first week of January when we knew that all we needed was the sequence of the new coronavirus.'

On the 10th of January 2020, as discussed earlier, Holmes and Zhang shared the genetic sequence of the 'Wuhan virus' online, and Graham's team designed the Moderna vaccine in just one weekend. Years of work had prepared them for this moment.

By the 27th of January 2020 they had started trials in animals. By the 19th of February the good news began when a two-week study in mice found that the blood of vaccinated mice was producing antibodies to fight SARS-CoV-2. A senior researcher involved with the study noted, 'When we got the first results from the mice, and they had a great antibody response, it was so gratifying.' Bancel also said, 'I realized that if it worked, it would change medicine forever.'

On the 16th of March 2020, at 8 a.m. at the Kaiser Permanente Washington Health Research Institute in Seattle, Jennifer Haller, Operations Manager for a small tech start-up, became the first volunteer to receive the Moderna vaccine as a jab into her upper arm. This was only sixty-six days after the genetic sequence had been posted; the Phase 1 trial had been approved within days. When Haller showed up for her appointment, she didn't know that she would be the first human to receive it, nor that she would be widely recognized for her courage. In an interview a year later, she reflected that she took a deep breath, reminded herself that this was just like another flu shot, and that it was a chance to help to find a solution. She said, 'Life has handed me opportunities to see how strong I can be and this was one of those opportunities that I've been gifted. I am hopeful that my experience can help encourage others to see that within themselves.'

Haller was part of a larger plan to give different doses to four groups of fifteen adults aged 18 to 65. Two jabs would be given twenty-eight days apart, with months of follow-up to determine side effects and optimum dosage, and to screen for the level of antibodies that was generated. To help support the efforts, on the 1st of April 2020 country singer Dolly Parton announced a major donation to the Moderna trial through Vanderbilt University, and fifteen days later the US Department of Health and Human Services pledged $483 million and then $472 million to pay for the trial.

On the 9th of May 2020 Phase 1 results came in. The blood of eight volunteers, including Haller's, had been examined. When Vanderbilt researchers took antibodies from the volunteers' blood and tested them on infected cells in the lab, the virus stopped replicating. Even better, there were no major safety concerns. On the 18th of May 2020 Moderna released these early findings in a press release, causing their stock to rise 250 per cent from its value in December 2019.

Within days Moderna announced plans for a 600-volunteer Phase 2 trial to establish optimum dosage; it had also started to design a

Phase 3 trial with the most promising dose from Phase 1 involving 30,000 participants, to start on the 27th of August 2020. Bancel said, 'It's the first Phase 3 of a COVID-19 vaccine in the US; it's the first Phase 3 of an mRNA vaccine ever and it's the company's first Phase 3 as well.'

One of the challenges Moderna faced was in recruiting volunteers and ensuring there was racial and ethnic diversity in the trials. Moncef Slaoui, the Head of Operation Warp Speed (the US federal programme to fund and develop COVID-19 vaccines), noticed the company wasn't recruiting enough Black Americans. He said, 'We were shouting at each other on the phone – shouting in a respectful but very angry, very stressed way. There was a very big tension because we need to recruit very quickly and we need to recruit the right people. Frankly developing a vaccine not used in the population or in a fraction of the population is the same as having no vaccine. A vaccine on the shelf is absolutely useless.' The pause to recruit additional participants from non-white backgrounds lasted two weeks and resulted in a 50 per cent increase in Black people in the trial.

The decision to prioritize diversity and inclusion in their trials resulted in Moderna losing front-runner status in the US to Pfizer. While this was a setback in the race to get the vaccine out first, it was a win to ensure that health inequities were addressed and would help to overcome vaccine hesitancy among those in Black and non-white communities. In a survey of Black Americans released in March 2021 more than one third of all study participants agreed or strongly agreed that they would not get a COVID-19 vaccine. Reasons given were mistrust of government motivations for pushing vaccination and previous history of being treated badly by the health care system. But having vaccine companies being transparent about tackling diversity issues seriously seemed to help some Black Americans get over their hesitation. For example, Kevin Lloyd, a 57-year-old Black American, said in April 2021 after getting his first jab, 'What changed? . . . transparency of the government and the scientists and companies responsible for developing the vaccine.'

Seeing these positive results, Donald Trump questioned Bancel, looking him in the eye and asking, 'Can I count on you? Can you get it done?' 'Yes, sir,' was Bancel's response. Trump was excited, given his upcoming re-election, 'So we're going to have a vaccine very soon, maybe even before a special date. You know what date I'm talking about.'

The US election was held on 3 November 2020 with no vaccine for Trump to announce before it. Thirteen days later, on the 16th of November 2020, Moderna announced early results of the trial, with data showing it was over 90 per cent effective. One of the trial co-leads, Lawrence Corey, celebrated: 'That efficacy is just beautiful, and there's no question about the veracity of it either.' Not only were the trial results better than they could have hoped for, the trial leads had confidence in the findings.

Moderna moved quickly from a focus on adults to testing whether their vaccine would also work in younger age groups. On the 2nd of December 2020 Moderna registered a trial to test the vaccine on young people aged twelve to eighteen, and on the 18th of December, Moderna became the second vaccine authorized for emergency use by the FDA.

While the development of Moderna and Pfizer/BioNTech was excellent news for richer countries, they weren't developed with the rest of the world in mind. In the US, Pfizer cost $20 per dose, Moderna between $25 and $37 per dose, and J&J $10 per dose. The prices of these vaccines were far beyond what most low- and middle-income countries could pay, plus cold-chain issues made it impractical to deliver in low-income contexts. And, given their links to the US and EU governments respectively, early supply would go to these richer markets. Luckily, the UK was also racing ahead with a vaccine, but with the goal of creating an affordable one for the entire world.

## *AstraZeneca/Oxford*

Oxford University had long experience in vaccine development and started to race towards a vaccine using a vector-based technology. Oxford's technology was known as 'plug and play', because it took a common cold adenovirus that infects chimpanzees (ChAdOx1), modified it to avoid replication and then further engineered it to become the building block of a nearly universal vaccine.

ChAdOx1 was chosen, as it can generate a strong immune response, but, since it cannot replicate, it cannot cause infection. It had already been used safely in vaccine trials against TB, influenza and MERS, featuring thousands of subjects; and it was effective in Nipah trials with hamsters, Rift Valley fever trials with sheep, goats and cattle, and prostate cancer trials. It even reached Phase I trials for human malaria.

The team's ongoing research interest into ChAdOx1 was, in part, preparation for the so-called Disease X, one of nine priority diseases that WHO had identified as posing a major public health risk. Professor Sarah Gilbert led this research, and, after reading about the new illness in Wuhan on New Year's Day 2020, prepared for when the sequencing would be released. She said, 'We'd been planning for Disease X, we'd been waiting for Disease X, and I thought this could be it' and 'We were planning how [we could] go really quickly to have a vaccine in someone in the shortest possible time. We hadn't got the plan finished, but we did do pretty well.'

On the 10th of January 2020 the sequencing was released by Zhang and Holmes, and Gilbert's team started to use this genetic data to create a vaccine: 'The genetic code was a bit like the recipe we needed for creating a vaccine.' In March 2020 Gilbert's team made the first batch of vaccines, testing it on pigs at the Pirbright Institute, a vet facility in Surrey, and the results were promising. On the 22nd of March 2020 Oxford became the fourth team to deliver results from a large-scale Phase 2 trial.

Drumming up support for her team's work was a major challenge for Gilbert: 'Getting money was my main activity until April, just trying to persuade people to fund it now.' Gilbert, and her Oxford collaborator Professor Andrew Pollard, knew that they would have to have a commercial partner in order to begin production immediately and provide enough for the whole world, but that commercial partner would also have to be willing to do this at an affordable price. They had fruitless discussions with several manufacturers, until the 30th of April 2020, when AstraZeneca, a large pharmaceutical company, agreed to their terms for a partnership.

AstraZeneca's CEO, Pascal Soriot, says there was never any question of making money out of a vaccine when COVID-19 was taking lives: 'It only works if we do it at no profit, so we can bring it to as many people as possible around the world.' The profit would be made for both AZ and Oxford once the pandemic was over, if annual vaccination was needed. As reported in the *Telegraph* on the 7th of October 2020, Dr Louise Richardson, the Vice-Chancellor of Oxford University, did not want to repeat Oxford's mistake in the 1940s, when it failed to make money after scientists at the university worked out how to turn penicillin, discovered by Alexander Fleming, into a drug (antibiotics) that could cure bacterial infections. Her view resulted in Oxford's being part of the profit-making deal for future annual COVID-19 vaccination.

In July 2020 preliminary results from the human trial showed that the vaccine produced the desired immune response. Gilbert was optimistic about creating an effective vaccine, putting the chances of success at 80 per cent and surprising many in the field with her confidence that it would be developed and rolled out within months. When I heard her speak about this prospect, it became clear to me that there wouldn't be just one or two vaccines but multiple safe and effective ones. The challenge was for governments to buy time until the end of 2020; there was now a reason to keep holding out and suppressing COVID-19.

All looked to be on track – until the 6th of October 2020, when Phase 3 trials of the Oxford vaccine were halted after a participant developed a rare neurological condition. The participant was a UK woman who experienced symptoms consistent with a serious spinal inflammatory disorder called transverse myelitis. On examination, that participant was diagnosed with multiple sclerosis deemed to be unrelated to the vaccine. The trial resumed several days later.

Dr Francis Collins, of the NIH, reflected, 'To have a clinical hold, as has been placed on AstraZeneca as of yesterday because of a single serious adverse event, is not at all unprecedented. This certainly happens in any large-scale trial where you have tens of thousands of people invested in taking part; some of them may get ill and you always have to try to figure out: is it because of the vaccine, or were they going to get that illness [anyway]?'

Pfizer and Moderna announced their positive trial results in November 2020 and the Oxford/AZ team found this reassuring. One of the Oxford team members told *The Times*, 'That was monumental because it gave us a lot of hope that we would be able to get a vaccine and there would be a lot of vaccines, a family of vaccines. It was never really a race to produce a vaccine but a race against the virus.'

On the 23rd of November 2020 Oxford/AZ announced their trial results: effectiveness was around 70 per cent. However, they had an added advantage over Pfizer and Moderna in that their vaccine could be stored at ordinary fridge temperatures, making distribution worldwide simpler. On the basis of this data, on the 30th of December 2020 the UK approved the Oxford/AZ vaccine for use, with the first doses rolled out in Britain on the 4th of January 2021.

But the vaccine was not without controversy. First, several countries in Europe and elsewhere halted their use of the vaccine after a small number of people who received it developed life-threatening blood clots. This resulted in the UK changing its guidance: only those over forty would be offered the vaccine. Denmark permanently stopped the use of AZ. At the time of writing this book the US hasn't approved the vaccine, and the doses it had ordered were sent to Canada and Mexico.

Second, the European Commission was angered that AstraZeneca didn't honour its contractual supply commitments to them and prepared a legal case against the company for failing to deliver enough vaccines. Under the contract agreed, the company should have made 'best reasonable efforts' to deliver 300 million doses from December to June 2021; instead, it said on the 12th of March 2021 that it could deliver only one third of those. All twenty-seven member states of the EU backed the lawsuit, although it hampered their vaccine rollout and caused them to pivot to using Pfizer/BioNTech supplies instead.

European leaders also made pejorative comments, with French President Emmanuel Macron saying that the Oxford vaccine was 'quasi-effective' for people over sixty-five. This was firmly debunked by WHO, but created mistrust among European citizens, who did not then want to take the vaccine. Finally, as mentioned previously, results from a very small trial in South Africa found that the Oxford/AZ vaccine provided only 'minimal protection' against their new Beta variant, and so South Africa halted plans for a rollout of 1 million doses.

As reflected by the standing ovation given on Wimbledon Centre Court to Sarah Gilbert in June 2021, the Oxford/AZ vaccine saved tens of thousands of lives in Britain in 2021, and helped to weaken the link between cases and severe illness. It was also deliberately designed to be 'the world's vaccine': the partnership between Oxford and AstraZeneca specified that low-income countries would receive the vaccine on a not-for-profit basis, at roughly $3 per dose. In the end, as with most medical products, different countries paid different prices for the vaccines based on what they negotiated. In the US, AstraZeneca cost $4 per dose, with Pfizer costing $20 per dose, Moderna between $25 and $37 per dose, and J&J costing $10 per dose.

## *J&J/Janssen*

One of the challenges of all three of the above vaccines is that they required two doses, meaning people would have to return several weeks or months later. J&J /Janssen went down the route of a single-dose injection, which could last up to two years frozen and up to three months refrigerated before it expired. This is in contrast with the Pfizer and AstraZeneca vacines, which expire thirty days after thawing.

In Cambridge, Massachusetts, on the 10th of January 2020, team members from the Center for Virology and Vaccine Research (CVVR) at the Harvard Medical School came to a similar conclusion as teams in other parts of the world: there was a dire need to develop a vaccine because of the pandemic potential of SARS-CoV-2. Professor Dan Barouch, the Director of the CVVR, said, 'It had all the hallmarks of a virus that we thought might have pandemic potential.' Like the Oxford group, Harvard scientists had been working on viral vectors for several years, developing an adenovirus vector called

Ad26 that had been modified to create vaccines, and they subsequently conducted clinical trials for HIV, TB and Zika.

The key feature of Ad26 was that several genes had been removed, so that it could still insert itself into human cells but couldn't replicate and make a person ill. Pfizer, Moderna, AZ and J&J all followed the same principle of injecting genetic information into either a fatty envelope (mRNA) or into some kind of vector (like an adenovirus) that could carry it inside the body. However, J&J used DNA, which is more stable (in comparison with mRNA), resulting in the need for only one dose, instead of two jabs as with the other vaccines.

Dr Johan Van Hoof, Head of Vaccine Development at Janssen, had been pursuing this idea in parallel with Barouch, given their close collaboration for over a decade. Barouch called him and said, 'Johan, this is looking bad. I think we need to make a vaccine.' With Van Hoof replying, 'Absolutely. We do.'

Both teams started by creating a dozen or so DNA strands and injecting them directly into mice, assessing which triggered the strongest immune response. Based on this, they created seven vaccine candidates, putting the DNA into the Ad26 vector and testing it on rhesus monkeys.

The decision to test multiple vaccines in animals, in order to find the best candidate to take forward, delayed J&J's progress in comparison with the Moderna and AZ teams, both of which quickly moved to choosing vaccine candidates for human trials. J&J instead ran tests on fifty-two animals, trying variations of different DNA sequences and then exposing the animals to COVID-19. Reassured by success in animal trials, Janssen finalized their vaccine candidate on the 30th of March 2020.

In July 2020 J&J began Phase 1/2 trials, which were successful. In September 2020 Phase 3 trials were launched using just one dose. On the 29th of January 2021, J&J announced that Phase 3 trials had proven that the vaccine was safe and effective, with the FDA issuing an emergency use authorization in the US on the 27th of February 2021. Two days earlier, South Africa began use of J&J for their health care workers. The vaccine trials showed 72 per cent efficacy in the US, 68 per cent in Brazil and 64 per cent in South Africa.

Also in the spring of 2021 J&J started trials in children, and in pregnant women, with the hope that the results would be available before the start of the 2021/2 school year for children. However, after six women aged 18 to 48 developed a rare disorder involving blood clots linked to the vaccine, the FDA in the US called for an immediate pause in its use on the 13th of April 2020. After a temporary pause, the FDA recommended that the J&J rollout should continue. While acknowledging the risk of rare blood clots in adult women younger than fifty, the FDA said that the potential benefits outweighed the known and potential risks of catching and dying from COVID-19.

The J&J vaccine was an important one for global rollout, given that it is only one dose and its storage is relatively easy. On the 21st of May 2021 Gavi, the Vaccine Alliance, announced that it would purchase 200 million doses for low- and middle-income countries through COVAX, with additional discussions for 300 million more doses in 2022. Dr Seth Berkley, CEO of Gavi, noted, 'As a one-dose vaccine, the J&J vaccine has particular relevance for places with difficult infrastructure, making it a very important addition to the portfolio.'

## *Sputnik V*

Halfway across the world, Russian scientists were also working around the clock to create a vaccine. They developed two candidates: the first was called Sputnik V, and the second was called EpiVac-Corona.

Sputnik V is similar to the Oxford vaccine in that it uses an adenovirus vector to trigger the immune system. Two doses are required, spaced three weeks apart. In June 2020 Phase 1 and 2 clinical trials were launched in Russia, with the results announced on the 1st of August. Researchers reported that the vaccine induced a strong antibody and cellular immune response, that no participants had become infected with COVID-19 after being vaccinated, and that all the volunteers felt well with no unforeseen or unwanted side effects.

Ten days later the Russian Ministry of Health approved the

vaccine, making it the first vaccine in the world to be approved, and, under emergency rules, it could be used immediately to start vaccinating the public. Russian President Vladimir Putin said that Sputnik V worked 'effectively enough' and declared its approval to be a 'very important step for our country, and generally for the whole world'. In the *Lancet*, Russian scientists published their results on thirty-eight people, but didn't provide the raw data, leading academics in other countries to question the validity of their findings.

Russia rolled out the vaccine before Phase 3 trials had even begun. A Phase 3 trial is necessary to demonstrate protection against COVID-19 on a large enough scale to reveal harmful side effects missed by small studies. On the 17th of October 2020 a Phase 2/3 trial was launched in India.

Less than a month later, on the 11th of November, Russian scientists analysed the first preliminary evidence and estimated efficacy of the vaccine to be 92 per cent. This was later published on the 2nd of December 2020 in the *Lancet*. Inconsistent clinical trial data had scientists questioning the analyses and wondering if it had been manipulated. On the 22nd of May 2021 a group of international scientists highlighted concerns over patterns in the *Lancet* data consistent with data manipulation. For example, multiple independent requests for access to raw datasets went unanswered; there were data inconsistencies in the published trial; and safety data was only partial.

Concerns continued to be raised about the speed at which Sputnik V had been developed, and the lack of transparency and independent scrutiny over the trial results. This has led to some scepticism on the part of countries receiving large shipments of this vaccine as well as hesitancy in Russia itself over taking the vaccine.

In early 2021 several ambassadors reached out to me to ask about the scientific consensus regarding Sputnik V: was it indeed as safe and effective as the Russian government was publicly stating? Russia was heavily pushing distribution of the vaccine into 39 countries and expected to reach 27 more, but recipient governments were unsure whether to accept these vaccines at face value or whether to apply further independent scrutiny. This shows the importance of trust in the scientific process, and the necessity to ensure that political demands, in this case by the Russian government, did not pressurize scientists and affect the actual technical process of ensuring the vaccine was good enough.

## *CoronaVac and Sinopharm*

Similar to Russia, China raced ahead with its own vaccines, the two most prominent being CoronaVac and Sinopharm. In January 2020 Sinovac Biotech, a private Chinese company, began to develop an inactivated vaccine candidate called CoronaVac. It developed the vaccine by first growing large stocks of SARS-CoV-2 in monkey kidney cells. It then doused the viruses with beta-Propiolactone, a compound that disables coronaviruses by bonding to their genes. The inactivated coronaviruses could no longer replicate, but their proteins, including spike, remained intact.

Because the coronaviruses in CoronaVac are dead, they can be injected into the arm without causing COVID-19. Once inside the body, some of the inactivated viruses are swallowed up by a type of immune cell called an antigen-presenting cell. The antigen-presenting cell tears the coronavirus apart and displays some of its fragments on its surface. A helper T cell may detect a fragment, and, if that fragment fits into one of its surface proteins, the T cell becomes activated and can help to recruit other immune cells to respond to the vaccine.

In June 2020 a preprint of Phase 1/2 trials on 743 volunteers found no severe adverse effects, and the details published in November 2020 in *Lancet Infectious Diseases* showed a comparatively modest production of antibodies. In July, Sinovac launched its Phase 3 trials in Brazil, followed by others in Indonesia and Turkey. Based on the positive early results, in July 2020 the Chinese government gave CoronaVac an emergency approval for limited use. In October 2020 (prior to results from the Phase 3 trials) authorities in Jiaxing, China, announced they were giving CoronaVac to people in high-risk jobs such as medical workers, port inspectors and public service personnel.

In December 2020 trials in Brazil and Turkey showed that the vaccine could protect against COVID-19, but those trials delivered strikingly different results because they had been designed differently. In Brazil the efficacy against symptomatic or asymptomatic COVID-19 was 50 per cent; in Turkey the efficacy against COVID-19 with at least one symptom was 91.25 per cent. Based on this, on the 11th of January 2021 Indonesia gave the vaccine emergency authorization, with the Indonesian President receiving an injection of CoronaVac on live television two days later. On the 13th of January 2021 Turkey authorized CoronaVac, and on the 19th of January Brazil authorized the jab. On the 6th of February 2021 Sinovac announced that China had given CoronaVac conditional approval, and within months had increased its manufacturing capacity to 2 billion doses.

A second Chinese vaccine was Sinopharm, which used the same inactivated virus platform as CoronaVac. This one was created in January 2020 by the Beijing Institute of Biological Products, with clinical trials run by the state-owned company Sinopharm. In June 2020 researchers reported that the vaccine produced promising results in monkeys, and by 17th of June 2020 it was clear that there were no serious adverse reactions during Phase 1 and 2 trials.

On the 23rd of June 2020 China and Afghanistan signed a clinical cooperation agreement, marking the official launch of the world's first international COVID-19 clinical Phase 3 trial. In addition, in July 2020 a Phase 3 trial began in the UAE, with Sheikh Abdullah bin Mohammed Al Hamed, the Chairman of the Department of Health, Abu Dhabi, becoming the first person to be given the vaccine. On the 14th of September 2020 the UAE gave Sinopharm emergency approval for health workers, with the health minister saying the vaccine would be available for 'first-line-of-defence heroes, who are most at risk of catching COVID, protecting them from any danger that they may be exposed to due to the nature of their work'. And on the 9th of December 2020 the UAE gave its full approval for use, reporting 86 per cent efficacy in a press release.

In China itself, on the 29th of December 2020 Sinopharm announced 79.3 per cent efficacy and requested regulatory approval to become China's first approved vaccine for general public use. Two days later China's regulatory agency granted it conditional market approval, making it China's first COVID-19 vaccine. This was followed by approval in Europe: Hungary became the first EU member state to approve Sinopharm, on the 29th of January 2021. Globally, 43 million doses had been provided by the 20th of February 2021.

However, countries using predominantly Chinese vaccines, usually CoronaVac or Sinopharm, went on to experience major outbreaks, for example, the Seychelles, Chile, Bahrain and Mongolia, even after 50–68 per cent of the population had been vaccinated. Dr Jin Dongyan, a virologist at the University of Hong Kong, noted, 'If the vaccines are sufficiently good, we should not see this pattern. The Chinese have a responsibility to remedy this.' This has raised some scepticism about the effectiveness of the Chinese vaccines in stopping transmission as well as in stopping people becoming severely ill. It has also led regulatory authorities in the US, UK and EU to question the validity of the trial data and to not approve Chinese vaccines for use in their own populations.

## *Abandoned Vaccines*

Reading about all the successful vaccine efforts might make it seem like creating a vaccine is easy. That's far from the truth. Hundreds of vaccine projects started off, but only the ones highlighted above made it past the finish line. Other promising starts fell at various hurdles along the way. Here are just three examples of projects that had to be abandoned.

First, Imperial College researchers, led by Professor Robin Shattock, developed an RNA vaccine in early 2020, which boosted production of a viral protein to stimulate the immune system. They began Phase 1/2 trials on the 15th of June 2020, partnering with Morningside Ventures to manufacture and distribute the vaccine through a new company called VacEquity Global Health.

The vaccine was based on a new method using self-amplifying RNA to re-create the spike protein sequence, without the need for animal cells or human stem cells. Short strands of self-amplifying RNA are packaged into fat droplets; and, once injected into the muscle, the cells take up these tiny fat droplets and the RNA therein.

The self-amplifying RNA generates copies of itself, instructing the cells to make the SARS-CoV-2 protein in the cytoplasm (the solution within cells). The muscle then produces lots of spike protein but not the entire virus. Some of the proteins will land on the surfaces of muscle cells, stimulating antibody production. The plan was also to make it stable up to 40°C, which eliminated the need for a cold chain in delivery. On the 18th of December 2020 Imperial even collaborated with Enesi Pharma to formulate a solid version of the vaccine that could be implanted into the skin without a needle.

But after Pfizer, Moderna and AstraZeneca raced ahead and got approval, Shattock abandoned the planned vaccine development on the 27th of January 2021 and the planned Phase 3 trials. He explained, 'Although our first generation COVID-19 vaccine candidate is showing promise in early clinical development, the broader situation has changed with the rapid rollout of approved vaccines. It is not the right time to start a new efficacy trial for a further vaccine in the UK, with the emphasis rightly placed on mass vaccination in response to the rapid spread of the new variant.'

Rather than seeing their efforts as wasted, Imperial showed how yet another platform could produce a vaccine and take it successfully into human trials within months. This kind of learning is crucial for the next pandemic, when the race will begin again to create an effective vaccine.

A second promising, but abandoned, vaccine was one produced from a collaboration between the US company Merck, the Austrian firm Themis Bioscience and the Institut Pasteur. This vaccine used a variant of the measles virus to introduce inactive parts of SARS-CoV-2, mostly antigens, into the body, thus prompting antibody production. This type of virus-vector approach had been used in vaccine development programmes for SARS, chikungunya, MERS and Lassa fever.

But, again, this attempt was slower than the others, and the results also didn't show a high enough efficacy. The consortium was formed only in March 2020, with Phase 1 trials starting in August. Early trials found that the immune response generated was weaker than the one created following natural infection, as well as those being reported for other vaccines. On the 25th of January 2021 Merck announced it would discontinue development of this vaccine and instead focus on treatments, including two antivirals that could be used both in hospital and in outpatient settings for COVID-19. One of those drugs, molnupiravir, would show extremely promising results as an outpatient five-day oral treatment in early trials in 2021. In March 2021 Merck partnered with J&J to help to boost production of their vaccine instead.

In addition to its project with Themis, Merck partnered on a second viral vector vaccine, this one with IAVI (International AIDS Vaccine Initiative). Based on vesicular stomatitis viruses (VSV), it utilized the same approach as the one Merck successfully produced for the first approved vaccine for Ebola. This platform used an attenuated strain of a VSV, a common animal virus, modified to use certain proteins. This development came through a long-standing research programme on an rVSV-based HIV vaccine candidate and rVSV-based vaccines for other emerging infectious diseases, e.g., Lassa fever and Marburg virus disease. Merck and IAVI designed this COVID-19 vaccine as a pill, which would have made it easier to distribute than vials and syringes for injections and required fewer trained health workers for administration. This was part of the US government initiative Operation Warp Speed mentioned earlier and received more than $38 million in funding from the government to accelerate research.

On the 30th of September 2020 Merck registered a Phase 1 trial. The results were not good enough: the immune response was weaker than the one seen following natural infection or those produced by other COVID-19 vaccines. On the 25th of January 2021 the two groups stopped development of this COVID-19 vaccine, with Mark Feinberg from IAVI saying, 'These results are not the ones we hoped for, but we conduct studies in order to learn the answer to a question. In this case, the answer is clear: more work is needed.' They continued to collaborate with a view to seeing if the rVSV platform could be improved through changes in route of administration, or in the viral vector, or to the S-protein, all useful learning for the next pandemic.

Even when science seems to fail, as in the case of the abandoned vaccine candidates, it succeeds. These candidates might prove to be useful for the next pandemic, as those teams of scientists will continue to discover what did and didn't work and improve their knowledge on how best to trigger the body's immune response using a vaccine. This is a similar story to the one about failed mRNA vaccines against RSV, which then proved useful when it came to developing an effective vaccine for COVID-19. Such is the beauty of science: even failed attempts are a step towards more information and progress forward.

## *The Shift from Science to Politics*

With the development of several safe and effective vaccines, the first half of the battle was completed. This achievement is absolutely remarkable: a truly extraordinary moment. I think many of you will understand the joy associated with being fully vaccinated: I felt elation and appreciation when receiving my first, then second, Moderna jab and knowing that I would be better protected against severe illness from COVID-19, as well as less likely to pass on infection to others.

While we celebrate the several safe and effective vaccines that were created, as the stories above describe, these were the exceptions, not the rule. Most attempts failed, even those that initially seemed promising. Most efforts to create vaccines pre-COVID-19 estimated a timescale of five to seven years. Before COVID-19 there were no promising coronavirus vaccine candidates (for example, against SARS or MERS). This just shows what scientific efforts can achieve in a crisis and with access to sufficient resources and support.

On the 22nd of April 2020, in my weekly *Guardian* column, I reflected that, while the focus was on the scientific challenge of developing an effective vaccine or discovering a treatment, the other half of the battle was the implementation challenge of getting vaccines into arms.

Manufacturing enough doses and distributing these fairly across the world was indeed a major problem: from the start of the pandemic, each country pursued its own selfish national interest and its own strategy without linking into the wider picture. Global cooperation is fundamental to responding to, and ending, a pandemic, and the bulk of this responsibility fell on to WHO.

Why did global cooperation break down? How did global agencies try to respond to the increasing division in the world between rich and poor countries, and between the US and the EU and China and Russia? In the next chapter we will take a closer look at global cooperation, in particular the efforts of international agencies such as WHO and the World Bank in trying to bring governments across the world together and the challenges they faced, given difficult leaders, such as Trump, as well as how they tried to influence the distribution of limited commodities, such as vaccines. While breakdown in cooperation occurred among governments, the race for a vaccine showed how scientists continued to work collectively towards finding solutions, regardless of nationality or politics.

# 10. Cooperation Breaks Down

The COVID-19 pandemic strained international cooperation, as tensions between the US and China reached boiling point, and agencies such as WHO and the World Bank found themselves caught in the middle. The hope was that countries would come together during a global crisis and find coordinated ways to tackle the challenge, in this case, SARS-CoV-2, and move towards solutions. The reality of how countries behaved was far from that ideal.

In this chapter we take a closer look at the role of WHO in health emergencies, its response to COVID-19 from early 2020 until the summer of 2021, how it evolved and improved its response following Ebola in 2014, and the major challenges posed by the Trump administration over the course of the COVID-19 pandemic. I also take a closer look at WHO's advice around travel restrictions, and how that fared in the face of this deadly pandemic.

WHO worked alongside the Coalition for Epidemic Preparedness Innovations (CEPI) to facilitate vaccine sharing through the mechanism of COVAX. French President Macron convened the first vaccine summit to bring together these various efforts to ensure they were coordinated, and that the most promising two or three vaccine candidates would be manufactured in sufficient numbers and distributed fairly to the world. Despite pledges from rich countries to share, vaccinating the world continued to be a major challenge, as we will explore.

We will also take a closer look at the World Bank's involvement in pandemic preparedness and response. The World Bank is often forgotten as a key stakeholder, despite being the largest financial contributor to global health. Over the course of the COVID-19 crisis it played a crucial role in supporting poorer countries in their public responses. But WHO and the World Bank were unable to stop the breakdown of global cooperation as governments pursued their own short-term, selfish interests.

## *What WHO Can and Can't Do*

WHO was established in 1948 as the chief director and coordinator of international health work in the United Nations. Its roots were deeply embedded in outbreak response, and can be traced back to the first International Sanitary Conference in 1851, when sovereign states came together to agree on infectious disease regulations.

WHO's strengths are threefold and referred to as its technical, normative and convening efforts. Technically, it shares data among countries, including standards, guidelines and key information. Normatively, it has the unique ability to secure agreement to international law, such as the International Health Regulations that govern the reporting and response to health outbreaks, as well as to set norms through its Codes of Practice and Global Strategies. Its convening efforts centre on the World Health Assembly, which brings together governments across the world annually to agree on priorities, debate thorny issues such as access to essential medicines and pass resolutions for action.

While it can advise, support and encourage countries, it cannot go into countries to change policies, investigate the source of outbreaks or penalize bad behaviour. It is, at its heart, a member state organization that serves as an independent and neutral body.

With COVID-19, WHO attempted to bring these three roles together to address the pandemic. After China reported the new outbreak to WHO's country office on the 30th of December 2019, as we have seen, WHO quickly sent out a bulletin to other countries about a new respiratory pathogen. As more data emerged on the virus, it helped to develop test-kits that could be sent to parts of the world that lacked lab capacity, as well as to encourage data-sharing from China so that other countries could learn and adapt.

On the 11th of January 2020 daily press briefings began, headed by the strong technical team of Dr Mike Ryan, Dr Maria Van Kerkhove and Dr Sylvie Briand, and chaired by Director-General Dr Tedros. On the 30th of January 2020 the WHO Emergency Committee declared COVID-19 a Public Health Emergency of International Concern – which, as we learnt earlier, is the technical term for the highest legal alarm bell there is – and strongly warned countries that the outbreak was on the way and to use the time well. This was the moment for countries to prepare (or not). The WHO Mission to China returned on the 24th of February 2020, and Dr Bruce Aylward shared what he had learnt about how the disease spreads and whom it affects; he also detailed the policies that the Chinese had enacted to contain the outbreak.

The clear messages were: testing; contact tracing; isolation of carriers of the virus and their contacts; physical distancing as needed (lockdowns); protection of health workers through adequate PPE; and ramping up hospital capacity to treat all patients who needed care. Tedros said, 'Find, isolate, test and treat every case and trace every contact. Ready your hospitals, protect and train your health workers and let's all look out for each other because we need each other.' While many of the policies adopted by the Chinese government were questionable, such as forcibly removing people from their homes and the intense surveillance of their population, some key messages could be drawn out on how other countries could prepare to contain the outbreak as it arrived on their shores.

WHO was criticized at the time for praising China, but leaked information indicated that this was the price for getting China's cooperation. WHO struggled to get the Chinese government to share information, to agree to an international mission to collect data and to allow an independent investigation on the origin of SARS-CoV-2. From the 30th of January 2020 onwards WHO has continued to update its technical advice, continue its briefings and keep its eye on all parts of the world, when traditional donor countries in Europe and the US were consumed by their own domestic outbreaks.

In May 2020, at the annual World Health Assembly conducted virtually, WHO managed to get all countries (including a reticent US) to agree by consensus to a resolution on the equitable access to, and fair distribution of, all essential health technologies and products for COVID-19. As explained above, its job is technical advice, normative authority and convening action on the part of its member states, and, in my view, it adequately delivered on those roles over the course of the pandemic.

## *Was WHO Too Slow?*

WHO has been criticized by some scientists for its guidance on masks and airborne spread since March 2020. I am sympathetic to the position of WHO scientists, as all of us have had to evolve our positions on issues based on new evidence and a changing situation. Signs of a good scientist are an ability to reassess data and their position as evidence alters, a willingness to acknowledge mistakes or having given suboptimal advice, and openness to recognize scientific uncertainty and several different scenarios for the future. Humility and listening skills are also essential.

For example, I was asked by children on a BBC *Newsround* clip in February 2020 whether they should be worried about their pets getting coronavirus. At the time we didn't have good data on SARS-CoV-2 in dogs or cats, so I said that they shouldn't be worried. By June 2021, with numerous cats and dogs testing positive, the advice became to stay away from pets if COVID-19 positive. My answer would be different in June 2021 based on this new data.

WHO, as already mentioned, changed its stance on mask wearing over the course of the pandemic. On the 1st of April 2020 its official position was that it was still gathering evidence regarding masks: 'There is ongoing debate about the use of masks at the community level. WHO recommends the use of medical masks for people who are sick, and those caring for them. However, in these circumstances, masks are only effective when combined with other protective measures.'

This was repeated on the 6th of April 2020, when WHO became concerned that 'the mass use of medical masks by the general population could exacerbate the shortage of these specialized masks for the people who need them most', as Tedros said. It didn't comment on using cloth or home-made masks at this point.

On the 5th of June 2020 the guidance on masks was updated to say, 'In areas of widespread transmission WHO advises medical masks for all people working in clinical areas of a health facility, not only workers dealing with patients with COVID-19', and, most crucially, it finally recommended masks for the wider public. Tedros noted that day, 'In light of evolving evidence WHO advises that governments should encourage the wider public to wear masks where there is widespread transmission and physical distancing is difficult such as on public transport, in shops or in other confined or crowded environments.' Some criticized WHO for being too slow to recommend masks in crowded, indoor settings.

The other issue that it's been perceived to be slow about is recognizing airborne transmission. In a scientific brief published on the 27th of March 2020 WHO said there was insufficient evidence to suggest that the virus was airborne, and maintained that COVID-19 was mainly spreading through droplets and contaminated surfaces. This was despite the experience on the *Diamond Princess* cruise ship.

In July 2020, 239 clinicians and scientists published a commentary in the journal *Clinical Infectious Diseases*, urging the medical and public health authorities to acknowledge the risk of airborne transmission. On the 7th of July 2020 WHO first acknowledged that airborne transmission was a possibility, which was reiterated in a scientific brief two days later. Was it slow? Possibly – but, again, the scientists working there are also human and were doing their best to provide guidance as new data arose.

## *US Withdrawal from WHO*

Geopolitics also proved a problem for WHO. In a speech on the 7th of April 2020 former US President Trump attacked WHO for being too 'China-centric' in tackling the COVID-19 pandemic. In the same speech he mentioned freezing the US financial contribution to WHO. His exact words were: 'The WHO really blew it . . . Chinese officials ignored their reporting obligations to the WHO and pressured the WHO to mislead the world when the virus was first discovered by Chinese authorities. Countless lives have been taken and profound economic hardship has been inflicted around the globe.'

Trump pointed particularly at WHO advice around keeping trade and travel open as being problematic – an assertion I will come back to later in this chapter. In a 14th of April 2020 press briefing, he pointed the finger at WHO for why the virus was seeded across the US: '[WHO] called it wrong. They really, they missed the call. Fortunately I rejected their advice on keeping our borders open to China early on. Why did they give us such faulty recommendation?'

As the World Health Assembly started on the 18th of May 2020, Trump sent a letter to WHO, unhappy with what he saw as its close ties to China, and unhappy with WHO's acceptance of China's explanation for the coronavirus origin. He remarked, 'China has total control over the World Health Organization' and that 'The world is now suffering as a result of the malfeasance of the Chinese government.'

The letter mentioned that if WHO did not make substantive improvements – details of what these entailed were not specified – within thirty days, Trump would freeze funding and reconsider the US's membership of WHO. The letter outlined perceived mistakes made by WHO and claimed that other global health institutions, like the Global Fund to Fight HIV/TB and Malaria, and Gavi, the Vaccine Alliance, could do better than WHO. The language was clear: 'I cannot allow American taxpayer dollars to continue to finance an organization that, in its present state, is so clearly not serving America's interest.'

China responded angrily, saying the letter was slanderous, while the EU threw its weight behind WHO, urging all countries to support the organization: 'This is time for solidarity. It is not time for finger-pointing or undermining multilateral cooperation,' said a European Commission spokesperson.

Despite initially giving WHO thirty days, on the 29th of May, only ten days after threatening to stop funding to WHO, Trump announced that the US would terminate its membership: 'Because they have failed to make the requested and greatly needed reforms, we will today be terminating our relationship with WHO and directing those funds to other worldwide and deserving, urgent global public health needs.'

On the 6th of July 2020 the US formally moved to withdraw from

WHO. A senior US administration official said that the US government had engaged with WHO regarding detailed reforms, but that WHO had refused to act. The withdrawal of the US would have been financially catastrophic for the agency, given that the US is traditionally one of the largest contributors of extrabudgetary funds, which altogether make up 80 per cent of WHO's total budget. The US usually provides $400–$500 million per year to WHO, out of an annual budget of roughly $2.4 billion. This loss of income would have resulted in major cuts to health agendas, such as supporting vaccination campaigns against childhood diseases like measles, polio and neglected tropical diseases that aren't covered by other organizations. In addition, the US is typically an active participant in the World Health Assembly, sending a large delegation and engaging and leading on major health issues.

What were Trump's exact complaints? First, he claimed that WHO did not adequately obtain, vet and share information in a timely and transparent fashion. He said that WHO ignored reports of the virus spreading in Wuhan, published as early as December 2019, including a report in the *Lancet*. The journal, however, said there was no report published in December 2019 regarding an outbreak in Wuhan or elsewhere in China. A *Lancet* editorial noted, 'The allegations levelled against WHO in President Trump's letter are serious and damaging to efforts to strengthen international cooperation to control this pandemic. It is essential that any review of the global response is based on a factually accurate account of what took place in December and January.'

Trump's second complaint was that throughout January 2020 WHO continued to tweet and say that there was no human-to-human transmission, despite evidence to the contrary. WHO denied that it delayed making evidence regarding human-to-human transmission public at China's request. Trump said that Taiwan warned about human-to-human transmission of COVID-19 on the 31st of December 2019, but provided no evidence to support this claim. Taiwan only sent a letter to WHO noting the reported cluster of cases and asking for further details.

Trump further claimed that WHO is not independent of China,

and that this affected its impartiality as an international agency. Trump's attacks on China were a complete reversal from his position earlier in the pandemic. On the 24th of January 2020 Trump had said, 'China has been working very hard to contain the coronavirus. The United States greatly appreciates their efforts and transparency. It will all work out well. In particular, on the behalf of the American People, I want to thank President Xi!'

Trump's final assertion was that WHO had fought the US on travel restrictions. WHO's recommendation was not to impose travel restrictions, but it never publicly criticized the US government, or any other government, for this.

What were the implications of US withdrawal from WHO? Because Joe Biden was elected President in November 2020 and reversed Trump's withdrawal on his first day in office in January 2021, there were very few real consequences. In fact, President Biden worked to strengthen relations and support for the agency, and repair the damage that the Trump administration had inflicted. This caused tension with the Chinese government, which didn't like the influence the US government seemed to be having over the agency, particularly over WHO's push to have an independent investigation into the origin of SARS-CoV-2 and conduct a lab audit.

What if Trump had been re-elected? Theoretically, there are checks and balances within the US political system that prevent rash actions by any one person, or any one branch of government (Executive, Legislative and Judicial). In the larger trajectory of history Trump's words remain just that – words. And the work of WHO continued, as there was no other agency that could perform its three core functions during the pandemic.

## *Travel Restrictions*

Alongside face masks, travel restrictions must rank as one of the most controversial and debated policies during the pandemic. Most countries implemented travel restrictions at one point or another. While Vietnam fully closed its borders, New Zealand and Australia remained open, but everyone entering had to stay in a mandatory quarantine facility for fourteen days. In Europe some countries banned flights/trains and/or ferries from specific countries because of high local transmission in these places and, later in the pandemic, to keep variants out. Others implemented 'light-touch' self-isolation requirements upon entry, while outliers like Britain stayed fully open with no checks during their entire first wave.

The kinds of travel restrictions that were implemented in 2020 and in 2021 would have been unthinkable before COVID-19. In June 2020, 91 per cent of the world's population lived in countries with travel restrictions, and 39 per cent of people lived in countries whose borders were completely closed to non-citizens and non-residents. These measures carried huge economic and social costs: global tourism, trade, business, education and labour mobility rely on cross-border movement of people. The year 2019, for example, saw the travel and tourism sector generate 10 per cent of global GDP, which stood at $8.9 trillion. International movement has also been seen as a right and a freedom enjoyed by those in democracies.

WHO actively advised against travel restrictions, with Tedros tweeting on the 31st of January 2020: '@WHO does not recommend limiting trade & movement. Travel restrictions can cause more harm than good by hindering info-sharing & medical supply chains & harming economies. We urge countries & companies to make evidence-based, consistent decisions.' Advising against trade or travel restrictions was made clear in the revised 2005 International Health Regulations.

The revised IHR were designed to 'prevent, protect against, control and provide a public health response to the international spread of disease in ways that are commensurate with, and restricted to public health risks, and which avoid unnecessary interference with international traffic and trade'. The revised IHR moved away from a focus on national self-protection through border measures and towards a more collaborative method where countries are assisted to contain outbreaks.

WHO provided four key reasons why it advised against travel and/or trade restrictions. First, border controls were considered ineffective in keeping disease out, and at best could delay importation only slightly; previous research on SARS, Ebola and seasonal flu indicated that targeted restrictions merely delay infections, while leading to social and economic costs.

Second, if travel restrictions are put in place for people travelling from certain countries, this has an economic impact on those countries, which may lead to governments keeping outbreaks secret for fear of the cost of reporting.

Third, border closures can hamper outbreak response, as vital supplies and personnel may not be able to enter countries. And, fourth, border restrictions raise a danger to human rights and civil liberties. For example, families may be split apart and unable to reunite. In April 2021 Australia was asked by the UN Human Rights Committee to allow the return of its citizens from the US on the grounds of the International Covenant on Civil and Political Rights, which states that 'No one shall be arbitrarily deprived of the rights to enter their own country.'

Why did the IHR warn against travel restrictions on the ground, saying they do not work at stopping disease spread? Professor Larry Gostin, of Georgetown University, helped to write this international law on dealing with outbreaks and said that this just reflected the global health community's 'almost religious belief that travel restrictions are bad'. He added, 'I have now realized [April 2021] that our belief about travel restrictions was just that – a belief. It was evidence-free.' But it was also a belief rooted in progressive, human rights, which saw travel bans as linked to nationalism and xenophobia.

Ignoring WHO's advice, countries started to close their borders one after the other. It was an approach no one had anticipated or had evidence to support. Professor Karen Grépin, of the University of Hong Kong, said, 'No one [had] modelled a scenario in which borders would be shut.'

Travel restrictions were examined in a rapid review aimed at assessing the effectiveness of international travel-related control measures during the COVID-19 pandemic on infectious disease transmission. The review established that the existing evidence-base was weak and failed to provide any real insights other than the fact that generally border measures could delay entry of a virus. Most studies on travel restrictions used modelling (theoretical mathematics based on assumptions) instead of observational data collection (looking at what happens in real life when countries introduce these).

The heavy reliance of SAGE on modelling partially explains why the UK stayed completely open to international travel. On the 3rd of February 2020 and the 23rd of March 2020 travel restrictions were discussed by SAGE but seen as having a negligible effect. Modelling predicted that 'If the UK reduces imported infections by 50 per cent, this would maybe delay the onset of any epidemic in the UK by about 5 days; 75 per cent would maybe buy 10 additional days; 90 per cent maybe buys 15 additional days; 95 per cent maybe buys a month.' Travel bans were seen as only delaying the spread of an uncontrollable virus, and not worth the associated disruption and economic costs.

Even after SAGE became more open towards travel restrictions in June 2020, the UK government was politically opposed to this – a decision, Johnson's former Chief Adviser Dominic Cummings said, that was driven by ideology and concerns over Brexit more than by science. It was ironic that a government that ran for election on the promise of 'taking back control of our borders' was so reluctant to implement border measures when they were actually necessary – in a pandemic.

A stark contrast is Vietnam, which went for elimination of the virus through sealing itself off from the world. On the 3rd of January 2020 Vietnam increased disease control measures on its border with China, and several weeks later introduced a complete flight ban from China. A member of the Prime Minister's Economic Advisory Group in Vietnam noted, 'Two countries taking the quickest action are Taiwan and Vietnam – they shared the same reasons: geographical proximity to and distrust in China.' There were also rumours that Vietnamese hackers spied on the Chinese government to collect intelligence about COVID-19, which may have led to inside information that was unavailable to other countries. The Vietnamese government denied these allegations.

In March 2020 the Vietnamese government cancelled all in-bound commercial flights, making it impossible for anyone, including Vietnamese residents, to enter the country through an airport. Limited repatriation flights for Vietnamese citizens, diplomats and other officials were constantly run, and air travel from low-risk countries like Taiwan, South Korea and Japan was increased. At the time of writing those wanting to travel to Vietnam require a special government pass and must complete 14–21 days of state-monitored quarantine plus regular PCR tests. Anyone testing positive is isolated in hospital regardless of how they feel.

As of the 9th of May 2021, Vietnam's approach has led to fewer than fifty deaths and just a few thousand infections in a population of 96 million. One professor in the UK downplayed Vietnam's success in containment, arguing that this was down to previous immunity in the population and having 'lots of bats'. However, Indonesia has more bats per capita and had many more cases and deaths, so bats couldn't be the explanation. More likely it was their strong approach to containment, including stopping all international flights, thus giving the virus limited opportunities to seed across the country.

Vietnam also had strong public messaging about how to avoid getting COVID-19, a massive amount of testing and contact tracing, and strict isolation policies for those testing positive. This avoided the need for a national lockdown. As an adviser to the Vietnamese government said, 'The reality [is] that Vietnam does not have enough budget to sacrifice the economy and support businesses and individuals who had to cease operation.' A Zero COVID-19 policy allowed Vietnam to stay open for business domestically, resulting in the economy growing in 2020 when most other countries in the world were having contraction.

From the COVID-19 pandemic, three clear lessons about travel measures are clear. First, targeted ones, i.e., applying them to just one country, don't work, as the way people travel and move is more complex. For example, much of the introduction to the UK of SARS-CoV-2 was through France, Spain, Italy and Austria, while travel restrictions focused on Wuhan, China. This is also clear from the earlier discussion on the ski resort in Austria that spread cases throughout Europe. Even stopping direct flights between China and the US didn't work, as people just connected in Singapore, Frankfurt or other hubs to get to where they wanted to go. Travel restrictions need to be for all countries, or no countries. And those countries that did seal off and go for a Zero COVID-19 approach did better domestically in 2020 and the first half of 2021: they sacrificed international mobility for domestic freedom. A harsh trade-off but a necessary one in a pandemic before there was a vaccine or treatment.

Second, the lower the cases, the more effective border restrictions are. For example, Scotland in the summer of 2020 practically eliminated COVID-19, and so at this point the most important policy measure to take was to stop importation of new cases that could set off new chains of infection. Iceland's second wave was kicked off by two French tourists who did not comply with self-isolation (even after testing positive) and then travelled around the country, with cases taking off quickly, given a completely open economy with no brakes within the system to limit spread. Grépin said, 'This is the lesson about border measures that's changed. The value of border restrictions goes up the fewer cases you have.'

Third, it is difficult to know at an early stage when to 'overreact' and shut down travel, and when to keep flights going. For scientific advisers, the challenge is to know exactly when to react and implement harsh travel measures, in the light of how many potential pandemics there are a year. Do governments jump at every report of a new virus, an unexplained cluster, a more severe TB strain or a highly infectious RSV variant? Professor Mark Jit, of the London School of Hygiene and Tropical Medicine, said, 'The natural thing is to think "When we have a big problem, there are many COVID cases, that's the point when we need to start doing a lot of things." But for travel restrictions – these are the solution to stop the problem from happening in the first place. It seems obvious in retrospect, but it's very paradoxical.'

What some countries understood was quite simple really: the virus moves when people move. This has been recognized by states since the fourteenth century, when quarantine was introduced by the Venetians; and was behind the inception of the International Sanitary Conferences, which started in Paris in 1851, given concerns about movement of people along trade routes. No need for maths, theoretical modelling or complex guidance. It's basic infectious disease prevention and control.

For the next pandemic the real lesson about travel restrictions is to know how best to stop the export of a virus at the early stages of an outbreak from an affected region. In late December 2019 and into early January 2020 China could have closed Wuhan to the world once the severity of the outbreak and the difficulty in controlling spread became clear. China is a wealthy and able country, with enough resources, personal protective equipment, health workers, hospitals and logistical know-how. It could have managed the outbreak without external assistance.

But, instead of closing its borders and stopping wider spread, it let flights keep taking off, delayed sharing information about the virus and, once it had solved its internal problem and eliminated SARS-CoV-2, shut its borders to the world. At the risk of sounding Trumpian, there are reasons to be frustrated by the decisions of the Chinese government, and the devastation for the world that these caused – memories remain of bodies being burned in the streets of India, and the children who will never return to school in Malawi, and the increasing number of homeless people in New York City suffering outside in the cold.

The responsibility of affected countries not to export viruses must be looked at more closely in the future, including how WHO can do more to create incentives (such as financial assistance from the World Bank) for countries to report and then close off. Member states need to look again at the travel and trade guidance in the International Health Regulations, and how this might need to be changed based on key learnings from the COVID-19 pandemic. Governments need better advice and direction on how and when to shut their borders, either for exporting or importing disease.

## *CEPI*

One of the most important international partnerships in COVID-19 was the Coalition for Epidemic Preparedness Innovations (CEPI), launched at the World Economic Forum in Davos in 2017. CEPI is a global partnership between public, private, philanthropic and civil society groups and was established to develop vaccines to stop future epidemics.

The idea for CEPI was seeded in 2015 in an article in the *New England Journal of Medicine* about the challenges in vaccine development. During the 2014 West Africa Ebola outbreak, when a vaccine was necessary, none was approved or ready to go. This was despite several promising Ebola vaccines on the shelf that had made it to Phase 1 trials. None of these had been further developed, because there was no incentive for companies to complete testing and produce a vaccine for a disease that had no regular market. Expedited trials in the field led to emergency approval, but this market failure was evident to the experts working there.

The problem is that the prevailing pharma business model focused on vaccines with large market potential (ideally rich countries), rather than on vaccines for some infectious diseases for which the market is too small or might be non-existent in the future (if a particular pathogen never spreads), thus making it hard to justify the capital investment for development and the necessary risk. Yet, for the next pandemic, scientists already need to have investments in place in order to do the research and development and be ready to run at the first sign of an outbreak. And here lies the challenge: how to convince vaccine manufacturers (and their shareholders) to devote the necessary resources to research, development and production when their business models don't support this?

Identifying this problem, experts proposed an international vaccine-development fund, CEPI, which helps to shoulder some of the risk in business terms, thereby incentivizing companies to continue to develop vaccines. It was given initial funding of $460 million from the governments of Germany, Japan and Norway, as well as the Bill and Melinda Gates Foundation and the Wellcome Trust. The initial targets were MERS, Lassa and Nipah viruses, which were all identified as having pandemic potential, with the aim of having two potential vaccines ready for each of the viruses quite soon. It also provided support for research on Ebola, Marburg and Zika viruses. Between 2017 and 2019 CEPI started projects to develop seventeen vaccine candidates against those priority pathogens, and created three vaccine manufacturing platforms to produce and test vaccine candidates.

When COVID-19 emerged, CEPI became a crucial stakeholder and invested in a range of vaccines, including different types of technologies, as it was impossible at the start to predict which ones would be successful in the early stages of development. It invested in scaling up manufacturing, development, clinical trials and licensing. For example, giving $4.9 million to the Institut Pasteur for the measles vector vaccine (which was abandoned, as discussed in the previous chapter), $1 million to Moderna for the mRNA platform and $384 million to Oxford for their vaccine. CEPI also played a big role in helping to expedite COVID-19 vaccine candidates through animal and human trials by the end of March 2020.

Apart from vaccine development, CEPI established a global network of labs to centralize testing and enable comparison of immune responses generated by COVID-19 vaccines. It also was a key stakeholder, alongside WHO, in COVAX, which was created to ensure equitable access to COVID-19 vaccines.

## *COVAX*

Alongside the race for a vaccine was the recognition that developing an effective one was only half the battle. Having adequate supply and distributing it equally across the world would be just as challenging, if not more. Recognizing this, in April 2020 the Access to COVID-19 Tools (ACT) Accelerator was set up to speed up the development, production and equitable access to COVID-19 diagnostics, treatments and vaccines. Within that, COVAX was the vaccine pillar that aimed to ensure fair allocation of vaccines as well as support procurement and delivery at scale. It was co-led by CEPI, Gavi, the Vaccine Alliance and WHO, with UNICEF helping with delivery, making use of its experience of doing on-the-ground work in low- and middle-income countries.

The target was to have enough vaccines donated to COVAX to protect at least 20 per cent of each country's population, starting with those most at risk, such as health and social care workers and vulnerable groups. This was estimated to be at least 2 billion doses by the end of 2021 (at a cost of $5.7 billion), half of which would go to lower-income countries and the other half to countries that would otherwise have limited or no access to COVID-19 vaccines. As of February 2021, 92 countries qualified as lower-income, while 98 additional self-financing countries opted to join this mechanism.

The basic idea of COVAX was to share the risks and benefits of developing and producing vaccines across countries by pooling financial resources to develop vaccines, purchasing these at scale and investing in manufacturing ahead of vaccine approval so that distribution could occur quickly. Early on COVAX monitored the vaccine landscape to identify promising candidates, worked with manufacturers to start production and used collective purchasing power to negotiate reasonable prices from manufacturers, for which all participating countries were eligible.

As of the 11th of June 2021 six vaccines had been given emergency use authorization by COVAX: AstraZeneca, Pfizer, J&J, Moderna, Sinopharm and CoronaVac. However, only Pfizer and AstraZeneca had delivered vaccines to the facility. Participating countries such as the US and UK pledged to donate money, as well as provide surplus doses from their own vaccine supplies once their own populations had been covered.

## *Vaccine Diplomacy*

While Western countries worked through COVAX, Russia and China developed their own vaccine programmes, likening them to the space race, in that they would not only protect their own populations but extend their nations' influence on the world stage. Their vaccine assistance to other countries, even though commercial (i.e., countries had to buy it; it wasn't gifted), came with economic or political strings attached, or as a reward to proven reliable past partners. For example, Russia started discussions with Bolivia about access to mines producing rare earth minerals and nuclear projects shortly after delivering a batch of Sputnik V, and Pakistan was offered doses in return for its approval of projects linked to the China Belt and Road Initiative (a huge infrastructure project that would stretch from East Asia to Europe). We can presume that China and Russia will expect recipient countries to back them at the UN, facilitate access to natural resources, fast-track the approval of investment projects, expedite trade and even become more open to buying defence equipment or 5G technology.

This worried Western countries. Charles Michel, President of the European Commission, stated, 'We should not let ourselves be misled by China and Russia, both regimes with less desirable values than ours, as they organize highly limited but widely publicized operations to supply vaccines to others.'

Western countries like the US and UK and those of the EU were largely absent from this vaccine diplomacy, given the intense political pressure to vaccinate their own populations first as well as their choice to largely work through COVAX, a multilateral sharing mechanism, rather than donate or sell vaccines directly to countries. Heiko Maas, Germany's Foreign Minister, insisted COVAX must deliver for countries: 'Our multilateral solutions must succeed, if we don't want to lose our ground to those who argue that authoritarian regimes are better at dealing with a crisis like this.'

## *Vaccine Hoarding by the Western World*

As trials started to show promising results, richer countries began to pre-order vaccines in the summer of 2020 – not only to ensure access to the first doses but also to show manufacturers what a large market would be awaiting them. This was quite an exceptional move by countries; the UK, for example, had never before ordered a drug or vaccine before it was approved.

Pre-arrangement led to high-income countries, which make up 19 per cent of the global population, collectively purchasing 54 per cent, or 4.6 billion doses, of global vaccine supply by the 17th of March 2021; 33 per cent of doses were purchased by low- and middle-income countries; and 13 per cent by COVAX. Globally enough vaccine doses were purchased to cover more than 80 per cent of adults everywhere; however, this was unequally purchased. High-income countries could vaccinate 245 per cent of their populations, while low- and middle-income countries could vaccinate only a third of theirs. Richer countries bought multiple doses of different vaccines, as they weren't sure which ones would ultimately be the most effective. The EU, US and Canada pre-purchased doses for up to six times their populations, and paid much less for these. For example, the EU paid $2.15 for each dose of AstraZeneca, while South Africa had to pay $5.25 per dose.

Dr Tedros of WHO warned against this 'vaccine nationalism' early on, stating: 'We need to prevent vaccine nationalism. Whilst there is a wish amongst leaders to protect their own people first, the response to this pandemic has to be collective.' Yet his pleas fell on deaf ears.

By the 18th of January 2021, 39 million doses had been given in high-income countries, while only 25, in total, doses were provided in low-income countries. A paltry 25 – which is outrageous. Several months later the US had administered doses to almost half of its population; the UK to more than half; and Canada and the EU to approximately one third. By contrast, in Ethiopia, Nigeria and South Sudan 0.1 per cent–0.9 per cent of their population had been vaccinated, with the highest vaccination rate on their continent being 3.6 per cent in South Africa. Additional problems arose when India, which manufactured the main supply of vaccines for the world, went into a large second wave, resulting in the government banning vaccine exports in order to support its own population.

Was this situation unexpected? No. Rich countries have always looked out for their own interests, and it would be difficult to see any national leader telling their citizens that after a year of restrictions and suffering, with vaccines as the main tool to lift lockdown, that doses would instead be going abroad. Rich countries are usually happy to help the world, after they have taken what they need. And, to be fair, there is an argument for putting out the fire in your own house before trying to help others.

Resentment started to grow against Western countries for hoarding vaccines. Aleksandar Vučić, the President of Serbia, said, 'Today it's harder to get the vaccines than nuclear weapons', and went on to compare the hoarding of vaccines by high-income countries to the sinking of the *Titanic*, where everyone wanted a lifeboat only for themselves.

Tedros continued to call out this moral failing, saying, 'Vaccine nationalism, where a handful of nations have taken the lion's share, is morally indefensible and an ineffective public health strategy against a respiratory virus that is mutating quickly and becoming increasingly effective at moving from human to human. At this stage in the pandemic, the fact that millions of health and care workers have still not been vaccinated is abhorrent.'

People like Dr Jeremy Farrar, Director of the Wellcome Trust, and myself tried to raise the issue that vaccinating the world was an issue of self-interest too for rich countries. Letting the virus spread uncontrolled would lead, as it did, to variants, and these variants could potentially undermine the gains made in richer countries. Rich countries would be safe only when everyone was safe in the world. In June 2021 I presented a BBC World Service documentary called *Vaccinating the World*, which laid out the arguments and obstacles involved in protecting people in all parts of the world.

But, as new variants arrived that challenged even the most advanced rollouts in Israel and the UK, the focus turned to boosters – that is a third shot of a vaccine to increase the immune response of those vaccinated. As the UK, Israel and Germany started planning and delivering boosters, many low-income countries had not even started to vaccinate health care workers and the most vulnerable. Referring to what he perceived as greed in richer countries, Ryan, of WHO's Health Emergencies Programme, said, 'There are some people who want to have their cake and eat it, and then they want to make some more cake and eat it too.'

On the other hand, as I reported in the BBC documentary, governments were under pressure to lift restrictions and get their economies going again. The clear path to doing this was vaccinating their people as quickly as possible and keeping hospitalizations low and within health care capacity. How do governments balance taking care of their electorate (who will punish leaders who do not deliver quickly enough) with helping people half a world away?

The other contentious issue was around vaccinating children, especially once a safe and effective vaccine (Pfizer) was given emergency approval for those twelve and older. While the US, Israel, France, Italy and Germany raced ahead to vaccinate teens, the UK's main decision-making expert body on vaccines, the JCVI (Joint Committee on Vaccination and Immunisation), dithered. Various JCVI members did press interviews noting their reluctance to vaccinate children: one argued natural infection was safer than the side effects of Pfizer, while another said it was morally indefensible to vaccinate children in the UK before health workers abroad. Again, the UK was an outlier and waited and watched, instead of realizing this was an emergency situation that required swift action, especially as Delta took off in younger age groups.

This is when I got targeted by anti-vaxxers after doing a short clip for the children's programme BBC *Newsround*, which was shown in schools. I spoke about the Pfizer vaccine and answered questions sent in by teens. I recorded it in three minutes on my phone, in between meetings, and didn't think much about it. I always enjoy doing *Newsround* and talking to children. However, this triggered anger among several parents with large social media followings, who felt their child was being exposed to propaganda at school and wanted me to stop talking about children and vaccinations.

I have always argued that this is a voluntary choice for those parents and children who want to be protected and have assessed the situation for themselves. It is not mandatory or forcibly given. On the flip side, in July 2021, while JCVI were deliberating about vaccinating 12–16-year-olds in the UK, the most common question asked when I got stopped in the street was about when children would be vaccinated. I asked one ten-year-old why he wanted to be vaccinated and he said, 'To protect myself from COVID.' He had realized he would get COVID-19 eventually but wanted a vaccine first.

Returning to vaccine nationalism, Africa was left behind. On the 30th of May 2021 only 1 per cent of the 1.3 billion vaccines administered worldwide had been given to Africa. This was less than 2 per cent of their population. Adequate vaccine coverage of even 20 per cent of their adult population would take until late 2022 or early 2023 according to some projections. John Nkengasong, the Director of the Africa CDC, warned Western countries that this could trigger social unrest and a backlash, saying, 'It would be extremely terrible if the world watches Africa not receiving vaccines and sees the global North getting the vaccines.'

Nkengasong had tried early on to get bulk orders from African countries, but struggled to compete against richer countries. His main lesson was that Africa must be self-sufficient and not rely on Western charity or deals. In an interview for our BBC documentary, he said:

> We have to look at this crisis in the face and ask ourselves the question – as Africans – and ask what future we want for our continent. We need to draw some lessons . . . we have to have continental manufacturing capabilities . . . we import 99 per cent of our vaccines, and only manufacture 1 per cent. If we continue to rely on that dynamic, I would characterize that as unacceptable going forward . . . you cannot do that if you don't strengthen your own research and public health institutions across the continent. We have to have that. Research hubs that are capable of doing R&D in areas of drugs, diagnostics and vaccines.

Regional manufacturing capacity is one of the major lessons as we prepare for the next pandemic.

And I remain sceptical of any talks about better agreements, or resolutions, or deals to make sharing more fair. Vaccine nationalism is not a new phenomenon: the same was true in the 2009 swine flu pandemic, and before then with smallpox and polio vaccines soon after they were developed in the twentieth century. The same was true during the 2014 Ebola outbreak with the experimental drug ZMapp. To expect rich countries to share at their own expense is to repeat a cycle of good words, paper agreements, and then disappointment and little real action. But to get low- and middle-income countries self-sufficient requires financing and building infrastructure, and the agency responsible for helping them with that is the World Bank.

## *The Lender to the World: The World Bank*

While WHO played a central and the most public role on the global stage during the pandemic, the World Bank had an equally important part in supporting low- and middle-income countries in their public health and economic response. What is the World Bank? It was established in 1944 to help to finance post-Second World War European reconstruction and recovery. When the US Marshall Plan kicked in to fulfil that role, the Bank pivoted to focusing on development in low- and middle-income countries. Our research team in Edinburgh has a large grant from the Wellcome Trust to analyse the role of the Bank in global health over the past forty years. And if you are keen to learn more about the World Bank and WHO, I co-wrote a technical book with Chelsea Clinton called *Governing Global Health*, which takes an in-depth look at these two agencies: their history, operations and influence.

For a more accessible take: the Bank is influential within countries because of its long-standing relations with ministers of finance; its technical support to countries based on the promise of a loan package towards programmes or policies; its 'revolving door' of people who move in and out of the Bank to positions in ministries of finance and health; and, finally, its combination of public health with economic concerns over productivity, growth and cost-effectiveness.

Unlike WHO, the World Bank provides countries with money to implement the advice it gives. From its first entry into health in the early 1970s until today the Bank has become one of the largest and most influential health funders worldwide. The presidency of Jim Yong Kim (2012–19), a medical doctor and anthropologist who argued for access to health care being a basic human right, ensured that health became a top priority for the organization.

The US, including under former President Trump, has had a cosy relationship with the World Bank, with the US always appointing the President and being the only country to have veto power over decisions of the Executive Board of Directors. The US has used threats to reduce, or withhold, funding from one of the arms of the

Bank in order to demand changes in policy for the entire institution. It has also used the Bank to host pet initiatives of the US government, like former 'First Daughter' Ivanka Trump's fund for women entrepreneurs in low- and middle-income countries.

After the 2014 Ebola outbreak was brought under control, President Jim Kim said 'never again' and committed the World Bank to create a financial mechanism to support global pandemic preparedness. Within months the Pandemic Emergency Financing Facility (PEF) was established, which would serve as a health insurance scheme for the world's poorest countries and for qualified responding agencies.

The core element was an 'insurance window' that would provide cover up to $500 million for infrequent, severe health pandemics. In an outbreak, part of the money it holds is paid out if the following criteria are met: a country must be affected by a specific kind of pathogen, including orthomyxoviridae (e.g., new influenza virus A, B and C); coronaviridae (e.g., SARS and MERS); filoviridae (e.g., Ebola, Marburg); and other zoonotic diseases (e.g., Crimean-Congo, Rift Valley and Lassa). Moreover, the size of an outbreak, measured in number of cases or of deaths, must be considerable (2,000 confirmed cases worldwide for influenza); outbreak growth must be fast (an increase of confirmed cases from 2,000 to 5,000 within a month); and spread of the outbreak must be broad (two or more countries must be affected).

Investors found these 'pandemic bonds' incredibly appealing and bought them up quickly. It was a calculated gamble: if a health pandemic breaks out with the insurance window becoming operative, they would lose part of their invested capital. Yet if the insurance window was not paid out, they would get the full principal back after three years. While investors hold on to the bonds and risk losing their money, donors like the German government pay them a reasonably good annual interest. The Bank marketed this widely, but scientists expressed concern over the 300-plus-page prospectus for the bonds and the secrecy surrounding the trigger for payments, which was held by a small Boston risk assessment firm. The risk assessment firm said their formula for triggering payments was 'proprietary'.

While creating these 'pandemic bonds' has necessitated dozens of reports and conferences and the spending of much time and money, they did not pay out during the 2018 Ebola outbreak in the Democratic Republic of Congo because the outbreak did not affect enough countries.

With COVID-19 the bonds did not pay out until June 2020, months after WHO had declared a global health emergency. The amount was a paltry few hundred million dollars for over seventy poor countries. By that time the Bank had already used its traditional financing mechanisms to create a $20 billion fund to support low- and middle-income countries' health sector efforts to prepare.

The story of 'pandemic bonds' are a tale of good intentions by a public sector leader going awry and being captured by investors looking to make a profit. The investors constructed the terms of play so that they would always win. And therein lies a cautionary tale about favouring new, shiny marketing mechanisms over the hard slog of building core public health structures.

Coming back to the Bank's core response to COVID-19, it eventually made available $160 billion in loans to countries affected by health or economic shocks, with an additional $50 million on highly concessional loans or grants to the poorest countries. This was for relief, restructuring and then recovery. Four pillars underpinned the Bank's response.

The first aim of funding was to save lives and strengthen the public health system through investing in disease monitoring and prevention, especially in low-income countries and those affected by conflict. For example, Haiti was granted $20 million to build up testing and contact tracing, as well as for lab equipment and PPE for health workers.

The second aim of funding was to scale up social protection to help families and businesses affected by lockdown measures, increased prices and unexpected medical bills from COVID-19. For example, in India, to help those unable to earn due to lockdown measures, the Bank issued a loan to help to facilitate cash transfers; and food benefits were provided for the poor and vulnerable.

The third aim was to try to ensure continued growth and job

creation and to stabilize the larger economy in the event of contraction. This is through a Bank funding window called the International Financing Corporation, which directly supports firms and businesses. For example, in Nigeria, the Bank provided a combined $200 million loan to three banks, which could then support small and medium enterprises.

The final aim focused on strengthening policies, institutions and investments for rebuilding post-COVID-19; for example, in helping countries deal with public debt and financial management better. In Mongolia the Bank and the EU partnered to support the country in reprioritizing of funding.

Alongside these aims funds were provided to help countries to purchase urgent medical equipment and health care supplies, such as masks, ventilators and test-kits; the Bank even facilitated a procurement programme that allowed it to negotiate directly with medical suppliers on behalf of low- and middle-income governments.

On the 30th of June 2021 the Bank announced more than $4 billion to help to purchase vaccines for fifty-one developing countries. The vaccine finance package is flexible and can be used to acquire doses through COVAX or other sources, and for vaccine deployment such as training health care workers, developing data and information systems, and communication and outreach campaigns. Alongside the financing, the Bank has pushed governments with excess vaccine supplies to release surplus doses as soon as possible to developing countries.

Taking a closer look at Senegal provides a country-level snapshot of how the Bank supported COVID-19 response. On the 2nd of April 2020 the Bank approved a $20 million credit to support Senegal's prevention, preparedness and response to the COVID-19 outbreak. Nathan Belete, the World Bank Country Director for Senegal, said, 'Senegal has built its response to COVID-19 on successful experiences in containing disease outbreaks in recent years through timely identification and response. The Bank is confident that the project will be implemented efficiently and in close coordination with all relevant partners and stakeholders.' Additionally, $4 million was provided to strengthen the overall health system and $1.5 million through the PEF.

On the 2nd of June 2021 the Bank approved $134 million to support effective access and deployment of vaccines, which would enable 9.5 million people to be vaccinated, about 55 per cent of Senegal's population. The Senegalese Health Minister said that this funding would support 'the government of Senegal's efforts to procure vaccines to improve availability and increase geographic accessibility to populations' as well as support outreach. The World Bank Task Team Leader, Djibrilla Karamoko, said that this communications campaign is 'part of a broader social engagement and mobilization strategy to address vaccine hesitancy. The content will be tailored and targeted to specific priority groups to increase acceptance for the vaccine among the population.'

This is similar to how the Bank supported Mongolia. On the 2nd of April 2020 the Bank approved $26.9 million for the Mongolia COVID-19 Emergency Response and Health System Preparedness Project to meet emergency needs. While community transmission was still not evident, the funding was used to train health care workers, prepare hospitals with adequate supplies and build diagnostic capacity. Andrei Mikhnev, the World Bank Country Manager for Mongolia, said, 'This emergency operation will not only provide immediate support to address the COVID-19 pandemic, but will also benefit Mongolia's health sector in the longer term to become more resilient to future health emergencies.' On the 11th of February 2021 another $50.7 million was provided to help to acquire and distribute vaccines in order to reach 60 per cent of the population.

You probably haven't heard about the role of the Bank during the COVID-19 pandemic. While WHO was at the forefront of press briefings and leading technical and normative guidance to the pandemic, the World Bank had the financial power to help governments respond with key policies, whether through building up health systems and testing, putting in place economic packages to support lockdown measures, or in acquiring and distributing vaccines. And the strength of the Bank was not in flashy new mechanisms like the PEF but in its core work of talking to ministries of finance, finding out what their needs were and then finding the funds to give to them to support response.

While the Bank's core response of providing finance to low- and middle-income governments was solid, it perhaps could have done more in its advocacy for the interests of these countries. For example, the Bank was silent on lifting the patent protection of pharmaceutical companies through an emergency intellectual property waiver to enable generic manufacturers to produce vaccines. As noted above, supply was a major problem and the 'charity' model of relying on rich countries to donate doses failed. What was needed was more production at regional hubs.

Second, the Bank could have publicly and loudly insisted on pharmaceutical companies having reduced, but still juicy, profit margins. When have companies made enough money? When have they made enough profit for their shareholders, and can turn to serving the needs of people across the world and saving lives? What other agency could have spoken up loudly about these issues on the international stage? WHO did speak about these issues, but their words fell on deaf ears. The Bank's additional voice would have carried more weight.

## *Future International Cooperation*

We now face a world in which cooperation is increasingly needed to manage interconnected health threats, but in which the rise of populism and nationalism makes this increasingly unlikely. COVID-19 has only accentuated this, as countries have taken care of their self-interest first, and, if anything, will increasingly move manufacturing and supply chains within their borders and within their control.

Out of global crises usually comes a reshuffling of the world order towards peace, cooperation and solidarity. The UN system itself came out of the destruction and horror of the Second World War, when world leaders came together to agree on a new set of multilateral institutions that would maintain international peace and security. COVID-19 has also caused shockwaves throughout the world and will, I hope, lead to reflections on what could be changed before the next pandemic, and on inequality and values. If all countries had worked together closely, instead of competing, would the pandemic have ended sooner? That's the question that will be asked for years and possibly decades to come.

# 11. Healing and Recovery

In an alternative universe a new virus emerges in China. The country quickly identifies the pathogen, closes its borders (to both incoming and outgoing traffic), launches an unprecedented campaign to eradicate the virus, and therefore manages to ensure that very few cases leave the country. The other countries that do report cases, such as South Korea, Taiwan and Singapore, rapidly identify those who are infected, trace the people they have contacted, isolate the carriers of the virus and contain its spread. Through this three-pronged strategy of test, trace and isolate, eradication is successful. Humanity is saved.

In reality SARS-CoV-2 escaped the public health interventions of the Chinese government and spread across the world. As other governments fumbled in their early responses, the virus spread undetected and quickly through communities, infecting many people and hospitalizing and killing some. Even WHO had to acknowledge that the genie was out of the bottle by mid-April 2020. Given where we are now, in August 2021 as the book goes to print, how could a pandemic like COVID-19 end? What will the rubble left behind look like, and how do we build a better future?

## *COVID's Legacy*

When the virus first emerged, infectious disease modellers obsessed over the case fatality rate, the percentage of people who die after acquiring the infection. Seasonal flu, for example, kills about 0.1 per cent of people who get it, and largely those who are elderly (over seventy) and those who are quite young (under five years of age). Contrast this with the Ebola virus, which kills 70 per cent of people who get it and are not treated; or even SARS, which is 11 per cent; or MERS, which is 35 per cent. The WHO Mission to China estimated the equivalent fatality rate due to COVID-19 would be 2–3 per cent, but this was revised downwards after emerging evidence about asymptomatic carriers. The likely number is between 0.5 per cent and 1.5 per cent depending on access to medical care.

While these seem like small numbers, at a population level this works out to over 5 million people dying in the US, close to a million in the UK and close to 20 million in India. But the real picture is more complex, because COVID-19 deaths are concentrated in the elderly and in younger individuals with comorbidities such as asthma, heart disease, suppressed immune systems, diabetes and hypertension. Children, as noted in Chapter 6, are largely unaffected, while younger people under the age of thirty are at higher risk of dying from drowning or getting hit by a car.

One of the biggest surprises of risk of death in certain countries like the US, UK and Sweden was its close association to being of Black or minority ethnic origin. In the US, African-Americans were nine times more likely to die of COVID-19 than their white counterparts, and in the UK people from South Asian backgrounds were 20 per cent more likely to die than white people. While recognizing this as a problem, governments struggled to understand why exactly these groups were at higher risk, and the even harder question was how exactly to create appropriate policies to address it? When the suggestion was made to take all ethnic minority staff off COVID-19 wards in the UK, managers reacted that the NHS would collapse without their participation in the workforce.

Existing inequalities within society having to do with socioeconomic status, race and occupation also became clear. As data has increasingly been analysed on who is most affected by COVID-19, three clear risk factors for death were: coming from a deprived background; being from a racial or ethnic minority group; and having an occupation such as cleaner, security guard, taxi driver, social care or health care worker. Often all three of those factors combined in those regarded as key workers, who had to continue to show up for their jobs (as they couldn't work remotely) and do them regardless of safety and appropriate equipment. Instead of being the great equalizer, COVID-19 pulled back the veil and revealed one set of rules for elites and another for essential workers, whether it was in housing (having enough rooms, windows, bathrooms, space in which to isolate), the ability to work from home, access to quality education for children, access to early testing and treatment, or involvement in key decisions over government policy.

## *Governments' Dilemma*

The age profile and relatively low mortality rate of COVID-19 compared with SARS and MERS created a dilemma for governments on which strategy to choose. If this was a disease killing younger people in large numbers, like MERS, the strategy was obviously elimination. No government could have let a disease that kills a third of people spread. However, because the disease largely affected elderly individuals, and had a relatively low case fatality rate compared with other infectious diseases, a misalignment was created between countries, which ranged from full elimination (like New Zealand) to repeated suppression with lockdowns (like in Britain) to trying to 'live with the virus' (like Sweden attempted, and what ultimately happened in Brazil). Each country was pursuing its own strategy, and there was no coordination on a global plan.

There were no silver bullets, clever models or easy answers for governments on what to do when coronavirus emerged. And there were relatively few choices. The first and most difficult path was to contain the virus through a programme of mass testing, contact tracing and isolating, and work towards elimination. But this required a huge effort, as witnessed in Taiwan, New Zealand and China. It required building a large infrastructure to monitor cases of the virus and identify hotspots, ensuring the system ran efficiently, providing adequate PPE to everyone who needed it and deploying strict border controls to vet who was entering the country.

The second path was far simpler, and the one taken by the UK, most of Europe and the US. It involved slowing the spread of the virus by largely using lockdowns, with the government issuing guidance on how much physical distancing would be required. But the side effects of this path were very costly: it ruined these economies, strained health and social care systems, and created social unrest. This was already clear at the start of the pandemic. On the 25th of March 2020 I tweeted: 'We will be stuck in an endless cycle of lockdown/release for next 18 months, if we do not start mass testing, tracing, & isolating those who are carriers of the virus while pursuing rapid research for antiviral treatment of vaccine. This is the message the public needs to hear.'

The third and easiest path available to governments was the one chosen by Sweden: a hands-off approach. The plan was for the virus to sweep across the population, with society and the economy remaining open. As discussed in Chapter 7, Sweden tried this approach but pivoted to look much like the rest of Europe when health care collapse was imminent.

Early decisions about which strategy to pursue were made in an initial vacuum of knowledge about COVID-19. The biggest gaps in knowledge related to long-term health complications for survivors, how long immunity would last, if at all, how many people were truly susceptible and the risk of mutations. Some leaders chose a risk-averse approach – gather as much information on the disease as possible before making crucial decisions – while others rushed in and were willing to take a gamble in order to preserve economic objectives.

For example, the focus on deaths from COVID-19 missed another toll of the virus: the long-term damage it can cause to the lungs, heart, kidneys, brain and even blood vessels among those who recover. Long COVID. We still do not understand the full and devastating consequences of this.

Another hope, floated by doctors in Italy in summer 2020, was that COVID-19 would grow weaker as it spread and through viral mutation become a milder infection more like the common cold. However, improved survival rates for those in hospital with COVID-19 were most likely due to patients going into hospital earlier in their disease trajectory as well as better clinical management of critically ill patients through the use of drugs, oxygen and ventilators. And, in fact, as Chapter 8 showed, the rise of more transmissible and more severe variants meant evolution was not on our side. The virus tried – as all viruses do – to outrace our efforts to contain and manage it.

Immunity was a looming question from the start, including whether having the virus triggered antibodies, and, if so, how long these could be protective for, as well as how many people had already been exposed to the virus unknowingly – which could mean there was more 'immunity' in a population than expected. At this point in time we still don't know how long antibodies provide protection for, and how likely reinfection is from the various variants circulating.

We were able to get more insights into exposure with the development of antibody testing. Early serology (antibody) tests emerged in April–May 2020 that indicated only 17 per cent of Londoners and 5 per cent of the UK population had been exposed by that point. These studies fit with those from France, Spain and Sweden. This is far off the estimated threshold for herd immunity initially set at 60 per cent for the original SARS-CoV-2 but moved up to 98 per cent with the Delta variant. Some scientists, like Andrew Pollard of Oxford University, don't think we can ever reach a herd immunity threshold, given that people can get COVID-19 multiple times and transmit it even after being fully vaccinated. He thinks we will be living with COVID-19 for the foreseeable future as it circulates among humans.

However, the picture is more complex. Several studies emerged in May and June 2020 reporting that individuals who had never been exposed to SARS-CoV-2 still had T cells – another key component of the body's immune response that focuses on attacking specific foreign particles – reacting to the virus. The authors suggested that having previous exposure to common cold coronaviruses could mean the body is able to detect the protein in SARS-CoV-2 and thus mount an immune response. And even if antibodies to SARS-CoV-2 fade, other parts of the immune system might be longer lasting, like T and B cells (another part of the immune system that fights bacteria and viruses by making antibodies specific to each pathogen), meaning we shouldn't read too much into serology testing.

## *Strategies under Uncertainty*

Governments had to make their strategies ahead of several potential future scenarios. The first was that an accessible and affordable vaccine would become available within the next twelve to eighteen months, and they just needed to hold out. Or antiviral therapies could become available to treat COVID-19 to make it a mild and manageable illness, so the population could gradually and safely build up natural herd immunity. Science needed time to deliver solutions. Those countries that bought time were vindicated when safe and effective vaccines became available, as well as more data on clinical management and treatments.

Or, failing a vaccine or therapeutic, governments that had the capacity could embark on a resource-intensive and gruelling campaign to eliminate the virus, particularly in the light of its impact on all parts of the body, making it a similar campaign to those for smallpox and polio. The country-by-country elimination ending would fit with how SARS and MERS outbreaks were controlled. The reason they never became global pandemics was because of active case finding, tracing of contacts and isolation of all those with the virus to ensure chains of infection were broken early. SARS and MERS also made infected individuals much sicker, meaning that they would often be in hospital or at home, and not out in bars/workplaces/churches circulating and spreading the virus to others. Presymptomatic transmission was also minimal.

Failing that, in a repeat of the 1918 flu pandemic, the virus could rip through the world, causing tens of millions of deaths, in a Darwinian culling of the elderly and vulnerable, and an individual gamble for those exposed to the virus. This unfortunately is what happened in most parts of the world, particularly in South Asia, Latin America and Sub-Saharan Africa.

## *It's the Virus Killing the Economy, Not Lockdown Alone*

The COVID-19 pandemic revealed governments doing a fragile balancing act between the interests of public health, society and the economy, all while trying to ensure minimum harm from COVID-19, as well as reducing harm from the measures taken to address COVID-19 such as lockdowns, and from a reduction in health services for other conditions. Throughout the pandemic a false dichotomous argument pitting public health against economic success emerged. In fact, one common argument against stringent public health measures like lockdown was the potential damage it inflicted on the national economy. Former US Treasury Secretary Steve Mnuchin said in June 2020, 'I think we've learned that if you shut down the economy, you're going to create more damage.'

It is incorrect to say that the entire loss of economic growth and job losses are a primary consequence of social-distancing measures, rather than of the virus itself. A study of forty-five countries' economic performance showed that those countries that succeeded (during their first wave) in suppressing COVID-19 had a smaller loss in private expenditure (consumer spending) than those that had a more relaxed approach to COVID-19 spread. This indicates that consumers themselves changed their behaviour (i.e., locked themselves down) to avoid COVID-19.

Not taking strict public health measures to protect the national economy during the pandemic was a short-sighted policy; in the long run a brief closure and temporal subsidization has proven to be more cost-beneficial than opening the economy during the pandemic.

Although New Zealand experienced an annual contraction in real gross domestic product (GDP) of 6.1 per cent, this is much lower than that of other comparable countries, and in Taiwan a net change in GDP of 0 per cent was maintained. These are the success stories of those countries that didn't pit economy against pandemic. Furthermore, compare this with the economic performance of the US – a country that 'prioritized' the economy – where net GDP losses from COVID-19 are estimated to range from $3.2 trillion (14.8 per cent) to $4.8 trillion (23 per cent). Effective public health measures, if implemented, can reduce these financial costs significantly. Contrary to the false – yet common – dichotomy, protecting the health of the people is equivalent to protecting the wealth of the people. Similar analyses have shown that this was also the case in the 1918 influenza pandemic.

A *Lancet* analysis in 2021 of OECD countries showed that those governments that pursued elimination strategies in 2020 did better in terms of lives saved, more civil liberties and freedoms maintained and better economic performance. This is a counter-intuitive finding to the mainstream media narrative that containment harms the economy. It was the virus circulating that was killing the economy, not the restrictions alone.

## *The Complacency of the Western World*

One of the starkest observations is how rich countries, particularly the US and UK, floundered while poorer countries such as Senegal, Ghana, Thailand, Vietnam and the state of Kerala in India did incredibly well. Even poorer countries in Europe like Greece and the Czech Republic could see the virus approaching and had the humility to listen to WHO advice on how best to prepare. Prior to COVID-19, multiple rankings of pandemic preparedness put the US in the top spot with the UK close behind, or vice versa. These rankings looked at capacity rather than at less tangible qualities like political will, leadership, trust in government, the capability of senior politicians, and willingness to learn quickly from other countries and evolve responses to new information.

Richer countries falsely believed they could treat their way through the outbreak. Deputy Chief Medical Officer in England, Jenny Harries, said in a press briefing that she believed WHO advice on testing, which was to test for all cases including in the community, was not suited to the UK, before praising the NHS's ability to offer world-class care: 'The clue for WHO is in its title. It is a World Health Organization and it is addressing all countries across the world with entirely different health infrastructures and particularly public health infrastructures. So the point is that they are addressing every country, including low- and middle-income countries.'

Similarly, as we have seen, Trump admitted that he tried to 'slow the testing down' to downplay the extent of the problem, and, through this active denial strategy, hoped that states and cities would manage and that their ICU capacity would not be breached.

However, both these countries soon realized that if 20 per cent of people require hospitalization and a third of those require ICU care (as estimated by WHO in January 2020), no health system in the world could cope with such numbers.

One of the consequences is that two of the previous leaders in global health, the US and UK, became tragic stories in the eyes of the rest of the world, which watched the unfolding of the pandemic with disbelief and pity. Countries like Mongolia, Ghana and South Africa have had to listen for years to US and UK health experts flying in and advising them on how best to run their health systems. Citizens and scientists in these countries started to wonder whether Western countries did indeed know best and where future leadership would come from in global health.

## *Messaging/Trust/Communication*

Clear and evidence-based communication during an outbreak is critical to building trust with the public and ensuring adherence to public health measures and successful containment. Most importantly, governments need to tell their people what they are trying to achieve, and how and when they will get there. Some leaders seem to have got the balance right, for example, in New Zealand, South Korea, Taiwan and Senegal, while others have struggled, for example, in the US and the UK.

As the pandemic has unfolded, knowledge about the virus, how to manage it and the interventions available to us have rapidly evolved. Some governments have been good at communicating uncertainty and necessary changes in strategy when better options have become clear. For instance, in New Zealand, government communicated that it was adopting an elimination strategy and it was also upfront about the uncertainty involved with any new virus. When the development of effective vaccinations allowed a change in strategy, New Zealand pivoted. It told the public that mass vaccination was the ultimate exit strategy. This wasn't a 'U-turn' but a sound approach of moving as more information and data came in.

In the US and the UK it has at times been unclear what success would look like, how this is measured, and what approach was being adopted: elimination, suppression or mitigation? In the US the Trump administration regularly ignored scientific evidence, trying to pretend COVID-19 wasn't a problem. In the UK sometimes questions about why policies were being adopted were met with boasts of having 'world-beating' approaches and 'following the science'.

One of the key differences between Scotland and England was around messaging, with First Minister Nicola Sturgeon, alongside Chief Medical Officer Gregor Smith and National Clinical Director Jason Leitch, providing constant guidance and reassurance to the public and trying to create a sense of collective responsibility. By contrast, the UK government would often tell newspapers and journalists what it was planning before it had made decisions, partly to see what the public response would be; and then, once it had the results of this polling, it would decide how best to move forward. Governments need the consent of their electorate, and yet they also need to make tough decisions for the good of their people, based on expert advice. In fact, the best way to get the support of the people is to make tough decisions, based on expert advice, and lead from the front – as Jacinda Ardern did in New Zealand.

I took the role of messaging and communications seriously and tried to help people feel, through my public outreach, that their concerns were important and that I would continue to provide impartial, neutral analysis to the best of my ability. It was often frustrating, for example, watching holiday travel in summer 2020 and knowing what the consequences would be. I wrote about this in the *New York Times*: 'We will pay for summer holidays with winter lockdowns.' And trying to prepare the public for a difficult winter with my *Scotsman* article: 'The next four months might be the hardest of your life.'

But together with the difficult messages was continued optimism, hope and reflection on the importance of finding joy each day, even if for a walk to #getoutside. I wanted to give people the idea that we have to look with optimism to tomorrow and not collapse under the burden of today. I was asked in an interview in March 2021 with Professor Topol of Stanford University if I had received any pushback for this sunny outlook on life. I responded that, yes, of course I'd had flak on Twitter. There was specifically a group of 'mean girls' who had said I was full of unicorn crap and rainbows. My reaction is: well, if that's the worst that people can say about me, then I'll take it. I bet you unicorn crap probably tastes like bubble gum. And one of Scotland's national symbols is the unicorn, an emblem of purity, innocence and power. Not a bad thing to be full of (the unicorn, not the crap).

One hobby I enjoyed tweeting about was otter-spotting. This began when I was sitting on a bench in the park on the 1st of December 2020, and a friend walked by and said he was off to see the otters. I had never seen an otter before and found this intriguing, so tweeted that day: 'Apparently there are wild otters wandering around #Edinburgh. My December goal is to spot them.' I later had to clarify that I meant the animal. Not the definition in the urban dictionary: 'An otter is a gay man who is very hairy all over his body, but is smaller in frame and weighs considerably less than a bear.' The playful, fun and elusive otter in question lived at the Dunsapie Loch at Arthur's Seat in the middle of the city, and from that point on I spent many winter days paying the otter a visit, sharing videos when I did catch him or her eating and out of the water, and encouraging others to get outside, enjoy what was locally available to them and find happiness in the relatively mundane (a single otter is hardly a Taylor Swift concert).

The otter soon left the Loch mysteriously but not before being featured in a BBC nature documentary – I like to think I helped make that little otter famous, and now, wherever he or she is, they're the talk of the town. My otter adventures continued on a trip to the north of Scotland in May 2021 (when domestic travel restrictions finally lifted), where I visited the International Otter Survival Fund, and I was regularly sent videos and photos of otters as well as books and otter-themed cards. Again, a little spark of brightness during a difficult time for Scotland, and for the world.

On the flip side, my tweets also got me into trouble. One Sunday morning in early February 2021 I shared a picture of a vegan chocolate cake I had baked following a Nigella Lawson recipe. This received several angry responses accusing me of promoting unhealthy eating and obesity, including complaints to Edinburgh University. When a senior colleague asked me why I felt the need to share a picture of cake, I responded, 'We millennials like to share. Sorry!' However, Nigella wrote to tell me that she thought the cake looked nice. Which is the highest culinary compliment there is.

How did I cope through a challenging work situation? I think by recognizing that this wasn't going to last forever, that vaccines would be arriving soon and that they would provide a breakthrough solution – and that, however hard it was, most people had it much harder than me. The children trapped in abusive homes. Those who had lost their jobs and couldn't pay their rent. And the families who lost loved ones to COVID-19, way before their time.

I also exercised compulsively to deal with the intense pressure of trying to ensure I was providing the best scientific evidence and guidance possible to governments and to the public. Did I get everything right? Of course not. But I hope I evolved in my analysis and thinking, as scientists are supposed to do as new data comes to light, and that I got more right than I got wrong.

While others got lockdown puppies, I also decided to get a pet. In November 2020 my lockdown tortoise arrived, providing slow, relaxed and zen companionship during a time when days seemed to fly due to the workload, and I was running from meeting to deadline to meeting trying to keep up with demands. Tortoises are remarkably undemanding, fasting regularly and hibernating for months in the winter either outside in a shed or inside the fridge. The perfect pet for workaholics like me.

Perhaps the most difficult personal moment was losing my grandfather in Chennai, after not being able to see him for two years; I couldn't fly to India to join my grandmother at his funeral. For elderly individuals, the situation was difficult. My grandparents, both in their late eighties, had had an active social life and kept physically and mentally fit by seeing others, working on office projects, going for walks at the beach and even lifting weights at the gym. The lockdown in India took all of this away and imprisoned them in their home. While important at their age and state of frailty to avoid getting COVID-19, the loneliness and lack of stimulation also took its toll on their health. How does one choose to spend the last years of life knowing that interactions, which bring joy, would lead to infection and likely death, while at that age avoiding interaction is a mental death sentence?

Those countries that were able to pursue Zero COVID-19 in 2020 avoided this situation. In an interview with the BBC in November 2020 I was asked whether, given most deaths to COVID-19 were in those over sixty-five, this was an 'acceptable loss'. It was upsetting to think of my grandparents, and all the inspiring seventy- and eighty-year-olds that I knew, who should have a chance to live life and socialize, while also being protected from serious illness and death. All people, whatever their age, ethnicity, gender, abilities – or whatever else – can lead rich, valuable lives. Loss of life is not 'acceptable'.

## *Have Diversity at the Table*

Scientists rely on peer review and transparency to ensure that their theories, methods and results are accurate. And scientists became key advisers to governments, helping to shape their responses in a new COVID-19 world. This happened in countries such as the US, UK, France, Germany, Hong Kong, Singapore, China and New Zealand.

Modelling somehow became the crystal ball that governments relied on. When governments wanted quick answers, they took modelled projections as certainty and lost sight of other crucial information sources. Other equally valuable sources of information included case-study analysis from other countries, talking to frontline health staff and patient groups, and policy documents and historical analyses of previous novel outbreaks. The models themselves are basically advanced statistics and mathematics applied to infectious disease spread. They are a technical tool to present different scenarios under specific assumptions, but deciding which model to follow and which factors to include is a political and moral choice.

Although better than relying on intuition or flying completely blind into a crisis, this reliance on modelling led to several missteps. For example, early models in the UK did not consider the effects of mass 'test, trace, isolate' strategies, or potential staff shortages in hospitals, or that care home workers might move between several different homes during the course of a day. Including these might have led to an earlier focus on testing capacity, appropriate PPE for frontline workers, and preventing hospitals and care homes from becoming hotspots of transmission themselves.

Contrasting the US, Germany, the UK and New Zealand's use of models illustrates this. In the US, former President Trump was unhappy with the predictions emerging from Harvard, so he looked for alternative modellers who were showing fewer deaths if the virus were to spread; and found one in the University of Washington. It mattered little to him that the Harvard projections ended up being more accurate than those from the University of Washington. He sought and found the science to support what he wanted to hear.

In Germany the authorities considered modelled predictions, but they also learnt from analyses of South Korea's successful strategy of mass testing, tracing and isolation. The UK completely relied on SAGE, whose models falsely put forward either global eradication or natural herd immunity as the only options, then added a third option of suppression through cycles of lockdown and release. These models completely neglected what other countries were doing to suppress their epidemics without lockdowns.

New Zealand advisers started with models based on pandemic flu, like the UK, but upon reading the WHO Mission Report to China quickly pivoted to basing their models on SARS and towards elimination. Prime Minister Ardern made a political decision that there would be no level of acceptable death from COVID-19 and that she would never consider 'herd immunity' given the loss of life it entailed.

What the examples above point to is that, when it comes to policy, scientific advice must not only rely on modelling but also ensure representation from diverse backgrounds and experiences, to help to flesh out multiple options that can be presented to political leaders. Also, mathematical models do not include value systems or morals. For example, a model might suggest that allowing 95 per cent of people to continue life as normal while 5 per cent become critically ill is a suitable path forward. This is when scientific advice is just that – advice – and leaders need to consider the values, needs and preferences of their population when deciding how to build a strategy going forward. Scientists also need to be reflective and learn from this pandemic. Several lessons are already apparent, as discussed in this book. First, there will be more dangerous outbreaks of infectious diseases in our lifetimes, given that spillover events have increased in frequency. But the global community must adapt to these and will need to learn how to react faster and better. Second, we cannot address risks to public health without addressing the inequalities that determine the unevenness of death and infection rates across different parts of the population. Third, populist governments and leaders pose a threat to global health and security and have largely neglected science and evidence. Fourth, scientific advice to policy-makers must ensure representation from diverse backgrounds and areas of expertise, and should be complemented by a clear outline of the values and ethics driving the advice, such as a 'child first' lens. Fifth, just measuring pandemic preparedness based on country wealth and capacity misses out the crucial role of leadership, decision-making and population-level behaviours. Finally, the economy and health are two sides of the same coin, and those countries, like Taiwan and Denmark, that controlled the virus had a steadier recovery.

## *Preparing for the Next Pandemic*

One of the most frequent questions I get is, will there be another pandemic? Of course there will. The world will face a virus similar to SARS-CoV-2, and, while we cannot prevent these challenges emerging, we can change how we respond and learn from past mistakes. How much suffering can be prevented the next time around?

As we reflect on the past eighteen months, the words that come to mind are 'never again'. Never again should millions of lives be lost to a new virus, a number that would have been unthinkable before COVID-19, when we thought infectious diseases that wiped out millions were consigned to history. Never again should children be taken out of school, unable to learn in classrooms or socialize with their peers, and sometimes even forced into marriages, employment and caring roles because of the need to earn for their family or the loss of caregivers to COVID-19. Never again should the unemployment rate soar, as small businesses close and larger ones make sweeping redundancies. Never again should over a hundred thousand health workers die, many of whom lacked adequate protective equipment, clothing and priority access to vaccines. Never again should we endure months and months of lockdowns, death and pain.

As countries now look towards rebuilding and healing, they must also think longer term about the mistakes made, the gaps exploited by the virus, and how best to prepare and respond the next time. We have already seen in this book that countries that learnt from their previous experiences in SARS, MERS and Ebola did better with COVID-19.

What will be the cause of the next pandemic? WHO has identified priority diseases such as Crimean-Congo haemorrhagic fever, Ebola, Marburg, Lassa fever, MERS, SARS, Nipah and Zika; some of these you may recognize, others you may not. Scientists can make educated guesses about disease types we are already familiar with, but there could be others that we don't yet know about: 'Disease X'. What particularly concerns me is MERS mutating into a form that is more transmissible and that then has pandemic potential.

Dr John-Arne Røttingen, a global health expert from Norway, said, 'History tells us that it is likely the next big outbreak will be something we have not seen before' and 'It may seem strange to be adding an "X" but the point is to make sure we prepare and plan flexibly in terms of vaccines and diagnostic tests. We want to see "plug and play" platforms developed which will work for any, or a wide number, of diseases; systems that will allow users to create countermeasures at speed.' What he means is tests, vaccines and systems of public health response that are built, and then can be deployed quickly once a specific pathogen is 'plugged' into that platform.

And pandemic influenza is always lurking as well. Early 2021 saw reports from Russia of the two first humans infected with avian influenza (H5N8) at a poultry plant, a stark warning for governments to prepare and to work together to build on COVID-19 structures and establish robust response systems. These include mass investment in testing, protocols for rapid research in hospitals, processes for fast-tracking public health research in communities and approval of vaccines, and, in some cases, strengthening local public health response structures such as testing sites, contact tracing and isolation facilities.

Throughout 2021 I have been Vice-Chair of the National Academies of Sciences, Engineering, and Medicine commission tasked with advising the US Department of Health and Human Services on how to prepare best for a future influenza pandemic, with a specific focus on vaccines. The first thing to clarify is the difference between seasonal and pandemic flu. Seasonal flu already causes roughly half a million deaths worldwide and largely affects young children under the age of five, pregnant women and elderly individuals. Annual flu epidemics occur worldwide every winter.

By contrast, pandemic flu is a global outbreak of a new influenza virus (likely from an animal) against which we have no previous immunity. Because it is a new virus with no immunity built up in the population, even healthy and young adults could be at risk of becoming seriously ill and dying. The 1918 flu pandemic, for example, infected one third of the world's population and killed 50 to 100 million people worldwide out of a total population of 1.8 billion. The death rate for 15–34-year-olds was twenty times higher in 1918 than in previous years. Since then pandemic flu has been seen as unstoppable, with countries' pandemic flu planning built around rapidly building enough hospitals to ensure no one dies from being unable to access care.

Five clear lessons have emerged from the COVID-19 pandemic on how best to prepare for the next one. First, the biggest public health risk that we face is an animal virus jumping to humans. Every time a virus circulates among animals, particularly bats, rodents, livestock and birds, and then comes into contact with humans, there's a chance that one of those viruses will infect humans and lead to human-to-human transmission. If that virus spreads through breathing or droplets of moisture, it becomes extremely difficult to stop.

Ever increasing international travel and global trade have connected the world and created new opportunities for diseases to spread. This is why we need global cooperation and surveillance to identify disease risks. We must invest in a Global Virus Surveillance Network (almost like a weather service) to scan for new pathogens of concern, as well as identify spillover risks (when animals and humans come in contact) and mitigate them. Currently the process is more bottom-up, with either countries reporting to WHO that they have an outbreak, or scientists and doctors in a country reporting to a website called ProMed, the largest publicly available system for the global reporting of infectious disease outbreaks. We need more.

Second, governments must invest in the resources necessary to rapidly sequence new viruses. It is not only new pathogens that pose a problem of course, but, as we have seen with SARS-CoV-2, new variants can emerge as a result of uncontrolled spread, such as happened in England, Brazil, India and South Africa. Most countries do not have the capacity or facilities to sequence the genomes of virus samples to detect new strains or even new pathogens. The UK, US, Denmark and a select few others are outliers in being able to do the amount of sequencing they've done over the course of the COVID-19 pandemic.

Why is sequencing important? It is crucial in developing testing and vaccines within weeks and manufacturing and distributing enough doses within months. Three scientific tools through a pandemic are testing, therapeutics and vaccines. While the COVID-19 vaccine development has been remarkable, as described in Chapter 9, we need to be even faster next time to avoid the disease, death and restrictions COVID-19 has brought. One hundred days has been cited by scientists such as BioNTech CEO Uğur Şahin as a realistic target for getting some kind of scientific breakthrough, most likely a vaccine, ready for mass distribution. With COVID-19 it took just under a year, 365 days. But developing the vaccine was only half the battle, as we have learnt.

Third, countries should now be thinking about their manufacturing capacity regionally, so as not to rely on other parts of the world, as Africa has seen first hand, as well as coordinating worldwide to have these hubs ready to mass produce. Getting enough supply to the world is not only about IP waivers, or tech transfer, or donations, or building more factories. It is about all of that, together. And African countries are tired of relying on the goodwill of rich countries to donate doses. Rich countries will never give valuable doses overseas if their citizens are waiting. Not even a pandemic treaty can overcome issues of sovereignty and state self-interest.

Instead the challenge is to create regional hubs with enough supply, so that it is about mass market production everywhere, and not just charitable donations of leftovers from the West. Countries will need financial and technical support to do this, which points to the key role of the World Bank in providing funding for infrastructure

Fourth, given that we know that a pandemic influenza is on the horizon, we should already be focusing on a universal influenza vaccine (as the next big push after COVID-19 efforts). Our current flu vaccines are given annually, with experts predicting which strain will become the most common. In some years these guesses are off, resulting in a vaccine that isn't effective for the circulating strain. To boost effectiveness against all strains, our ambition should be for a one-dose universal vaccine that provides immunity over several years and against various influenza strains. The first human clinical trial of a universal flu vaccine has been completed using a new technology that mixes different pieces of flu strains; it was successful. We must accelerate this progress now.

Finally, governments need to carefully examine their 'flu pandemic' playbook on how we respond to any incoming acute respiratory pathogens based on what we've learnt from the COVID-19 experience. Much of the early COVID-19 debate fixated on whether it was like flu, or like SARS, or something entirely different. It would have been more productive to focus less on 'Is it flu?' and more on 'What can be done to stop, or at least delay, spread?'

In the process of managing COVID-19, countries unintentionally eliminated seasonal flu and other respiratory infections. For countries the question is: should governments ever move into the mitigation phase (where they accept spread) before a vaccine is rolled out, rather than containment? How many could have survived COVID-19 had we found a way to buy time and delay infections until a life-saving vaccine could have been given to them? Can we plan for pan-coronavirus and pan-influenza vaccines so that we have the scientific solutions in place? And then can we focus on rapid manufacturing and distribution? How can we build on our seasonal flu infrastructure and move towards a global pandemic flu response infrastructure?

As I also describe in this book, one of the lessons from COVID-19 is that science holds the solutions for the future. It feels like magic to have vaccines or treatments, but these reflect years of research, training and building on the work of the scientists who came before. When Sabin's polio vaccine was approved in 1961, it was met with universal celebration and relief. In 1963 mass vaccinations became widespread against measles, mumps, rubella and rabies, saving millions of lives, first in the US and Europe and then across the world. Antibiotic development, starting with Fleming's discovery of penicillin in 1928, made life-threatening medical procedures such as C-sections and surgeries safe. They currently treat most bacterial infections and stop people dying who are having organ transplantation, urinary tract infections and even cancer treatment. And, in the case of the COVID-19, the leaps made in mRNA technologies mean that promising malaria, HIV and influenza mRNA vaccines are being developed; they could completely change how we manage these infectious diseases in the future.

And it's not just vaccine development. The COVID-19 response relied on evolutionary biology, genetic sequencing, public health, epidemiology, behavioural science, governance and financing, engineering and mathematical modelling. Now should be the time that governments are investing in science. Instead, in 2021, the UK government announced major cuts to research budgets funding global health research across the world. This included a 70 per cent reduction in the 2021/2 overseas aid budget that funded collaborative projects with research institutions in low- and middle-income countries. Now is the time to be stepping up to invest in science so that when the next pandemic arrives we are ready with the words 'never again' burned into our minds. And now is the time to be mentoring and cultivating the next generation of children to pursue scientific careers. Let's hope that our children are inspired by the way scientists, like US-voice-of-reason Dr Tony Fauci and Professor Sarah Gilbert (now featured as a new doll by Barbie), have been celebrated (and rightly so), and that they aren't dissuaded by the appalling way these very same scientific superheroes have been harassed and belittled by a few loud voices in government, media and public. I am hopeful that they will.

## *A Brighter Future*

For humanity, the challenge now is how to take the open wounds that COVID-19 has exposed and build a better world moving forward. Wealth was indeed the best shielding strategy not only from COVID-19 but from the response to it as well: lockdown exposing the plight of the poor in overcrowded housing compared with the country estates of the rich. Will countries learn from their mistakes and rebuild society in a way that is more resilient and more equal? Will governments cooperate to think of how to do things better, or turn inwards and become more selfish? How do we take the legacy of COVID-19 and improve life for our children? That is the challenge for leaders who must bring their people together when uncertainty and division are rife. I am optimistic that we can and will do better in the months and years ahead. Recognizing these inequalities and the horrendous impact that they have on our health is the first step; the next is designing policies to start to address them.

Where does this leave us scientists and advisers? A mix of emotions is my sense of what colleagues have experienced: relief, burn-out, frustration, guilt and exhaustion after sprinting for two years. Yet all we can do is to keep doing our jobs, including designing and executing high-quality research projects, analysing evidence and contributing expertise without fear or favour, and hoping that political leaders listen and give weight to this advice and seek long-term solutions, not quick, easy and ineffective short-term ones. And, at the time of finishing this book, I still haven't had COVID-19, although I will probably eat my words later. No one knows how long it will take for COVID-19 to recede into the background in all parts of the world – possibly 2023 or 2024. But, whenever this may be, you will find me paddle-boarding off the coast of an island in the Pacific Ocean, because, I don't know about you, but I need a rest.

# Afterword

I'm writing this on the 6th of January 2022, but, sadly, I'm not in the Pacific on my paddle-board. I'm in freezing and dark Scotland, and have just turned off the television after the Prime Minister announced that the worst is behind us and that the UK government's plan is to 'ride out the wave' of Christmas cases from 'Omicron', and to see how it goes. Omicron, a new variant first identified in South Africa, has arrived and is now the dominant strain in most places able to do the sequencing to identify it. The numbers are astonishing: 1 in 15 people positive in England, and 1 in 20 in Scotland. Similar numbers are being reported in Manhattan, New York, and in those places with the ability to detect cases. I presume these numbers also to be true in India, South Africa, Brazil and other countries, although the surveillance systems don't exist to confirm this.

I was asked by the BBC to comment right after Johnson's announcement but declined. I wasn't sure what he and the Chief Medical Officer (now Sir) Chris Whitty were going to say, and I would have needed time to think and process it all. So what *do* I think?

The world is still struggling with this wily virus. New Delhi has gone back into lockdown, and the US has surpassed one million daily confirmed cases. Even the UK has abandoned PCR testing, as systems had become overwhelmed, and asked people to test at home, which, while quicker, means that the sequencing of genomes and any tracking of cases gets lost. Some even think that now is the time to abandon testing as a whole and move towards treating COVID-19 like the common cold. In fact, South Africa has asked those who are asymptomatic (but test positive) to return to their jobs and not isolate any longer, given that isolation policies for healthcare staff have caused such widespread disruption in hospitals.

As the world tries to figure out how to 'live with COVID-19', global eradication seems a far-off dream. Unfortunately, no country has yet figured out how to do this well over time without crashing health services, stifling social lives, impacting the economy, or having widespread disruption in one way or another, be that in travel or hospitality – not to mention mass deaths. Countries that did well in 2020 and into 2021, like South Korea, Hong Kong, New Zealand and Senegal, now struggle with imported variants that have become harder and harder to contain and manage.

China is now the only country left on earth pursuing Zero COVID, locking down millions of people to catch a handful of cases. Part of this is because its home-grown vaccines (Sinovac and Sinopharm) lack effectiveness in protecting against new variants such as Omicron and even Delta, with an internal push to move towards mRNA vaccines like Moderna and Pfizer. Australia and New Zealand have opened up their borders using a combination of vaccination and testing, and are effecting a transition towards 'living with COVID-19' but in a strictly managed way. Tennis is again making headlines, with the main story being Novak Djokovic, World Number One and notorious anti-vaxxer, being stuck in a border isolation facility because he didn't comply with the Australian Open guidance that players should be fully vaccinated. Djokovic claims he has a medical exemption, as he tested positive for the virus a few weeks back (16 December 2021). Australian authorities have strictly enforced the rules, trying to show that neither wealth nor celebrity can buy special privileges. The Australian Prime Minister Scott Morrison tweeted, 'Mr Djokovic's visa has been cancelled. Rules are rules, especially when it comes to our borders. No one is above these rules. Our strong border policies have been critical to Australia having one of the lowest death rates in the world from COVID-19, we are continuing to be vigilant.' These stories of 'one rule for us, another for them' have continued to play out across the globe. In the UK, for example, photographs of Boris Johnson at what is seemingly a garden party have surfaced, from a time during lockdown when such mixing was prohibited for the public at large. (Johnson & Co. claim that this was a 'work' meeting – we don't have cheese and wine at most of my work meetings.)

And, while vaccines have delivered on their promise to stop severe infection and people dying, those vaccinated can still get infected and pass on COVID-19 to others. Plus getting COVID-19 once doesn't mean you won't get it again, possibly in a different form. This means achieving any kind of 'herd immunity' static state isn't possible; the virus will continue to circulate, finding hosts to infect and continuing its spread. So the scientific community has a new goal: next generation broad-spectrum vaccines that give sterilizing immunity, which means those who are vaccinated cannot get infected or transmit to others. This would be better than having to vaccinate the population every four to six months (so-called 'boosters').

No, we still don't know where SARS-CoV-2 came from. Six more months haven't presented any new clues as to an animal host in China, nor has the Chinese government opened its labs for independent scrutiny. The discussion around 'lab leak' has moved from outlandish conspiracy theory to reasonable hypothesis needing examination, given an original animal reservoir for SARS-CoV-2 has yet to be identified almost two years later. In the meantime, SARS-CoV-2 loves animals and has found an excellent host in wild deer in the US, as well as in hippos, tigers, rats, and house cats and dogs around the world.

MERS has flared up again in the Middle East, and this also keeps me up at night. Saudi Arabia has reported 4 additional MERS cases this week in January 2022, which makes 17 total cases in that country in 2021. This is a disease with a case fatality rate of 35 per cent, and which has human-to-human transmission potential through respiratory mechanisms. Luckily, outbreaks have been managed quickly, but a version that could take off into a global pandemic is only a few mutations away.

If a pandemic-ready version of MERS started to spread, would it be treated the same way as COVID-19? Would thousands of people show up at protests because they had read on Facebook that MERS was a hoax? One would hope that such carnage would be recognized as such. Unfortunately, with COVID-19, we have seen the line between facts and lies disintegrate. Expertise and training have given way to influencers on YouTube or Facebook who have figured out that misinformation sells, pushing their clickbait to hundreds of thousands of people. This misinformation is often worryingly organized, making the management of any pandemic incredibly challenging for governments and experts.

I don't have a good answer on how we can overcome the flood of absolute rubbish on the internet that is costing lives (and causing lockdowns). Will the big tech companies change their policies? Will governments force them to do so? Even after all this time, as a scientist, the best answer for me seems to be that the scientific community must continue to engage with the public, talking and writing as accessibly and patiently as we can, in the hope that people choose reliable sources for their information.

Over the past month, my energy has been spent on schools policy and trying to make the argument that we now have the tools to keep schools open. This is now being recognized globally: for example, New York City former Mayor Bill de Blasio and his replacement Eric Adams gave a 'Stay Safe, Stay Open' briefing in December 2021 in which I was invited to give expert remarks on schools and COVID-19. (Although it did make me chuckle to hear de Blasio introduce me via Zoom: 'He is a world-class expert in global health and his research on schools is influential.' I guess I'll never escape being misgendered as a man.)

Omicron is one of the most infectious viruses in human history. And, while vaccines have blunted the severe impact on health, it is creating widespread disruption as a result of the sheer number of people infected and having to be off work to isolate. Flights and trains have been cancelled because of staff shortages. Hospitals are missing a significant portion of their workforce. The same applies to border control, supermarkets, schools, police forces and even Apple Stores. Omicron has now turned into a disease that challenges how we keep society functioning. And, while some point to its being 'milder' – the hospitalization rates are estimated to be half of those of the Delta variant – it is still hospitalizing and killing people, especially those who are unvaccinated.

With almost two years of 'pandemic memory' behind them, the public has also become more adept at avoiding COVID-19 (and understanding the severity of the disease), which means even if governments don't shut the hospitality sector, footfall has been significantly reduced. People have locked themselves down and changed their behaviour. This is more economically devastating than a legal lockdown that comes with financial support and benefits. Businesses are starved week by week rather than supported. This comes back to the point I repeatedly tried to make through the pandemic: it is the virus causing economic harm rather than restrictions alone.

While some say we should adapt normal social relations and mixing for the foreseeable future, I struggle with this line of thought. Humans are social: we need to hug, talk, dance, sing, kiss and be around others. We're not bears or rhinos or other solitary creatures. We like seeing each other's faces. And we know that a sense of community and connection are vital to wellbeing too. A holistic approach to public health is vital, and this includes not just people's mental health but also their ability to pay rent, feed their family, stay warm through the winter and have a meaningful role in society, be that going to church or being part of a glee club. For a certain period of time, altering these made sense, so that we could avoid preventable illness and deaths; allow vaccines to be created, trialled and distributed in 2020 and into 2021; allow clinicians to better understand how to treat COVID-19; and allow a better understanding of transmission and risk.

But it would be naive to say that this time was 'bought' with no cost. One has only to look at children and young people: they are the generation who have largely paid the price for this time in terms of their mental health, employment opportunities and educational future. Childhood obesity has increased, especially among the poorest quintile. Over the next year I want to research physical activity and children in greater depth, and hopefully engage with governments about this topic. While I haven't had time to finish the personal-training certification that I started in my free time, I continue to incorporate fitness, exercise, getting outside and green space into our research work on public health, wellbeing and pandemic recovery.

As we start to understand the full harms to society and the economy of lockdowns, is there another universe out there somewhere

where certain countries could have avoided lockdown? Was there a way to isolate 'vulnerable people' and just let others get it and acquire natural immunity? Could other paths have been taken early on? These reflections are important, although we have to be clear on the facts. The core of our response to COVID-19 was about 'flattening the curve': keeping hospitalizations within healthcare capacity and trying to provide care to everyone who needs it. Vaccines and antivirals helped with this problem. Shouting about how harmful lockdown is didn't, and still doesn't.

Personally, I've taken a step back from media work. I spoke up in the first place because I felt there was an absence of voices and expertise in the public space to communicate what was happening, why it was happening and where it was going. Now a multitude of scientists are engaged in daily briefings, and the landscape has been transformed. If someone else can say what I'd say, I'd prefer they be the ones to say it. Not because I'm afraid of the backlash or the heat that comes with saying difficult truths – as I've mentioned, this is vital in tackling misinformation – but because a democracy and society are richer when there is a multitude of voices and perspectives – especially expert voices. You've heard enough from me.

The scientists I respect the most have evolved their position on, and analysis of, this pandemic based on emerging data and new knowledge. That is the core of science: thinking about a problem differently based on new information. Unfortunately, perhaps out of fear of being seen to be wrong – especially when they might have defended that position for a long time – COVID-19 'camps' have become even more hard-line, with people remaining terrified of COVID-19, locking themselves away and waiting for it all to end, or continuing to talk about it as if we've overreacted to plain old flu. There's a middle path: one in which we evolve our management and response based on new data and tools. Omicron in the highly vaccinated population of 2022 is different from SARS-CoV-2 in the population of 2020, which had no prior immunity through vaccination or infection (a 'naive' population). Most sensible people should understand this.

The question has really changed from 'When does it end?' to 'How do we manage and live alongside this virus?'; and I'm hopeful. As we saw earlier in the book, we now have safe and effective vaccines, new oral antivirals and rapid lateral-flow tests – the last are nifty little kits that look like pregnancy tests and allow people to swab their noses and throats at home to figure out if they're infectious, all within minutes. This was the stuff of dreams in early 2020. It is now reality.

Science continues to push ahead and make breakthroughs to transform COVID-19 from a deadly virus to a manageable one. Will it fade into the background and become just like other manageable infectious diseases? I hope so. But we're nowhere near that now. And, in fact, as Omicron is more concentrated in the upper respiratory tract, rather than in the lungs – as we read about earlier in the book – we are seeing an increase in paediatric admissions in under-fives. Unfortunately, this means that in poorer countries it could transform into a major childhood killer. COVID-19 could become a disease of children similar to childhood pneumonia, croup and RSV. Childhood pneumonia, for example, is already a major challenge in countries that don't have good health systems, and lack the ability to provide oxygen to those children suffering from breathing difficulties.

And perhaps that's what COVID-19 is becoming: something similar to other global health problems like measles, malaria and even tuberculosis – a disease we've managed in richer countries using tools likes vaccines and medicines, but that remains a major disrupter of normal life in poorer countries. That is the story of global health: richer countries race ahead and solve the major challenges of infectious diseases, and poorer countries are left behind.

Vaccine inequality has also triggered major resentment towards the West, which hoarded vaccines and then continued to buy up supply with plans to vaccinate three, four, five . . . who knows how many times, as well as 'boost' people with the mRNA vaccines that have come out the 'winner' in the vaccine race. Maybe Pfizer will start offering a reward card with a free coffee after the seventh jab. The selfishness of the West won't easily be forgotten. Will the world continue to see Britain and the US as generous and able donors in global health? Can WHO pick itself up and rebuild for the next pandemic, given its entanglements in fights between the Chinese and US governments? How will countries engage with China in the months and years to come if it becomes clear that the government knows more than it has let on thus far? Will COVID-19 be the spark for the Third World War? . . . These are the questions that loom in international diplomacy.

As for me, have I had COVID-19 yet? No, I haven't, which feels somehow impossible. In a population of 5.5 million in Scotland, one million people have tested positive (and this is just confirmed cases). I've been testing several times a week, and, while I cautiously avoid crowded spaces, and wear masks on public transport and in shops, I continue to go to the gym and to hot yoga and to see friends outside or in small groups. I've found a sustainable way to live alongside COVID-19 for now. But, as with everything else in a pandemic, this could change in the next few days. If we ever run into each other – perhaps on our paddle-boards in the Pacific Ocean; or, more realistically in the short term, in Wardie Bay in Edinburgh – I will let you know what happened.

# Acknowledgements

A book is a labour of love and time, and far from the product of just one person. This is my chance to thank all those who supported me during the COVID-19 crisis and made this writing project possible. First, my family and friends. You know who you are. Thank you for your love, warmth and making each day bright and exciting. Second, my research team at Edinburgh University, particularly Genevie Fernandes, Lorna Thompson and my two research assistants for this book, Jay Patel and Maartje Kletter. I'd also like to thank the team at the Global Health Governance Programme: Joey Brooke, Adriel Chen, Jonny Grey, Lois King, Gwenetta Curry, Ines Hassan and Felix Stein. I'm grateful to my colleagues at Edinburgh University; within the Scottish government civil service and the UK Cabinet Office civil service; and at the Wellcome Trust for supporting our team's research. Also a shout-out to all those I've learned from on various expert advisory groups, such as the Royal Society DELVE initiative, *Lancet* commissions, and Scottish and UK government advisory groups. This book wouldn't have been written without my agent, Andrew Gordon, and my editor, Connor Brown (as well as Assallah Tahir). And finally my thanks to all of you reading this book, including those who supported me either online or in real life through a roller-coaster of two years. I hope I'll have a chance to meet you in person someday – I'm kind of fed up with all this virtual stuff.

# Bibliography

*Web links are correct at the time of going to press.*

## *Introduction: Warnings of a Global Pandemic*


Byatnal, A., 'Will we ever know the real death toll of the pandemic?', *National Geographic*, 2021 https://www.nationalgeographic.com/science/article/will-we-ever-know-the-real-death-toll-of-the-pandemic

*Economist*, 'Tracking COVID-19 excess deaths across countries', 2021 https://www.economist.com/graphic-detail/coronavirus-excess-deaths-tracker

Evans, D., and Over, M., 'The economic impact of COVID-19 in low- and middle-income countries', Center for Global Development, 2020 https://www.cgdev.org/blog/economic-impact-covid-19-low-and-middle-income-countries

Georgieva, K., 'Confronting the crisis: priorities for the global economy', International Monetary Fund, 2020 https://www.imf.org/en/News/Articles/2020/04/07/sp040920-SMs2020-Curtain-Raiser

International Labour Organization (ILO), 'ILO Monitor: COVID-19 and the world of work', ILO, 2021 https://www.ilo.org/global/topics/coronavirus/impacts-and-responses/WCMS_767028/lang--en/index.htm

International Monetary Fund, 'Transcript of Kristalina Georgieva's participation in the World Health Organization Press Briefing', 2020 https://www.imf.org/en/News/Articles/2020/04/03/tr040320-transcript-kristalina-georgieva-participation-world-health-organization-press-briefing

Karlinsky, A., and Kobak, D., 'The World Mortality Dataset: tracking excess mortality across countries during the COVID-19 pandemic', medRxiv, 2021 https://www.medrxiv.org/content/10.1101/2021.01.27.21250604v1

Kharas, H., 'What to do about the coming debt crisis in developing countries', Brookings, 2020 https://www.brookings.edu/blog/future-development/2020/04/13/what-to-do-about-the-coming-debt-crisis-in-developing-countries/

Laker, C., Yonzan, N., Mahler, D. G., et al., 'Updated estimates of the impact of COVID-19 on global poverty: looking back at 2020 and the outlook for 2021', World Bank Blogs, 2021 https://blogs.worldbank.org/opendata/updated-estimates-impact-covid-19-global-poverty-looking-back-2020-and-outlook-2021

Murdock, J., 'Bill Gates feels "terrible" about coronavirus pandemic, wishes he had done more to warn about danger', *Newsweek*, 2020 https://www.newsweek.com/bill-gates-interview-coronavirus-covid19-pandemic-warnings-trump-administration-1503385

Organisation for Economic Co-operation and Development, 'The impact of the coronavirus (COVID-19) crisis on development finance', 2020 https://www.oecd.org/coronavirus/policy-responses/the-impact-of-the-coronavirus-covid-19-crisis-on-development-finance-9de00b3b/

Prasad, R., 'Coronavirus: do excess deaths suggest mortality crossed one million?', *Hindu Times*, 2021 https://www.thehindu.com/sci-tech/science/coronavirus-do-excess-deaths-suggest-mortality-crossed-one-million/article34860615.ece

United Nations Conference of Trade and Development, 'COVID-19 is a matter of life and debt, global deal needed', 2020 https://unctad.org/news/covid-19-matter-life-and-debt-global-deal-needed

United Nations Conference of Trade and Development, 'Global foreign direct investment fell by 42 per cent in 2020, outlook remains weak', 2021 https://unctad.org/news/global-foreign-direct-investment-fell-42-2020-outlook-remains-weak

United Nations Development Programme, 'COVID-19: Looming crisis in developing countries threatens to devastate economies and ramp up inequality', 2020 https://www.undp.org/press-releases/covid-19-looming-crisis-developing-countries-threatens-devastate-economies-and-ramp

US Global Leadership Coalition, 'COVID-19 brief: impact on the economies of developing countries. United States Global Leadership Coalition', 2021 https://www.usglc.org/coronavirus/economies-of-developing-countries/

World Bank Group, 'World Bank country and lending groups', 2021 https://datahelpdesk.worldbank.org/knowledgebase/articles/906519-world-bank-country-and-lending-groups

World Health Organization, 'The true death toll of COVID-19: estimating global excess mortality', 2021 https://www.who.int/data/stories/the-true-death-toll-of-covid-19-estimating-global-excess-mortality

Yong, E., 'Our pandemic summer', *Atlantic*, 2020 https://www.theatlantic.com/health/archive/2020/04/pandemic-summer-coronavirus-reopening-back-normal/609940/?utm_content=edit-promo&utm_campaign=the-atlantic&utm_term=2020-04-14T19per cent3A52per cent3A58&utm_source=twitter&utm_medium=social

Yoshida, N., Narayan, A., and Wu, H., 'How COVID-19 affects households in poorest countries – insights from phone surveys', World Bank Blogs, 2020 https://blogs.worldbank.org/voices/how-covid-19-affects-households-poorest-countries-insights-phone-surveys

## 1. Spillover

Aguirre, A. A., Catherina, R., Frye, H., et al., 'Illicit wildlife trade, wet markets and COVID-19: preventing future pandemics', *World Medical and Health Policy*, 2020 https://onlinelibrary.wiley.com/doi/full/10.1002/wmh3.348

Andersen, K. G., Rambaut, A., Lipkin, W. I., et al., 'The proximal origin of SARS-CoV-2', *Nature Medicine*, 2020 https://doi.org/10.1038/s41591-020-0820-9

Anthony, S. J., Johnson, C. K., Greig, D. J., et al., 'Global patterns in coronavirus diversity', *Virus Evolution*, 2017 https://academic.oup.com/ve/article/3/1/vex012/3866407

Baker, Nicholson, 'The lab-leak hypothesis', *Intelligencer*, 2021 https://nymag.com/intelligencer/article/coronavirus-lab-escape-theory.html

Banerjee, A., Doxey, A. C., Mossman, K., et al., 'Unraveling the zoonotic origin and transmission of SARS-CoV-2', *Trends in Ecology and Evolution*, 2021 https://www.cell.com/trends/ecology-evolution/fulltext/S0169-5347(20)30348-7?_returnURL=https%3A%2F%2Flinkinghub.elsevier.com%2Fretrieve%2Fpii%2FS0169534720303487%3Fshowall%3Dtrue

Barber, M. R., Guan, Y., Magor, K. E., et al., 'The role of animal surveillance in influenza preparedness: the consequence of inapparent infection in ducks and pigs', *Influenza and Other Respiratory Viruses*, 2011 https://www.researchgate.net/publication/51487875_The_role_of_animal_surveillance_in_influenza_preparedness_the_consequence_of_inapparent_infection_in_ducks_and_pigs

Barnett, T., and Fournié, G., 'Zoonoses and wet markets: beyond technical interventions', *Lancet Planet Health*, 2021 https://doi.org/10.1016/S2542-5196(20)30294-1

*Beijing News*, 'Dialogue "rumor" was reprimanded. Doctor: I'm reminding everyone to pay attention to prevention', 2020 https://web.archive.org/web/20200206144253/http://www.bjnews.com.cn/feature/2020/01/31/682076.html

Boni, M. F., Lemey, P., Jiang, X., et al., 'Evolutionary origins of the SARS-CoV-2 sarbecovirus lineage responsible for the COVID-19 pandemic', *Nature Microbiology*, 2020 https://doi.org/10.1038/s41564-020-0771-4

Brown, C., 'Review: *Spillover: animal infection and the next human pandemic*', *Emerging Infectious Diseases*, 2013 https://doi.org/10.3201/eid1902.121694

Burkhi, T., 'The origin of SARS-CoV-2', *Lancet Infectious Diseases*, 2020 https://www.ncbi.nlm.nih.gov/pmc/articles/PMC7449661/

Collins, F. S., 'Statement on funding pause on certain types of gain-of-function research', National Institutes of Health, 2014 https://www.nih.gov/about-nih/who-we-are/nih-director/statements/statement-funding-pause-certain-types-gain-function-research

Collins, F., 'Genomic study points to natural origin of COVID-19', National Institutes of Health, 2020 https://directorsblog.nih.gov/2020/03/26/genomic-research-points-to-natural-origin-of-covid-19/

Cyranoski, D., 'Bat cave solves mystery of deadly SARS virus – and suggests new outbreak could occur', *Nature*, 2017 https://doi.org/10.1038/d41586-017-07766-9

Cyranoski, D., 'Inside the Chinese lab poised to study world's most dangerous pathogens', *Nature*, 2017 https://www.nature.com/news/inside-the-chinese-lab-poised-to-study-world-s-most-dangerous-pathogens-1.21487

Dawood, F. S., Jain, S., Finelli, L., et al., 'Emergence of a novel swine-origin influenza A (H1N1) virus in humans', *New England Journal of Medicine*, 2009 https://www.nejm.org/doi/pdf/10.1056/NEJMoa0903810

de Oliveira Padilha, M. A., de Oliveira Melo, J., Romano, G., et al., 'Comparison of malaria incidence rates and socioeconomic-environmental factors between the states of Acre and Rondônia: a spatio-temporal modelling study', *Malaria Journal*, 2019 https://doi.org/10.1186/s12936-019-2938-0

Dwyer, D., 'I was on the WHO's mission to China, here's what we found', 2021 https://www.theguardian.com/commentisfree/2021/feb/22/i-was-on-the-whos-covid-mission-to-china-heres-what-we-found

Eurogroup for Animals, 'Dr Anthony Fauci demands global shutdown of wet markets', 2020, https://www.eurogroupforanimals.org/news/dr-anthony-fauci-demands-global-shutdown-wet-markets

Forgey, Q., ' "Shut down those things right away": calls to close "wet markets" ramp up pressure on China', *Politico*, 2020 https://www.politico.com/news/2020/04/03/anthony-fauci-foreign-wet-markets-shutdown-162975

Ge, X.-Y., Li, J.-Y., Yang, X.-L., et al., 'Isolation and characterization of a bat SARS-like coronavirus that uses the ACE2 receptor', *Nature*, 2013 https://doi.org/10.1038/nature12711

Gostin, L. O., Habibi, R., and Meier, B. M., 'Has global health law risen to meet the COVID-19 challenge? Revisiting the International Health Regulations to prepare for future threats', *Journal of Law, Medicine and Ethics*, 2020 https://doi.org/10.1177%2F1073110520935354

Green, A., 'Li Wenliang: obituary', 2020, *Lancet*, https://www.thelancet.com/pdfs/journals/lancet/PIIS0140-6736(20)30382-2.pdf

Green, M. H., 'How a microbe becomes a pandemic: a new story of the Black Death', *Lancet Microbe*, 2020 https://doi.org/10.1016/S2666-5247(20)30176-2

Greenfield, P., 'Ban wildlife markets to avert pandemics, says UN biodiversity chief', *Guardian*, 2020 https://www.theguardian.com/world/2020/apr/06/ban-live-animal-markets-pandemics-un-biodiversity-chief-age-of-extinction

Habibi, R., Burci, G. L., de Campos, T. C., et al., ' "Do not violate the International Health Regulations during the COVID-19 outbreak" ', *Lancet*, 2020 https://doi.org/10.1016/S0140-6736(20)30373-1

Hu, B., Zeng, L.-P., Yang, X.-L., et al., 'Discovery of a rich gene pool of bat SARS-related coronaviruses provides new insights into the origin of SARS coronavirus', *PLoS Pathogens*, 2017 https://doi.org/10.1371/journal.ppat.1006698

Jee, Y., 'WHO International Health Regulations Emergency Committee for the COVID-19 outbreak', *Epidemiological Health*, 2020 https://doi.org/10.4178/epih.e2020013

Johnson, C. K., et al., 'Spillover and pandemic properties of zoonotic viruses with high host plasticity', *Scientific Reports*, 2015 https://doi.org/10.1038/srep14830

Jones, K., Patel, N., Levy, M., et al., 'Global trends in emerging infectious diseases', *Nature*, 2018 https://doi.org/10.1038/nature06536

Judson, S. D., Fischer, R., Judson, A., et al., 'Ecological contexts of index cases and spillover events of different Ebola viruses', *PLoS Pathogens*, 2016 https://doi.org/10.1371/journal.ppat.1005780

Keele, B. F., Van Heuverswyn, F., Li, Y.-Y., et al., 'Chimpanzee reservoirs of pandemic and nonpandemic HIV-1', *Science*, 2006 https://doi.org/10.1126/science.1126531

Li, W.-D., Shi, Z.-L., Yu, M., et al., 'Bats are natural reservoirs of SARS-like coronaviruses', *Science*, 2005 https://www.science.org/doi/10.1126/science.1118391

Liu, S.-L., Saif, L. J., Weiss, S. R., et al., 'No credible evidence supporting claims of the laboratory engineering of SARS-CoV-2', *Emerging Microbes and Infections*, 2020 https://www.tandfonline.com/doi/full/10.1080/22221751.2020.1733440

Ma, J., and Mai, J., 'Death of coronavirus doctor Li Wenliang becomes catalyst for "freedom of speech" demands in China', *South China Morning Post*, 2020 https://www.scmp.com/news/china/politics/article/3049606/coronavirus-doctors-death-becomes-catalyst-freedom-speech

MacDonald, A. J., and Mordecai, E. A., 'Amazon deforestation drives malaria transmission, and malaria burden reduces forest clearing', *Proceedings of the National Academy of Sciences*, 2019 https://doi.org/10.1073/pnas.1905315116

MacLean, O. E., Lytras, S., Weaver, S., et al., 'Natural selection in the evolution of SARS-CoV-2 in bats, not humans, created a highly capable human pathogen', bioRxiv, 2020 https://doi.org/10.1101/2020.05.28.122366

Mallapaty, S., 'Where did COVID come from? Five mysteries that remain', *Nature*, 2021 https://media.nature.com/original/magazine-assets/d41586-021-00502-4/d41586-021-00502-4.pdf

Marshall, M., 'COVID-19 – a blessing for pangolins?', *Guardian*, 2020 https://www.theguardian.com/environment/2020/apr/18/covid-19-a-blessing-for-pangolins

Menachery, V., Yount, B., Debbink, K., et al., 'A SARS-like cluster of circulating bat coronaviruses shows potential for human emergence', *Nature Medicine*, 2015 https://doi.org/10.1038/nm.3985

Metzl, J., 'How to hold Beijing accountable for the coronavirus', *Wall Street Journal*, 2020 https://www.wsj.com/articles/how-to-hold-beijing-accountable-for-the-coronavirus-11595976973

Morse, S. S., Mazet, J. A., Woolhouse, M., et al., 'Prediction and prevention of the next pandemic zoonosis', *Lancet*, 2012 https://doi.org/10.1016/S0140-6736(12)61684-5

Mullen, L., Potter, C., Gostin, L. O., et al., 'An analysis of International Health Regulations Emergency Committees and Public Health Emergency of International Concern Designations', *BMJ Global Health*, 2020 https://gh.bmj.com/content/5/6/e002502

Nadimpalli, M. L., and Pickering, A. J., 'A call for global monitoring of WASH in wet markets', *Lancet Planetary Health*, 2020 https://doi.org/10.1016/S2542-5196(20)30204-7

*New York Times*, 'He warned of coronavirus. Here's what he told us before he died', 2020 https://www.nytimes.com/2020/02/07/world/asia/Li-Wenliang-china-coronavirus.html

Newey, S., 'All hypotheses remain on the table in COVID-19 origins investigation, says WHO', *Telegraph*, 2021 https://www.telegraph.co.uk/global-health/science-and-disease/hypotheses-remain-table-covid-19-origins-investigation-says/

Newey, S., 'WHO mission a "really big step forward" to understand COVID-19 origins, team members say', *Telegraph*, 2021 https://www.telegraph.co.uk/global-health/science-and-disease/exclusive-mission-really-big-step-forward-understand-covid-19/

Nichol, S. T., Arikawa, J., and Kawaoka, Y., 'Emerging viral diseases', *Proceedings of the National Academy of Sciences*, 2000 https://doi.org/10.1073/pnas.210382297

Nie, J.-B., and Elliott, C., 'Humiliating whistle-blowers: Li Wenliang, the response to COVID-19, and the call for a decent society', *Journal of Bioethical Inquiry*, 2020 https://doi.org/10.1007/s11673-020-09990-x

Niewijk, G., 'Controversy aside, why the source of COVID-19 matters', *Genetic Engineering and Biotechnology News*, 2020 https://www.genengnews.com/insights/controversy-aside-why-the-source-of-covid-19-matters/

Norwegian Institute of Public Health, 'Urbanization and preparedness for outbreaks with high-impact respiratory pathogens', 2020 https://www.fhi.no/en/publ/2020/urbanization-and-preparedness-for-outbreaks-with-high-impact-respiratory-pa/

Palmer, J., 'Wuhan gets its first virus martyr', *Foreign Policy*, 2020 https://foreignpolicy.com/2020/02/06/li-wenliang-coronavirus-lies-wuhan-gets-its-first-virus-martyr/

Petrikova, I., Cole, J., and Farlow, A., 'COVID-19, wet markets and planetary health', *Lancet Planetary Health*, 2020 https://doi.org/10.1016/S2542-5196(20)30122-4

Pilling, D., 'Pandemics and the paradox of human progress', *Financial Times* 2021 https://www.ft.com/content/6916622e-ab7c-4e05-a8cc-7e70152cd8eb

Plowright, R. K., Field, H. E., Smith C., et al., 'Reproduction and nutritional stress are risk factors for Hendra virus infection in little red flying foxes (*Pteropus scapulatus*)', *Proceedings: Biological Sciences*, 2008 https://doi.org/10.1098/rspb.2007.1260

Pulliam, J. R. C., Epstein, J. H., Dushoff, J., et al., 'Agricultural intensification, priming for persistence and the emergence of Nipah virus: a lethal bat-borne zoonosis', *Journal of the Royal Society Interface*, 2012 https://doi.org/10.1098/rsif.2011.0223

Qifeng, Z., Gan, J., and Xun, L., 'Eating habits in south China driving endangered animals to extinction', *China Dialogue*, 2012 https://chinadialogue.net/en/nature/5506-eating-habits-in-south-china-driving-endangered-animals-to-extinction/

Rasmussen, A. L., 'On the origins of SARS-CoV-2', *Nature Medicine*, 2021 https://doi.org/10.1038/s41591-020-01205-5

Richardson, V., 'Lindsey Graham calls on China to shut down "absolutely disgusting" wet markets', *Washington Times*, 2020 https://www.washingtontimes.com/news/2020/mar/31/lindsey-graham-calls-china-shut-down-absolutely-di/

Sallard, E., Halloy, J., Casane, D., et al., 'Tracing the origins of SARS-COV-2 in coronavirus phylogenies: a review', *Environmental Chemistry Letters*, 2021 https://doi.org/10.1007/s10311-020-01151-1

Samuel, S., 'The coronavirus likely came from China's wet markets. They're reopening anyway', *Vox*, 2020 https://www.vox.com/future-perfect/2020/4/15/21219222/coronavirus-china-ban-wet-markets-reopening

*Science Daily*, 'The link between virus spillover, wildlife extinction and the environment', 2020 https://www.sciencedaily.com/releases/2020/04/200407215653.htm

Smith, N., '"The West was not humble enough": Taiwan's health minister on why COVID brought Britain to its knees', *Telegraph*, 2021 https://www.telegraph.co.uk/global-health/science-and-disease/west-not-humble-enough-taiwans-health-minister-covid-brought/

Su, A., 'A doctor was arrested for warning China about the coronavirus. Then he died of it', *Los Angeles Times*, 2020 https://www.latimes.com/world-nation/story/2020-02-06/coronavirus-china-xi-li-wenliang

Taylor, A. L., Habibi, R., Burci, G. L., et al., 'Solidarity in the wake of COVID-19: reimagining the International Health Regulations', *Lancet*, 2020 https://www.thelancet.com/pdfs/journals/lancet/PIIS0140-6736(20)31417-3.pdf

Wan, Y., Shang, J., Graham, R., et al., 'Receptor recognition by the novel coronavirus from Wuhan: an analysis based on decade-long structural studies of SARS coronavirus', *Journal of Virology*, 2020 https://doi.org/10.1128/JVI.00127-20

*Washington Post*, 'Experts debunk fringe theory linking China's coronavirus to weapons research', 2020 https://www.washingtonpost.com/world/2020/01/29/experts-debunk-fringe-theory-linking-chinas-coronavirus-weapons-research/?fbclid=IwAR1Xvt6WAMTFnAROnbRx1KBNOTqj5m2bG6tQaYy4NwegmsmBb8KEXQ_uDCQ

Watts, N., Amann, M., Arnell, N., et al., 'The 2020 report of the *Lancet* countdown on health and climate change: responding to converging crises', *Lancet*, 2021 https://doi.org/10.1016/S0140-6736(20)32290-X

Webster, R., 'Wet markets – a continuing source of severe acute respiratory syndrome and influenza?', *Lancet*, 2004 https://doi.org/10.1016/S0140-6736(03)15329-9

Wilson, K., Halabi, S., Gostin, L. O., 'The International Health Regulations (2005), the threat of populism and the COVID-19 pandemic', *Global Health* 2020 https://doi.org/10.1186/s12992-020-00600-4

Woo, P.-C., Lau, S.-K., and Yuen, K.-Y., 'Infectious diseases emerging from Chinese wet markets: zoonotic origins of severe respiratory viral infections', *Current Opinion in Infectious Diseases*, 2006 https://journals.lww.com/co-infectiousdiseases/Abstract/2006/10000/Infectious_diseases_emerging_from_Chinese.2.aspx

World Health Organization, 'COVID-19 virtual press conference transcript – 12 February 2021', 2021 https://www.who.int/publications/m/item/covid-19-virtual-press-conference-transcript---12-february-2021

World Health Organization, 'Episode 21. COVID-19 – origins of the SARS-CoV-2 virus', 2021 https://www.who.int/emergencies/diseases/novel-coronavirus-2019/media-resources/science-in-5/episode-21---covid-19---origins-of-the-sars-cov-2-virus

World Health Organization, 'International health regulations committees and expert roster', 2020 https://www.who.int/teams/ihr/ihr-emergency-committees

World Health Organization, 'International Health Regulations', 2005 https://www.globalhealthrights.org/wp-content/uploads/2013/10/International-Health-Regulations-2005.pdf

Wu, F., Zhao, S., Yu, B., et al., 'A new coronavirus associated with human respiratory disease in China', *Nature*, 2020 https://doi.org/10.1038/s41586-020-2008-3

Yang, L., Wu, Z., Ren, X., et al., 'Novel SARS-like betacoronaviruses in bats, China, 2011', *Emerging Infectious Diseases*, 2013 https://wwwnc.cdc.gov/eid/article/19/6/12-1648_article

Yu, S., and Liu, X., 'Coronavirus piles pressure on China's exotic animal trade', *Financial Times*, 2020 https://www.ft.com/content/74f1b26e-53c0-11ea-90ad-25e377c0ee1f

Yuan, J., Tang, X., Yang, Z., et al., 'Enhanced disinfection and regular closure of wet markets reduced the risk of avian influenza A virus transmission', *Clinical Infectious Diseases*, 2014 https://doi.org/10.1093/cid/cit951

Zhang, Q., Zhang, H., Gao, J., et al., 'A serological survey of SARS-CoV-2 in cat in Wuhan', *Emerging Microbes and Infections*, 2020 https://doi.org/10.1080/22221751.2020.1817796

Zhang, T., Wu, Q., Zhang, Z., 'Pangolin homology associated with 2019-nCoV', bioRxiv, 2020 https://doi.org/10.1101/2020.02.19.950253

Zhong, S., Crang, M., and Zeng, G., 'Constructing freshness: the vitality of wet markets in urban China', *Agriculture and Human Values*, 2020 https://doi.org/10.1007/s10460-019-09987-2

Zhou, P., Yang, X.-L., Wang, X.-G., et al., 'A pneumonia outbreak associated with a new coronavirus of probable bat origin', *Nature*, 2020 https://doi.org/10.1038/s41586-020-2012-7

Zimmer, K., 'Deforestation is leading to more infectious diseases in humans', *National Geographic*, 2019 https://www.nationalgeographic.com/science/article/deforestation-leading-to-more-infectious-diseases-in-humans

Zorocostas, Z., 'WHO team begins COVID-19 origin investigation', *Lancet*, 2021 https://www.ncbi.nlm.nih.gov/pmc/articles/PMC7906677/

## 2. China's Hammer

BBC News, 'China COVID-19: nearly 500,000 in Wuhan may have had virus, says study', 2020 https://www.bbc.co.uk/news/world-asia-china-55481397

Branswell, H., 'A disease detective is thrust on to the frontlines of WHO's COVID-19 response', STAT News, 2020 https://www.statnews.com/2020/07/08/a-disease-detective-is-thrust-onto-the-front-lines-of-whos-covid-19-response/

Burki, T., 'China's successful control of COVID-19', *Lancet Infectious Diseases*, 2020 https://www.thelancet.com/journals/laninf/article/PIIS1473-3099(20)30800-8/fulltext

Chan, J. F.-W., Yuan, S., Kok, K.-H., et al., 'A familial cluster of pneumonia associated with the 2019 novel coronavirus indicating person-to-person transmission: a study of a family cluster', *Lancet*, 2020 https://doi.org/10.1016/S0140-6736(20)30154-9

Duan, S., Zhou, M., Zhang, W., et al., 'Seroprevalence and asymptomatic carrier status of SARS-CoV-2 in Wuhan City and other places of China', *PLoS Neglected Tropical Diseases*, 2021 https://doi.org/10.1080/15387216.2020.1762103

Gill, V., 'Coronavirus: virus provides leaps in scientific understanding', BBC News, 2021 https://www.bbc.co.uk/news/science-environment-55565284

Gunia, A., 'China's draconian lockdown is getting credit for slowing coronavirus. Would it work anywhere else?', *Time*, 2020 https://time.com/5796425/china-coronavirus-lockdown/

Holshue, M., DeBolt, C., Lindquist, S., et al., 'First case of 2019 novel coronavirus in the United States', *New England Journal of Medicine*, 2020 https://www.nejm.org/doi/full/10.1056/NEJMoa2001191

Huang, C., Wang, T., Li, X., et al., 'Clinical features of patients infected with 2019 novel coronavirus in Wuhan, China', *Lancet*, 2020 https://doi.org/10.1016/S0140-6736(20)30183-5

Imai, N., Dorigatti, I., Cori, A., et al., 'Report 1: estimating the potential total number of novel coronavirus cases in Wuhan City, China', Imperial College, London, COVID-19 Response Team, 2020 https://www.imperial.ac.uk/media/imperial-college/medicine/sph/ide/gida-fellowships/Imperial-College-COVID19-epidemic-size-17-01-2020.pdf

Kang, D., 'The shunned: people from virus-hit city tracked, quarantine', Associated Press, 2020https://apnews.com/article/travel-wuhan-asia-pacific-beijing-weekend-reads-7f7336d2ed099936bd59bf8cb7f43756

Kuo, L., and Yang, L., '"Liberation" as Wuhan's coronavirus lockdown ends after seventy-six days', *Guardian*, 2020 https://www.theguardian.com/world/2020/apr/07/liberation-as-wuhans-coronavirus-lockdown-ends-after-76-days

Lau, H., Khosrawipour, V., Kocbach, P., et al., 'The positive impact of lockdown in Wuhan on containing the COVID-19 outbreak in China', *Journal of Travel Medicine*, 2020 https://doi.org/10.1093/jtm/taaa037

Li, Q., Guan, X., Wu, P., et al., 'Early transmission dynamics in Wuhan, China, of novel coronavirus-infected pneumonia', *New England Journal of Medicine*, https://www.nejm.org/doi/full/10.1056/nejmoa2001316

Lu, R., Zhao, X., Li, J., et al., 'Genomic characterisation and epidemiology of 2019 novel coronavirus: implications for virus origins and receptor binding', *Lancet*, 2020 https://doi.org/10.1016/S0140-6736(20)30251-8

Qian, Y., and Hanser, A., 'How did Wuhan residents cope with a 76-day lockdown?', *Chinese Sociological Review*, 2021 https://doi.org/10.1080/21620555.2020.1820319

Read, J. M., Bridgen, J. R., Cummings, D. A., et al., 'Novel coronavirus 2019-nCoV: early estimation of epidemiological parameters and epidemic predictions', medRxiv, 2020 https://www.medrxiv.org/content/10.1101/2020.01.23.20018549v2

Ren, X., 'Pandemic and lockdown: a territorial approach to COVID-19 in China, Italy and the United States', *Eurasian Geography and Economics*, 2020 https://doi.org/10.1080/15387216.2020.1762103

Rigby, J., '"No evidence for COVID-19 lab leak" but plenty for natural origins, leading virus-hunters conclude', *Telegraph*, 2021 https://www.telegraph.co.uk/global-health/science-and-disease/no-evidence-covid-19-lab-leak-plenty-natural-origins-leading/

Sanche, S., Lin, Y. T., Xu, C., et al., 'The novel coronavirus, 2019-nCoV, is highly contagious and more infectious than initially estimated', medRxiv, 2020 https://www.medrxiv.org/content/10.1101/2020.02.07.20021154v1

Senior, K., 'Recent Singapore SARS case a laboratory accident', *Lancet Infectious Diseases*, 2003 https://doi.org/10.1016/S1473-3099(03)00815-6

Tang, B., Wang, X., Li, Q., et al., 'Estimation of the transmission risk of the 2019-nCoV and its implication for public health interventions', *Journal of Clinical Medicine*, 2020 https://doi.org/10.3390/jcm9020462

Tian, H., Liu, Y., Li, Y., et al., 'Early evaluation of transmission control measures in response to the 2019 novel coronavirus outbreak in China', medRxiv, 2020 https://www.medrxiv.org/content/10.1101/2020.01.30.20019844v3.full.pdf

World Health Organization, 'Novel coronavirus (2019-nCoV) situation report – 3', 2020 https://www.who.int/docs/default-source/Coronaviruse/situation-reports/20200123-sitrep-3-2019-ncov.pdf

World Health Organization, 'WHO lists two COVID-19 tests for emergency use', 2020 https://www.who.int/news/item/07-04-2020-who-lists-two-covid-19-tests-for-emergency-use

Wu, F., Zhao, S., Yu, B., 'A new coronavirus associated with human respiratory disease in China', *Nature*, 2020 https://doi.org/10.1038/s41586-020-2008-3

Wu, J. T., Leung, K., and Leung, G. M., 'Nowcasting and forecasting the potential domestic and international spread of the 2019-nCoV outbreak originating in Wuhan, China: a modelling study', *Lancet*, 2020 https://www.thelancet.com/journals/lancet/article/PIIS0140-6736(20)30260-9/fulltext

Yang, Y., Lu, Q., Liu, M., et al., 'Epidemiological and clinical features of the 2019 novel coronavirus outbreak in China', medRxiv, 2020 https://www.medrxiv.org/content/10.1101/2020.02.10.20021675v2

Yuan, Z., Xiao, Y., Dai, Z., et al., 'Modelling the effects of Wuhan's lockdown during COVID-19, China', *Bulletin of the World Health Organization*, 2020 https://doi.org/10.2471/BLT.20.254045

Zhang, R., Liu, H., Li, F., et al., 'Transmission and epidemiological characteristics of severe acute respiratory syndrome coronavirus 2 (SARS-CoV-2) infected pneumonia (COVID-19): preliminarily evidence obtained in comparison with 2003-SARS', medRxiv, 2020 https://www.medrxiv.org/content/10.1101/2020.01.30.20019836v4

Zhou, P., Yang, X. L., Wang, X. G., et al., 'A pneumonia outbreak associated with a new coronavirus of probable bat origin', *Nature*, 2020 https://doi.org/10.1038/s41586-020-2012-7

Zhu, N., Zhang, D., Wang, W., et al., 'A novel coronavirus from patients with pneumonia in China, 2019', *New England Journal of Medicine*, 2020 https://www.nejm.org/doi/full/10.1056/nejmoa2001017

## *3. South Korea, West Africa and the* Diamond Princess

Adepoju, P., 'Africa's struggle with inadequate COVID-19 testing', *Lancet Microbe*, 2020 https://doi.org/10.1016/S2666-5247(20)30014-8

Africa CDC, 'Africa CDC establishes continent-wide task force to respond to global coronavirus epidemic', 2020 https://africacdc.org/news-item/africa-cdc-establishes-continent-wide-task-force-to-respond-to-global-coronavirus-epidemic/

Africa CDC, 'Africa CDC guidance for assessment, monitoring and movement restrictions of people at risk for COVID-19 in Africa', 2020 https://africacdc.org/download/africa-cdc-guidance-for-assessment-monitoring-and-movement-restrictions-of-people-at-risk-for-covid-19-in-africa/

Africa CDC, 'Africa CDC official launch', 2020 web.archive.org/web/20180630023917/https:/au.int/en/newsevents/20170131/africa-cdc-official-launch

Africa CDC, 'Africa50 supports COVID-19 response with US$300,000 grant to Africa CDC', 2020 https://africacdc.org/news-item/africa50-supports-covid-19-response-with-us300000-grant-to-africa-cdc/

Africa CDC, 'African Development Bank supports continental strategy on COVID-19 with US$27.33 million', 2020 https://africacdc.org/news-item/african-development-bank-supports-continental-strategy-on-covid-19-with-us27-33-million/

Africa CDC, 'Annual progress report', 2020 https://africacdc.org/download/annual-progress-report-2020/

Africa CDC, 'Governance', 2017 https://africacdc.org/governance/

Africa CDC, 'Report of an emergency meeting of Africa ministers of health on the COVID-19 outbreak', 2020 https://africacdc.org/download/report-of-an-emergency-meeting-of-africa-ministers-of-health-on-the-covid-19-outbreak/

Africa CDC, 'Report on a joint meeting of Africa ministers of health and finance on the progress and status in controlling the COVID-19 pandemic and its economic shocks', 2020 https://africacdc.org/download/report-on-a-joint-meeting-of-africa-ministers-of-health-and-finance-on-the-progress-and-status-in-controlling-the-covid-19-pandemic-and-its-economic-shocks/

Africa CDC, 'Team Europe supports Africa's COVID-19 response with 1.4 million test-kits donated by Germany and delivered by the EU Humanitarian Air Bridge', 2020 https://africacdc.org/news-item/team-europe-supports-africas-covid-19-response-with-1-4-million-test-kits-donated-by-germany-and-delivered-by-the-eu-humanitarian-air-bridge/

Africa CDC, 'The African Union Commission and Africa CDC launch partnership to accelerate COVID-19 testing: trace, test and track', 2020 https://africacdc.org/news-item/african-union-and-africa-centres-for-disease-control-and-prevention-launch-partnership-to-accelerate-covid-19-testing-trace-test-and-track/

Africa CDC, 'Wellcome and DFID support Africa COVID-19 continental response with €2.26 million', 2020 https://africacdc.org/news-item/wellcome-and-dfid-support-africa-covid-19-continental-response-with-e-2-26-million/

African Union, 'Press briefings on the coronavirus disease outbreak' (this live event holds all press briefings of Africa CDC on the coronavirus disease outbreak), 2020 https://www.facebook.com/watch/live/?v=253091766005286&ref=watch_permalink

Apuzzo, M., Rich, M., and Yaffe-Bellany, D., 'Failures on the *Diamond Princess* shadow another cruise ship outbreak', *New York Times*, 2020 https://www.nytimes.com/2020/03/08/world/asia/coronavirus-cruise-ship.html

Baker, A., 'These Ebola fighters helped halt an epidemic. Now, they're preparing to battle coronavirus', *Time*, 2020 https://time.com/5794329/africa-ebola-epidemic-coronavirus-liberia-covid-19/

Baraniuk, C., 'What the *Diamond Princess* taught the world about COVID-19', *BMJ*, 2020 https://doi.org/10.1136/bmj.m1632

BBC News, 'Coronavirus in Senegal: keeping COVID-19 at bay', 2020 https://www.bbc.co.uk/news/world- africa-54388340

Bedingfield, W., 'What the world can learn from South Korea's coronavirus strategy', *Wired*, 2020 https://www.wired.co.uk/article/south-korea-coronavirus

Bicker, L., 'South Korea on brink of nationwide virus outbreak, officials warn', BBC News, 2020 https://www.bbc.co.uk/news/world-asia-53888219

Biegon, E., 'China-funded Africa CDC headquarters' construction works launched', KBC, 2020 https://www.kbc.co.ke/china-funded-africa-cdc-hq-construction-works-launched/

Boseley, S., 'Make masks compulsory in public in UK, says virus expert', *Guardian*, 2020 https://www.theguardian.com/world/2020/jun/30/make-masks-compulsory-in-public-says-virus-expert-peter-piot

Boyce, M. R., and Katz, R., 'Community health workers and pandemic preparedness: current and prospective roles', *Frontiers in Public Health*, 2019 https://doi.org/10.3389/fpubh.2019.00062

Campbell, M., and Lee, H., 'There's still time to beat COVID without lockdowns', Bloomberg, 2020 https://www.bloomberg.com/features/2020-south-korea-covid-strategy/

Cha, V., and Kim, D., 'A timeline of South Korea's response to COVID-19', Center for Strategic and International Studies, 2020 https://www.csis.org/analysis/timeline-south-koreas-response-covid-19

Chakamba, R., 'How Senegal has set the standard on COVID-19', Devex, 2020 https://www.devex.com/news/how-senegal-has-set-the-standard-on-covid-19-98266

Chua, A. Q., Al Knawy, B., Grant, B., et al., 'How the lessons of previous epidemics helped successful countries fight COVID-19', *BMJ*, 2021 https://www.bmj.com/content/372/bmj.n486

Cooper, K., 'Prepared for the worst – how South Korea fought off COVID-19', BMA, 2020 https://www.bma.org.uk/news-and-opinion/prepared-for-the-worst-how-south-korea-fought-off-covid-19

Delamou, A., Sidibé, S., and Camara, A., 'Tackling the COVID-19 pandemic in West Africa: have we learned from Ebola in Guinea?', *Preventative Medicine Reports*, 2020 https://doi.org/10.1016/j.pmedr.2020.101206

*Dhaka Tribune*, 'What Sierra Leone can teach Bangladesh about beating coronavirus', 2020 https://www.dhakatribune.com/world/2020/04/20/what-sierra-leone-can-teach-bangladesh-about-beating-coronavirus

Dunlop, J., 'Ebola in Sierra Leone – the "Don't touch!" rule', Unicef, 2014 https://blogs.unicef.org/blog/ebola-in-sierra-leone-the-dont-touch-rule/

DW News, 'Coronavirus: the lessons to learn from Ebola', 2020 https://www.dw.com/en/coronavirus-the-lessons-to-learn-from-ebola/a-52759301

*Economist*, 'African countries are struggling to keep track of COVID-19', 2020 https://www.economist.com/middle-east-and-africa/2020/06/18/african-countries-are-struggling-to-keep-track-of-covid-19

*Egypt Today*, 'Egypt announces first coronavirus infection', 2020 https://www.egypt-today.com/Article/1/81641/Egypt-announces-first-Coronavirus-infection

Emery, J. C., Russell, T. W., Liu, Y., et al., 'The contribution of asymptomatic SARS-CoV-2 infections to transmission on the *Diamond Princess* cruise ship', eLife, 2020 https://elifesciences.org/articles/58699

Ferguson, N., 'Capturing human behaviour', *Nature*, 2007 https://doi.org/10.1038/446733a

Fernández-Luque, L., and Bau, T., 'Health and social media: perfect storm of information', *Healthcare Informatics Research*, 2015 https://doi.org/10.4258/hir.2015.21.2.67

*Foreign Policy*, 'The COVID-19 global response index', 2020 https://globalresponseindex.foreignpolicy.com

Friedhoff, K., 'South Koreans becoming more accepting of LGBTQ community', Chicago Council on Global Affairs, 2020 https://www.thechicagocouncil.org/commentary-and-analysis/blogs/south-koreans-becoming-more-accepting-lgbtq-community

GHS Index, 'Senegal', 2019 https://www.ghsindex.org/wp-content/uploads/2019/08/Senegal.pdf

Gunia, A., 'What South Korea's nightclub coronavirus outbreak can teach other countries as they reopen', *Time*, 2020 https://time.com/5834991/south-korea-coronavirus-nightclubs/

Harvard Global Health Institute, 'From Ebola to COVID-19: lessons in digital contact tracing in Sierra Leone', 2020 https://globalhealth.harvard.edu/from-ebola-to-covid-19-lessons-in-digital-contact-tracing-in-sierra-leone/

Hung, I. F., Cheng, V. C., Li, X., et al., 'SARS-CoV-2 shedding and seroconversion among passengers quarantined after disembarking a cruise ship: a case series', *Lancet Infectious Diseases*, 2020 https://doi.org/10.1016/S1473-3099(20)30364-9

Juhasz, A., 'Forty-four Americans on the *Diamond Princess* cruise ship diagnosed with coronavirus', NPR, 2020 https://www.npr.org/2020/02/16/806470340/americans-evacuate-diamond-princess-cruise-ship-amid-spread-of-infection

Kagbai, J. B., 'Sierra Leone is using lessons from Ebola to prepare for coronavirus', *The Conversation*, 2020 https://theconversation.com/sierra-leone-is-using-lessons-from-ebola-to-prepare-for-coronavirus-132443

Kang, C., Lee, J., Park, Y., et al., 'Coronavirus disease exposure and spread from nightclubs, South Korea', *Emerging Infectious Diseases*, 2020 https://wwwnc.cdc.gov/eid/article/26/10/20-2573_article

Kaseje, J., 'Why Sub-Saharan Africa needs a unique response to COVID-19?' World Economic Forum, 2020 https://www.weforum.org/agenda/2020/03/why-sub-saharan-africa-needs-a-unique-response-to-covid-19/

Kelly, A. H., Street, A., and Vernooi, E., 'COVID-19 laboratory preparedness in Africa: lessons can be learned from the Ebola outbreak', University of Edinburgh,

2020 https://blogs.ed.ac.uk/covid19perspectives/2020/04/09/covid-19-laboratory-preparedness-in-africa-lessons-can-be-learned-from-the-ebola-outbreak-write-dr-ann-h-kelly-eva-vernooij-and-dr-alice-street/

Kim, J., Ah-Reum, J., Min, P., et al., 'How South Korea responded to the COVID-19 outbreak in Daegu', *New England Journal of Medicine Catalyst Innovations in Care Delivery*, 2020 https://doi.org/10.1056/CAT.20.0159

Kim, J.-H., Ah-Rheum, J., Oh, S. J., et al., 'Emerging COVID-19 success story: South Korea learned the lessons of MERS', Our World in Data, 2020 https://ourworldindata.org/covid-exemplar-south-korea

Kim, N., '"More scary than coronavirus": South Korea's health alerts expose private lives', *Guardian*, 2020 https://www.theguardian.com/world/2020/mar/06/more-scary-than-coronavirus-south-koreas-health-alerts-expose-private-lives

Kim, N., 'South Korea struggles to contain new outbreak amid anti-gay backlash', *Guardian*, 2020 https://www.theguardian.com/world/2020/may/11/south-korea-struggles-to-contain-new-outbreak-amid-anti-lgbt-backlash

Lee, D., and Lee, J., 'Testing on the move: South Korea's rapid response to the COVID-19 pandemic', *Transportation Research Interdisciplinary Perspectives*, 2020 https://doi.org/10.1016/j.trip.2020.100111

Macola, I. G., 'Timeline: how COVID-19 spread aboard the *Diamond Princess* cruise ship', *Ship Technology*, 2020 https://www.ship-technology.com/features/timeline-covid-spread-aboard-diamond-princess-cruise-ship/

Magazi, I., 'Guinea: "Knowledge comes from experience". How the lessons learned from combating Ebola led to a quick response from the very start of the COVID-19 epidemic', World Bank Blogs, 2020 https://blogs.worldbank.org/nasikiliza/guinea-knowledge-comes-experience-how-lessons-learned-combating-ebola-led-quick-response

Makoni, M., 'Africa prepares for coronavirus', *Lancet*, 2020 https://www.thelancet.com/pdfs/journals/lancet/PIIS0140-6736(20)30355-X.pdf

Mallapaty, S., 'What the cruise-ship outbreaks reveal about COVID-19', *Nature*, 2020 https://www.nature.com/articles/d41586-020-00885-w

Martin, T., and Yoon, D., 'How South Korea successfully managed coronavirus', *Wall Street Journal*, 2020 https://www.wsj.com/articles/lessons-from-south-korea-on-how-to-manage-covid-11601044329

Mastercard Foundation, 'Mastercard Foundation and Private Sector Foundation Uganda launch COVID-19 economic recovery and resilience response program', 2020 https://mastercardfdn.org/mastercard-foundation-and-private-sector-foundation-uganda-launch-covid-19-economic-recovery-and-resilience-response-program/

Mawolo, A., 'COVID-19 and lessons from Ebola: how Senegal is confronting the challenge', *Down To Earth*, 2020 https://www.downtoearth.org.in/news/africa/covid-19-and-lessons-from-ebola-how-senegal-is-confronting-the-challenge-70201

Maxmem, A., 'Ebola prepared these countries for coronavirus – but now even they are floundering', *Nature*, 2020 https://www.nature.com/articles/d41586-020-02173-z

Mizumoto, K., and Chowell, G., 'Transmission potential of the novel coronavirus (COVID-19) onboard the *Diamond Princess* cruise ship', *Infectious Disease Modelling*, 2020 https://doi.org/10.1016/j.idm.2020.02.003

Mizumoto, K., Kagaya, K., Zarebski, A., et al., 'Estimating the asymptomatic proportion of coronavirus disease 2019 (COVID-19) cases on board the *Diamond Princess* cruise ship, Yokohama, Japan', 2020 *Eurosurveillance*, 2020 https://doi.org/10.2807/1560-7917.ES.2020.25.10.2000180

Mwangi, N., 'All African countries now have coronavirus lab testing capacity – WHO chief', 2020 https://africa.cgtn.com/2020/06/24/all-african-countries-now-have-coronavirus-lab-testing-capacity-who-chief/

Nakazawa, E., Ino, H., and Akabayashi, A., 'Chronology of COVID-19 cases on the *Diamond Princess* cruise ship and ethical considerations: a report from Japan', *Disaster Medicine and Public Health Preparedness*, 2020 https://doi.org/10.1017/dmp.2020.50

Ndjandjo, E., 'Senegal Muslims divided on mosques opening during pandemic', Voice of America, 2020 https://www.voanews.com/covid-19-pandemic/senegal-muslims-divided-mosques-opening-during-pandemic

Nkengasong, J. N., Maiyegun, O., and Moeti, M., 'Establishing the Africa Centres for Disease Control and Prevention: responding to Africa's health threats', *Lancet Global Health*, 2020 https://doi.org/10.1016/S2214-109X(17)30025-6

Normile, D., 'Scientist decries "completely chaotic" conditions on cruise ship Japan quarantined after viral outbreak', *Science*, 2020 https://www.sciencemag.org/news/2020/02/scientist-decries-completely-chaotic-conditions-cruise-ship-japan-quarantined-after

Oh, J., Lee, J. K., Schwarz, D., et al., 'National response to COVID-19 in the Republic of Korea and lessons learned for other countries', *Health Systems and Reform*, 2020 https://doi.org/10.1080/23288604.2020.1753464

Ondoe, P., Kebede, Y., Loembe, M. M., et al., 'COVID-19 testing in Africa: lessons learnt', *Lancet Microbe*, 2020 https://doi.org/10.1016/S2666-5247(20)30068-9

Our World in Data, 'South Korea: coronavirus pandemic country profile 2021', 2021 https://ourworldindata.org/coronavirus/country/south-korea

Perraudin, F., 'Coronavirus: Britons stuck on *Diamond Princess* to be flown home', *Guardian*, 2020 https://www.theguardian.com/world/2020/feb/20/diamond-princess-coronavirus-britons-stuck-to-be-flown-home

Petesch, C., 'African nations seek their own solutions in virus crisis', Associated Press, 2020 https:// apnews.com/article/66e8d6229ce8cfa535c3db2e821e7753

Pew Research Center, 'The age gap in religion around the world', 2018 https://www.pewforum.org/2018/06/13/the-age-gap-in-religion-around-the-world/

Peyton, N., 'Using lessons from Ebola, West Africa prepares remote villages for coronavirus', Reuters, 2020 https://www.reuters.com/article/us-health-coronavirus-africa-ebola-idUSKBN21C38B

Princess Cruises, 'News updates', 2020 https://www.princess.com/news/notices_and_advisories/notices/diamond-princess-update.html

Ratcliffe, R., and Fonbuena, C., 'Inside the cruise ship that became a coronavirus breeding ground', *Guardian*, 2020 https://www.theguardian.com/global-development/2020/mar/06/inside-the-cruise-ship-that-became-a-coronavirus-breeding-ground-diamond-princess

Reuters, 'Sierra Leone has confirmed its first case of coronavirus, President says', 2020 https://www.reuters.com/article/us-health-coronavirus-leone-idUSKBN21I1MY

Richards, P., 'Ebola and COVID-19 in Sierra Leone: comparative lessons of epidemics for society', *Journal of Global History*, 2020 https://www.cambridge.org/core/journals/journal-of-global-history/article/ebola-and-covid19-in-sierra-leone-comparative-lessons-of-epidemics-for-society/5672DE34C06149CDC142A38C2294EA6E

Richards, P., *Ebola: how a people's science helped end an epidemic*, Zed Books, 2016 http://dx.doi.org/10.5040/9781350219779

Sakurai, A., Sasaki, T., Kato, S., et al., 'Natural history of asymptomatic SARS-CoV-2 infection', *New England Journal of Medicine*, 2020 https://www.nejm.org/doi/full/10.1056/NEJMc2013020

Sall, M., 'Go early, go hard and keep it simple: how Senegal is staying ahead of the COVID-19 pandemic', World Economic Forum 2021 https://www.weforum.org/agenda/2021/02/senegal-covid- response-macky-sall/

Salyer, S. J., Maeda, J., Sembuche, S., et al., 'The first and second waves of the COVID-19 pandemic in Africa: a cross-sectional study', *Lancet*, 2021 https://www.thelancet.com/journals/lancet/article/PIIS0140-6736(21)00632-2/fulltext

Schumaker, E., 'Japanese expert who sneaked on to *Diamond Princess* cruise ship describes "zero infection control" for coronavirus', ABC News, 2020 https://abcnews.go.com/Health/japanese-expert-sneaked-diamond-princess-describes-infection-control/story?id=69071246

Scott, D., and Park, J., 'South Korea's COVID-19 success story started with failure', *Vox*, 2021 https://www.vox.com/22380161/south-korea-covid-19-coronavirus-pandemic-contact-tracing-testing

Search for Common Ground, 'Lessons from Guinea', 2020 https://www.sfcg.org/guineapostebola/

Shesgreen, D., 'Senegal's quiet COVID success: test results in twenty-four hours, temperature checks at every store, no fights over masks', *USA Today*, 2020 https://eu.usatoday.com/story/news/world/2020/09/06/covid-19-why-senegal-outpacing-us-tackling-pandemic/5659696002/

Soy, A., 'Coronavirus in Africa: five reasons why COVID-19 has been less deadly than elsewhere', BBC News, 2020 https://www.bbc.co.uk/news/world-africa-54418613

Srivatsa, S. I., and Mchenga, M., 'Economic impacts of and policy responses to the coronavirus pandemic: early evidence from Sierra Leone', Supporting Economic Transformation (SET), 2020 https://set.odi.org/wp-content/uploads/2020/04/Economic-impacts-of-and-policy-responses-to-the-coronavirus-pandemic-early-evidence-from-Sierra-Leone-1.pdf

Swift, R., '"Bureaucrats were in charge": Japanese doctor blasts ship quarantine', Reuters, 2020 https://www.reuters.com/article/us-china-health-japan-doctor-idUKKBN20D1MF

Swift, R., 'With stricken cruise ship, Japan draws criticism over coronavirus response', Reuters, 2020 https://www.reuters.com/article/us-china-health-japan-response-idUSKBN20D0HE

Sylvester, Squire J., Hann, K., Denisiuk, O., et al., 'The Ebola outbreak and staffing in public health facilities in rural Sierra Leone: who is left to do the job?', *Public Health Action*, 2017 https://doi.org/10.5588/pha.16.0089

UN News, 'Liberia: Ebola contact tracing lessons inform COVID-19 response', 2020 https://news.un.org/en/story/2020/04/1062582

United Nations, 'South Africa's COVID-19 testing capacity increased with sixty new mobile lab units launched', 2020 https://www.un.org/africarenewal/news/coronavirus/south-africa's-covid-19-testing-capacity-increased-60-new-mobile-lab-units-launched

United Nations, 'WHO press conference: update on the coronavirus (COVID-19)', 2020 https://www.unmultimedia.org/tv/unifeed/asset/2536/2536753/

US Embassy and Consulates in Japan, 'Updates on COVID-19 in Japan', 2020 https://jp.usembassy.gov/updates-on-diamond-princess-quarantine/

US Embassy Tokyo, 'Message to US citizen *Diamond Princess* passengers and crew', 2020 https://japan2.usembassy.gov/pdfs/alert-20200215-diamond-princess.pdf

Varagur, K., 'How Ebola helped Liberia prepare for COVID', *MIT Technology Review*, 2020 https://www.technologyreview.com/2020/08/19/1007139/liberia-kateh-covid-coronavirus-ebola/

Ventures Africa, 'Africa grappling with lack of COVID-19 test-kits', 2020 https://venturesafrica.com/africa-grappling-with-lack-of-covid-19-test-kits/

Werner, M., '"We clap if none die": COVID forces hard choices in Sierra Leone', *Guardian*, 2020 https://www.theguardian.com/global-development/2021/mar/18/childrens-healthcare-covid-forces-hard-choices-sierra-leone

Wiah, O., Subah, M., Varpilah, B., et al., 'Prevent, detect, respond: how community health workers can help in the fight against COVID-19, *BMJ* Blogs, 2020 https://blogs.bmj.com/bmj/2020/03/27/prevent-detect-respond-how-community-health-workers-can-help-fight-covid-19/

World Bank Group, 'In the face of coronavirus, African countries apply lessons from Ebola response', 2020 https://www.worldbank.org/en/news/feature/2020/04/03/in-the-face-of-coronavirus-african-countries-apply-lessons-from-ebola-response

World Health Organization, 'Listings of WHO's response to COVID-19', 2020 https://www.who.int/news/item/29-06-2020-covidtimeline

World Health Organization, 'Nigeria's polio infrastructure bolster COVID-19 response', 2020 https://www.afro.who.int/news/nigerias-polio-infrastructure-bolster-covid-19-response

World Health Organization, 'Sharing COVID-19 experiences: the Republic of Korea response', 2020 https://www.who.int/westernpacific/news-room/feature-stories/item/sharing-covid-19-experiences-the-republic-of-korea-response

World Health Organization, 'WHO demonstrates strategic commitment by supporting COVID-19 testing and laboratory capacity in Sierra Leone', 2020 https://www.afro.who.int/news/who-demonstrates-strategic-commitment-supporting-covid-19-testing-and-laboratory-capacity

Wright, L., Steptoe, A., and Fancourt, D., 'Predictors of self-reported adherence to COVID-19 guidelines. A longitudinal observational study of 51,600 UK adults', *Lancet Regional Health Europe*, 2021 https://doi.org/10.1016/j.lanepe.2021.100061

You, J., 'Lessons from South Korea's COVID-19 policy response', *American Review of Public Administration*, 2020 https://doi.org/10.1177%2F0275074020943708

YOUNA, 'How Ebola experience provided insights to fight against COVID-19 in Senegal', 2020 https://www.youna-ihs.com/blog-youna/how-ebola-experience-provided-insights-to-fight-against-covid-19-in-senegal

YouTube, 'Japan professor describes visit to "chaotic" virus ship', 2020 https://www.youtube.com/watch?v=xjSR_Pdy7PU

## *4. Europe's Tsunami*

Aledort, J. E., Lurie, N., Wasserman, J., et al., 'Non-pharmaceutical public health interventions for pandemic influenza: an evaluation of the evidence base', *BMC Public Health*, 2007 https://doi.org/10.1186/1471-2458-7-208

Althouse, B. M., Wenger, E. A., Miller, J. C., et al., 'Super-spreading events in the transmission dynamics of SARS-CoV-2: opportunities for interventions and control', *PLoS Biology*, 2020 https://doi.org/10.1371/journal.pbio.3000897

Amendola, A., Bianchi, S., Gori, M., et al., 'Evidence of SARS-CoV-2 RNA in an oropharyngeal swab specimen, Milan, Italy, early December 2019', *Emerging Infectious Diseases*, 2021 https://doi.org/10.3201/eid2702.204632

Apuzzo, M., Gebrekidan, S., and Kirkpatrick, D. D., 'How the world missed COVID-19's silent spread', *New York Times*, 2020 https://www.nytimes.com/2020/06/27/world/europe/coronavirus-spread-asymptomatic.html

Baeten, R., Spasova, S., Vanhercke, B., et al., 'Inequalities in access to healthcare: a study of national policies, 2018', European Commission, 2018 https://www.researchgate.net/publication/329128920_Inequalities_in_access_to_healthcare_A_study_of_national_policies_2018

Bamias, G., Lagou, S., Gizis, M., et al., 'The Greek response to COVID-19: a true success story from an IBD perspective', *Inflammatory Bowel Diseases*, 2020 https://doi.org/10.1093/ibd/izaa143

BBC News, 'Coronavirus: Venice Carnival closes as Italy imposes lockdown', 2020 https://www.bbc.co.uk/news/world-europe-51602007

Bourguignon, D., 'The precautionary principle: definitions, applications and governance', European Parliament, 2015 https://www.europarl.europa.eu/RegData/etudes/IDAN/2015/573876/EPRS_IDA(2015)573876_EN.pdf

Brainard, J. S., Jones, N., Lake, I., et al., 'Face masks and similar barriers to prevent respiratory illness such as COVID-19: a rapid systematic review', medRxiv, 2020 https://doi.org/10.1101/2020.04.01.20049528

Branswell, H., 'The fireman of global health: the WHO's emergencies chief is put to the test', STAT News, 2019 https://www.statnews.com/2019/10/30/the-fireman-of-global-health-the-whos-emergencies-chief-is-put-to-the-test/

Buonanno, P., Galletta, S., and Puca, M., 'Estimating the severity of COVID-19: evidence from the Italian epicenter', *PLoS One*, 2020 https://doi.org/10.1371/journal.pone.0239569

Centers for Disease Control and Prevention, 'History of 1918 Flu Pandemic', 2018 https://www.cdc.gov/flu/pandemic-resources/1918-commemoration/1918-pandemic-history.htm

Cheng, K. K., Lam, T. H., and Leung, C. C., 'Wearing face masks in the community during the COVID-19 pandemic: altruism and solidarity', *Lancet*, 2020 https://doi.org/10.1016/S0140-6736(20)30918-1

Correa-Martínez, C. L., Kampmeier, S., Kümpers, P., et al., 'A pandemic in times of global tourism: super-spreading and exportation of COVID-19 cases from a ski area in Austria', *Journal of Clinical Microbiology*, 2020 https://doi.org/10.1128/JCM.00588-20

Davies, P., 'Major "miscalculations" made, probe into COVID outbreak at Austrian ski resort finds', *Travel Weekly*, 2020 https://travelweekly.co.uk/articles/389038/major-miscalculations-made-probe-into-covid-outbreak-at-austrian-ski-resort-finds

Delauney, G., and Kallergis, K., 'Coronavirus: Greece and Croatia acted fast, now need to save summer', BBC News, 2020 https://www.bbc.co.uk/news/world-europe-52491204

DW News, 'Austria's Ischgl: a ski resort struggling to restore its image', 2021 https://www.dw.com/en/austrias-ischgl-a-ski-resort-struggling-to-restore-its-image/a-56842089

DW News, 'Commission blames Austrian government for poor handling of Ischgl coronavirus outbreak', 2020 https://www.dw.com/en/commission-blames-austrian-government-for-poor-handling-of-ischgl-coronavirus-outbreak/a-55251063

DW News, 'How Greece's crisis is helping it bend the COVID-19 curve', 2020 https://www.dw.com/en/how-greeces-crisis-is-helping-it-bend-the-covid-19-curve/a-53280532

Enimerosi, 'Supermarkets to close earlier during lockdown', 2020 https://enimerosi.com/details_en.php?id=51012

European Centre for Disease Prevention and Control, 'Infographic: using face masks in the community', 2020 https://www.ecdc.europa.eu/en/publications-data/infographic-using-face-masks-community

European Centre for Disease Prevention and Control, 'Using face masks in the community: first update – effectiveness in reducing transmission of COVID-19', 2021 https://www.ecdc.europa.eu/en/publications-data/using-face-masks-community-reducing-covid-19-transmission

European Observatory on Health Systems and Policies (EOHSP), 'Policy responses for the Czech Republic', 2020 https://www.covid19healthsystem.org/countries/czechrepublic/livinghit.aspx?Section=1.5%20Testing&Type=Section#9Testing-springandsummer2020

European Observatory on Health Systems and Policies (EOHSP), 'Policy responses for Greece', 2020 https://www.covid19healthsystem.org/countries/greece/livinghit.aspx?Section=1.5%20Testing&Type=Section

Eyres, H., 'How coronavirus has led to the return of the precautionary principle', 2020 https://www.newstatesman.com/politics/2020/04/how-coronavirus-has-led-return-precautionary-principle

Fouda, A., Mahmoudi, N., Moy, N., et al., 'The COVID-19 pandemic in Greece, Iceland, New Zealand, and Singapore: health policies and lessons learned', *Health Policy and Technology*, 2020 https://doi.org/10.1016/j.hlpt.2020.08.015

Fouskas, V. K., Gökay, B., and Vankovska, B., 'Crisis in the Eastern Mediterranean and COVID-19', *Journal of Balkan and Near Eastern Studies*, 2020 https://doi.org/10.1080/19448953.2020.1755154

GardaWorld, 'Greece: government confirms first COVID-19 case February 26', 2020 https://www.garda.com/crisis24/news-alerts/317701/greece-government-confirms-first-covid-19-case-february-26

GardaWorld, 'Italy: Towns in Lombardy and Veneto placed on lockdown February 22 /update 5', 2020 https://www.garda.com/crisis24/news-alerts/316651/italy-towns-in-lombardy-and-veneto-placed-on-lockdown-february-22-update-5

Genome Web, 'Austrian SARS-CoV-2 genomic study delivers details for superspreader events early in pandemic', 2020 https://www.genomeweb.com/sequencing/austrian-sars-cov-2-genomic-study-delivers-details-superspreader-events-early-pandemic#.YXhloxzTVdh

Giannopoulou, I., and Tsobanoglou, G. O., 'COVID-19 pandemic: challenges and opportunities for the Greek health care system', *Irish Journal of Psychological Medicine*, 2020 https://doi.org/10.1017/ipm.2020.35

Giuffrida, A., 'Why was Lombardy hit harder than Italy's other regions?', *Guardian*, 2020 https://www.theguardian.com/world/2020/may/29/why-was-lombardy-hit-harder-covid-19-than-italys-other-regions

Giuffrida, A., and Tondo, L., 'Italians struggle with "surreal" lockdown as coronavirus cases rise', *Guardian*, 2020 https://theguardian.com/world/2020/feb/24/italians-struggle-with-surreal-lockdown-as-coronavirus-cases-rise

Giuffrida, A., and Tondo, L., 'Leaked coronavirus plan to quarantine 16 million sparks chaos in Italy', *Guardian*, 2020 https://www.theguardian.com/world/2020/mar/08/leaked-coronavirus-plan-to-quarantine-16m-sparks-chaos-in-italy

Gollier, C., and Treich, N., 'The precautionary principle', *Encyclopedia of energy, natural resource and environmental economics*, 2013 https://www.sciencedirect.com/topics/earth-and-planetary-sciences/precautionary-principle

Good Governance Institute, 'The precautionary principle in action', 2021 https://www.good-governance.org.uk/publications/insights/the-precautionary-principle-in-action

Goumenou, M., Sarigiannis, D., Tsatsakis, A., et al., 'COVID-19 in northern Italy: an integrative overview of factors possibly influencing the sharp increase of the outbreak', *Molecular Medicine Reports*, 2020 https://www.spandidos-publications.com/10.3892/mmr.2020.11079

Government of the Czech Republic, 'The government requires the wearing of protective equipment and reserved time for pensioners to do their food shopping', 2020 https://www.vlada.cz/en/media-centrum/aktualne/the-government-has-decided-to-require-the-wearing-of-protective-equipment-and-reserved-time-for-senior-citizens-to-do-their-food-shopping-180465/

Grasselli, G., Pesenti, A., and Cecconi, M., 'Critical care utilization for the COVID-19 outbreak in Lombardy, Italy: early experience and forecast during an emergency response, *Journal of the American Medical Association*, 2020 https://jamanetwork.com/journals/jama/fullarticle/2763188

Greenhalgh, T., Schmid, M. B., Czypionka, T., et al., 'Face masks for the public during the COVID-19 crisis', *BMJ*, 2020 https://doi.org/10.1136/bmj.m1435

Guan, W.-J., Ni, Z.-Y., Hu, Y., et al., 'Clinical characteristics of coronavirus disease 2019 in China', *New England Journal of Medicine*, 2020 https://doi.org/10.1016/j.jemermed.2020.04.004

Henley, J., Connolly, K., Willsher, K., et al., 'France may follow Germany in making clinical masks mandatory', *Guardian*, 2021 https://www.theguardian.com/world/2021/jan/20/france-may-follow-germany-in-making-clinical-masks-mandatory

Hruby, D., 'How an Austrian ski resort helped coronavirus spread across Europe', CNN Travel, 2020 https://edition.cnn.com/2020/03/24/europe/austria-ski-resort-ischgl-coronavirus-intl/index.html

Huang, C., Wang, Y., Li, X., et al., 'Clinical features of patients infected with 2019 novel coronavirus in Wuhan, China', *Lancet*, 2020 https://doi.org/10.1016/S0140-6736(20)30183-5

Huggler, J., 'Austrian ski resort at heart of coronavirus "super-spreader event" to face first lawsuits', *Telegraph*, 2020 https://www.telegraph.co.uk/news/2020/08/26/austrian-ski-resort-heart-coronavirus-superspreader-event-face/

Hutt, D., 'Coronavirus: Czechs facing up to COVID-19 crisis by making masks mandatory', Euronews, 2020 https://www.euronews.com/2020/03/24/coronavirus-czechs-facing-up-to-covid-19-crisis-by-making-masks-mandatory

Johnson, M., and Ghiglione, D., 'Coronavirus in Italy: "When we heard about the lockdown we rushed to the station" ', *Financial Times*, 2020 https://www.ft.com/content/27458814-6159-11ea-a6cd-df28cc3c6a68

Knabl, L., Mitra, T., Kimpel, J., et al., 'High SARS-CoV-2 seroprevalence in children and adults in the Austrian ski resort Ischgl', *Communications Medicine*, 2020 https://doi.org/10.1038/s43856-021-00007-1

Kottasová, I., 'How the Czech Republic slipped into a COVID disaster, one misstep at a time', CNN, 2021 https://edition.cnn.com/2021/02/28/europe/czech-republic-coronavirus-disaster-intl/index.html

KPMG, 'Czech Republic – COVID-19: extended state of emergency and related measures', 2020 https://home.kpmg/xx/en/home/insights/2020/04/flash-alert-2020-186.html

Kreidl, P., Schmid, D., Maritschnik, S., et al., 'Emergence of coronavirus disease 2019 (COVID-19) in Austria', *Wiener klinische Wochenschrift*, 2020 https://doi.org/10.1007/s00508-020-01723-9

Kringos, D., Boerma, W., Bourgueil, Y., et al., 'The strength of primary care in Europe: an international comparative study', *British Journal of General Practice: The Journal of the Royal College of General Practitioners*, 2013 https://doi.org/10.3399/bjgp13X674422

La Rosa, G., Mancini, P., Ferraro, G. B., et al., 'SARS-CoV-2 has been circulating in northern Italy since December 2019: evidence from environmental monitoring', *Science of the Total Environment*, 2021 https://doi.org/10.1016/j.scitotenv.2020.141711

Lawler, D., 'Timeline: how Italy's coronavirus crisis became the world's deadliest', Axios, 2020 https://www.axios.com/italy-coronavirus-timeline-lockdown-deaths-cases-2adb0fc7-6ab5-4b7c-9a55-bc6897494dc6.html

Lee, H. K., Knabl, L., Pipperger, L., et al., 'Immune transcriptomes of highly exposed SARS-CoV-2 asymptomatic seropositive versus seronegative individuals from the Ischgl community', *Scientific Reports*, 2021 https://doi.org/10.1038/s41598-021-83110-6

Longman, J., 'Unlikely underdog Greece is beating coronavirus. But, like all ancient heroes, the test isn't over', ABC News, Reporter's Notebook, 2020 https://abcnews.go.com/International/underdog-greece-beating-coronavirus-ancient-heroes-test-isnt/story?id=70741588

Lopez, M. D., 'COVID-19, masks and the precautionary principle', *Gibraltar Chronicle*, 2020 https://www.chronicle.gi/covid-19-masks-and-the-precautionary-principle/

Lytras, T., Dellis, G., Flountzi, A., et al., 'High prevalence of SARS-CoV-2 infection in repatriation flights to Greece from three European countries', *Journal of Travel Medicine*, 2020 https://doi.org/10.1093/jtm/taaa054

Májek, O., Ngo, O., Jarkovský, J., et al., 'Modelling the first wave of the COVID-19 epidemic in the Czech Republic and the role of government interventions', medRxiv, 2020 https://doi.org/10.1101/2020.09.10.20192070

Maltezou, H. C., Papadima, K., Gkolfinopoulou, K., et al., 'Coronavirus disease 2019 pandemic in Greece, February 26–May 3, 2020: the first wave', *Travel Medicine and Infectious Disease*, 2021 https://doi.org/10.1016/j.tmaid.2021.102051

Matuschek, C., Moll, F., and Haussman, J., 'Face masks: benefits and risks during the COVID-19 crisis', *European Journal of Medical Research*, 2020 https://doi.org/10.1186/s40001-020-00430-5

Médecins sans Frontières, 'MSF steps up response to COVID-19 in Europe', 2020 https://prezly.msf.org.uk/msf-steps-up-response-to-covid-19-in-europe#

Ministero della Salute, 'COVID-19 – situazione in Italia', 2020 http://www.salute.gov.it/portale/nuovocoronavirus/dettaglioContenutiNuovoCoronavirus.jsp?area=nuovoCoronavirus&id=5351&lingua=italiano&menu=vuoto

Ministerstvo Zdravotnictví České Republiky, 'Doporučený postup pro testování pacientů COVID-19 prováděné v odběrových místech', 2020 https://koronavirus.mzcr.cz/ministerstvo-zdravotnictvi-vydalo-doporuceny-postup-pro-testovani-pacientu-a-stanovilo-kriteria-pro-propusteni-z-izolace/

Mitze, T., Kosfeld, R., Rode, J., et al., 'Face masks considerably reduce COVID-19 cases in Germany', *Proceedings of the National Academy of Sciences of the United States of America*, 2020 https://doi.org/10.1073/pnas.2015954117

Mugnai, G., and Bilato, C., 'COVID-19 in Italy: lesson from the Veneto Region', *European Journal of Internal Medicine*, 2020 https://doi.org/10.1016/j.ejim.2020.05.039

Muller, R., and Hovet, J., 'Czech government tightens lockdown, limits movement to fight COVID surge', Reuters, 2021 https://www.reuters.com/article/us-health-coronavirus-czech-idUSKBN2AQ23G

National Public Health Organization, Greece, 'Current state of COVID-19 outbreak in Greece and timeline of key containment events', 2020 https://eody.gov.gr/en/current-state-of-covid-19-outbreak-in-greece-and-timeline-of-key-containment-events/

*Neos Kosmos*, 'Greek government bans gathering of groups of ten in open spaces as coronavirus numbers rise to 418', 2020 https://neoskosmos.com/en/160738/greek-government-bans-gathering-of-groups-of-10-in-open-spaces-as-coronavirus-numbers-rise-to-418/

Odone, A., Delmonte, D., Scognamiglio, T., et al., 'COVID-19 deaths in Lombardy, Italy: data in context', *Lancet Public Health*, 2020 https://doi.org/10.1016/S2468-2667(20)30099-2

OECD/European Observatory on Health Systems and Policies, *Greece: country health profile*, OECD Publishing, 2017 http://doi.org/10.1787/9789264283404-en

Oltermann, P., and Hoyal, L., 'Could one Austrian village have infected much of Europe?', *Irish Times*, 2020 https://www.irishtimes.com/life-and-style/health-family/could-one-austrian-village-have-infected-much-of-europe-1.4348688

Paterlini, M., 'COVID-19: Italy has wasted the sacrifices of the first wave, say experts', *BMJ*, 2020 https://www.bmj.com/content/bmj/371/bmj.m4279.full.pdf

Percivalle, E., Cambiè, G., Cassaniti, I., et al., 'Prevalence of SARS-CoV-2 specific neutralising antibodies in blood donors from the Lodi Red Zone in Lombardy, Italy, as at 6 April 2020', *Eurosurveillance*, 2020 https://doi.org/10.2807/1560-7917.ES.2020.25.24.2001031

Poggioli, S., 'Prosecutors question Italy's top leaders over coronavirus response', NPR, 2020 https://www.npr.org/sections/coronavirus-live-updates/2020/06/12/876143187/prosecutors-question-italys-top-leaders-over-coronavirus-response?t=1617262978974

Popa, A., Genger, J. W., Nicholson, M. D., et al., 'Genomic epidemiology of superspreading events in Austria reveals mutational dynamics and transmission properties of SARS-CoV-2', *Science Translational Medicine*, 2020 https://www.science.org/doi/10.1126/scitranslmed.abe2555

Poula, V., 'Greece in the time of COVID-19: a chance to defend European ideals', London School of Economics Blogs, 2020 https://blogs.lse.ac.uk/greeceatlse/2020/04/21/greece-in-the-time-of-covid-19-a-chance-to-defend-european-ideals/

Queen Mary, University of London, 'Greece's COVID-19 response: not beyond reproach', 2021 https://www.qmul.ac.uk/law/news/responding-to-covid-19/items/greeces-covid-19-response-not-beyond-reproach.html

Redfield & Wilton Strategies, 'A detailed timeline of public opinion on the UK's mask policy', 2020 https://redfieldandwiltonstrategies.com/a-detailed-timeline-of-public-opinion-on-the-uks-mask-policy/

Relief Web, 'Czech Republic sent material aid to China to fight the coronavirus', 2020 https://reliefweb.int/report/china/czech-republic-sent-material-aid-china-fight-coronavirus

Reuters, 'Czech government orders people to wear masks in public against coronavirus', 2020 https://www.reuters.com/article/health-coronavirus-czech-covering-idUSL8N2BB4C6

Reuters, 'Greece reports two more coronavirus fatalities, bans all flights to Italy', 2020 https://www.reuters.com/article/us-health-coronavirus-greece-death/greece-reports-two-more-coronavirus-fatalities-suspends-all-flights-to-italy-idUSKBN2110H1

Ricci, P., and Sheng, H. X., 'Benefits and limitations of the precautionary principle', reference module in *Earth Systems and Environmental Sciences*, 2013 https://doi.org/10.1016/B978-0-12-409548-9.01935-7

Smith, H., 'How Greece is beating coronavirus despite a decade of debt', *Guardian*, 2020 https://www.theguardian.com/world/2020/apr/14/how-greece-is-beating-coronavirus-despite-a-decade-of-debt

Statista, 'How Ischgl set off a coronavirus avalanche', 2020 https://www.statista.com/chart/21699/coronavirus-outbreak-in-ischgl/

Tait, R., 'Czechs get to work making masks after government decree', *Guardian*, 2020 https://www.theguardian.com/world/2020/mar/30/czechs-get-to-work-making-masks-after-government-decree-coronavirus

Taleb, N. N., and Bar-Yam, Y., 'The UK's coronavirus policy may sound scientific, it isn't', *Guardian*, 2020 https://www.theguardian.com/commentisfree/2020/mar/25/uk-coronavirus-policy-scientific-dominic-cummings

Tidman, Z., 'Coronavirus: Czech Republic seizes more than 100,000 face masks sent by China to help Italy tackle spread', *Independent*, 2020 https://www.independent.co.uk/news/world/europe/coronavirus-face-masks-china-italy-czech-republic-latest-a9416711.html

Tountas, Y., Kyriopoulos, J., Lionis, C., et al., 'The new NHS: reform of the National Health System', 2020 https://www.dianeosis.org/wp-content/uploads/2020/ 02/health_system_final.pdf

Tubiana, M., 'The precautionary principle: advantages and risks', *Journal de Chirurgie*, 2001 https://www.researchgate.net/publication/12017376_The_precautionary_principle_advantages_and_risks

Usuelli M., 'The Lombardy region of Italy launches the first investigative COVID-19 commission', *Lancet*, 2020 https://doi.org/10.1016/S0140-6736(20)32154-1

van Daalen, K. R., Bajnoczki, C., Chowdhury, M., et al., 'Symptoms of a broken system: the gender gaps in COVID-19 decision-making', *BMJ Global Health*, 2020 http://dx.doi.org/10.1136/bmjgh-2020-003549

Wackerhage, H., Everett, R., Hefti, U., et al., 'SARS-CoV-2, COVID-19 and mountain sports: specific risks, their mitigation and recommendations for policy makers', preprint, 2020 https://www.mountaineering.scot/assets/contentfiles/pdf/2020-SARS-CoV-2-mountaineering-v26_2.pdf

Walker, M., 'COVID-19: what makes "waves" during a pandemic?', *MedPage Today*, 2020 https://www.medpagetoday.com/infectiousdisease/covid19/89599

Waxman, O., 'How does a pandemic end? Here's what we can learn from the 1918 flu', *Time*, 2020 https://time.com/5894403/how-the-1918-flu-pandemic-ended/

Wikipedia, 'COVID-19 pandemic in the Czech Republic', https://en.wikipedia.org/wiki/COVID-19_pandemic_in_the_Czech_Republic

Winter, J., tweet, 2020 https://twitter.com/winter_jakob/status/1259377668702830594

World Health Organization, 'Advice on the use of masks in the context of COVID-19: interim guidance', 2020 https://apps.who.int/iris/handle/10665/331693

World Health Organization, 'Mask use in the context of COVID-19', 2020 https://apps.who.int/iris/handle/10665/337199

YouTube, 'How to significantly slow coronavirus?', 2020 https://www.youtube.com/watch?v=HhNo_IOPOtU

## 5. *Britain, 'Herd Immunity' and 'Following the Science'*

ABC News, 'Ship passengers cruisy in swine flu quarantine', 2009 https://www.abc.net.au/news/2009-05-28/ship-passengers-cruisy-in-swine-flu-quarantine/1697442

Ahmed, S., Karim, M. M., Ross, A. G., et al., 'A five-day course of ivermectin for the treatment of COVID-19 may reduce the duration of illness', *International Journal of Infectious Diseases*, 2021 https://doi.org/10.1016/j.ijid.2020.11.191

Axfors, C. A., Schmitt, A. M., Janiaud, P., et al., 'Mortality outcomes with hydroxychloroquine and chloroquine in COVID-19: an international collaborative meta-analysis of randomized trials', *Nature Communications*, 2020 https://doi.org/10.1038/s41467-021-22446-z

BBC News, 'Belgian manhunt for heavily armed far-right soldier', 2021 https://www.bbc.co.uk/news/world-europe-57168576

BBC News, 'Coronavirus: Boris Johnson in "good spirits" in hospital', 2020 https://www.bbc.co.uk/news/uk-52180223

BBC News, 'Coronavirus: health workers on frontline to be tested in England', 2020 https://www.bbc.co.uk/news/health-52070199

BBC News, 'Coronavirus: Prime Minister Boris Johnson returns to work after illness', 2020 https://www.bbc.co.uk/newsround/52432524

BBC News, 'Coronavirus: Prime Minister Boris Johnson tests positive', 2020 https://www.bbc.co.uk/news/uk-52060791

Beigel, J. H., Tomashek, K. M., Dodd, L. E., et al., 'Remdesivir for the treatment of COVID-19: final report', *New England Journal of Medicine*, 2020 https://www.nejm.org/doi/full/10.1056/nejmoa2007764

Bland, A., 'The Cummings effect: study finds public faith was lost after aide's trip', *Guardian*, 2020 https://www.theguardian.com/politics/2020/aug/06/the-cummings-effect-study-finds-public-faith-was-lost-after-aides-trip

Boseley, S., 'Herd immunity: will the UK's coronavirus strategy work?', *Guardian*, 2020 https://www.theguardian.com/world/2020/mar/13/herd-immunity-will-the-uks-coronavirus-strategy-work

Bowcott, O., Carrell, S., and Teather, D., 'UK buys 60 million face masks as new cases of swine flu emerge', *Guardian*, 2009 https://www.theguardian.com/world/2009/apr/30/swine-flu-uk-masks-antiviral-drugs

Britton, A., 'UK was prepared for "wrong pandemic" former chief medical officer claims', *Independent*, 2020 https://www.independent.co.uk/news/uk/home-news/coronavirus-uk-prepared-for-wrong-pandemic-b1722321.html

Bryant, E., 'Final report confirms remdesivir benefits for COVID-19', National Institutes of Health, 2020 https://www.nih.gov/news-events/nih-research-matters/final-report-confirms-remdesivir-benefits-covid-19

Busby, M., and Stewart, H., 'Coronavirus: science chief defends UK plan from criticism', *Guardian*, 2020 https://www.theguardian.com/world/2020/mar/13/coronavirus-science-chief-defends-uk-measures-criticism-herd-immunity

Centers for Disease Control and Prevention, '2009 H1N1 flu ("swine flu") and you', 2010 https://www.cdc.gov/h1n1flu/qa.htm

Centers for Disease Control and Prevention, '2009 H1N1 pandemic (H1N1pdm09 virus)', 2019 https://www.cdc.gov/flu/pandemic-resources/2009-h1n1-pandemic.html

Centers for Disease Control and Prevention, 'The 2009 H1N1 pandemic: summary highlights, April 2009–April 2010', 2010 https://www.cdc.gov/h1n1flu/cdcresponse.htm

Chaccour, C., Casellas, A., Blanco-Di Matteo, A., et al., 'The effect of early treatment with ivermectin on viral load, symptoms and humoral response in patients with non-severe COVID-19: a pilot, double-blind, placebo-controlled, randomized clinical trial', *EClinical Medicine*, 2021 https://doi.org/10.1016/j.eclinm.2020.100720

Chalmers, V., 'Former medical tsar Dame Sally Davies admits Britain was ill-prepared for COVID-19 because PHE told her coronavirus could never reach the UK', *Daily Mail*, 2020 https://www.dailymail.co.uk/news/article-8944677/Former-medical-tsar-Dame-Sally-Davies-admits-Britain-ill-prepared-Covid-19.html

Cohen, D., and Carter, P., 'Feature conflicts of interest: WHO and the pandemic flu "conspiracies"', *BMJ*, 2010 https://www.bmj.com/content/bmj/340/7759/Feature.full.pdf

Conn, D., Lawrence, F., Lewis, P., et al., 'Revealed: the inside story of the UK's COVID-19 crisis', *Guardian*, 2020 https://www.theguardian.com/world/2020/apr/29/revealed-the-inside-story-of-uk-covid-19-coronavirus-crisis

Department of Health and Social Care, 'Coronavirus action plan: a guide to what you can expect across the UK', 2020 https://www.gov.uk/government/publications/coronavirus-action-plan/coronavirus-action-plan-a-guide-to-what-you-can-expect-across-the-uk#:~:text=The%20overall%20phases%20of%20our,long%20as%20is%20reasonably%20possible

DPA, 'Tensions escalate in Hong Kong's swine-flu hotel', *Taipei Times*, 2009 https://www.taipeitimes.com/News/world/archives/2009/05/03/2003442656

Edwardes, C., 'Jeremy Hunt on Boris, coronavirus and how to escape lockdown', *The Times*, 2020 https://www.thetimes.co.uk/article/jeremy-hunt-on-boris-coronavirus-and-how-to-escape-lockdown-stj6l50fs

Fineberg, H., 'Pandemic preparedness and response – lessons from the H1N1 influenza of 2009', *New England Journal of Medicine*, 2014 https://www.nejm.org/doi/full/10.1056/nejmra1208802

Gibbs, A., Armstrong, J., and Downie, J., 'From where did the 2009 "swine-origin" influenza A virus (H1N1) emerge?', *Virology Journal*, 2009 https://doi.org/10.1186/1743-422X-6-207

Gilead, 'Final results of National Institute of Allergy and Infectious Diseases' ACTT-1 trial published in *New England Journal of Medicine* expand clinical benefits of Veklury® (remdesivir) for the treatment of COVID-19', 2020 https://www.gilead.com/news-and-press/press-room/press-releases/2020/10/final-results-of-national-institute-of-allergy-and-infectious-diseases-actt-1-trial-published-in-new-england-journal-of-medicine-expand-clinical-bene

GOV.UK., 'Scientific Advisory Group for Emergencies (SAGE): about us', 2021 https://www.gov.uk/government/organisations/scientific-advisory-group-for-emergencies/about

Harding, L., Mason, R., Sabbagh, D., et al., 'Boris Johnson and coronavirus: the inside story of his illness', *Guardian*, 2020

Health Protection Agency, 'Changes to HPA pandemic flu media updates', 2009 https://web.archive.org/web/20091112202811/http://www.hpa.org.uk/web/HPAweb&HPAwebStandard/HPAweb_C/1246607819501

Health Protection Agency, 'Health protection report', 2009 https://web.archive.org/web/20090630234954/http://www.hpa.org.uk/hpr/archives/2009/news1809.htm

Health Protection Agency, 'Treatment approach announced for pandemic flu 2009', 2009 https://web.archive.org/web/20091112202811/http://www.hpa.org.uk/web/HPAweb&HPAwebStandard/HPAweb_C/1246607819501

Hsu, J., 'COVID-19: what now for remdesivir?', *BMJ*, 2020 https://doi.org/10.1136/bmj.m4457

Iacobucci, G., 'COVID-19: validity of key studies in doubt after leading journals issue expressions of concern', *BMJ*, 2020 https://doi.org/10.1136/bmj.m2224

Jick, H., MacLaughlin, D., Egger, P., et al., 'The United Kingdom 2009 swine flu outbreak as recorded in real time by general practitioners', *Epidemiology Research International*, 2011 https://doi.org/10.1155/2011/381597

Kupferschmidt, K., 'Remdesivir and interferon fall flat in WHO's megastudy of COVID-19 treatments', *Science*, 2020 https://www.science.org/content/article/remdesivir-and-interferon-fall-flat-who-s-megastudy-covid-19-treatments

Mason, R., 'Boris Johnson boasted of shaking hands on day SAGE warned not to', *Guardian*, 2020 https://www.theguardian.com/politics/2020/may/05/boris-johnson-boasted-of-shaking-hands-on-day-sage-warned-not-to

Mason, R., 'Coronavirus: just 2,000 NHS frontline workers tested so far', *Guardian*, 2020 https://www.theguardian.com/world/2020/apr/01/hancock-orders-all-spare-coronavirus-tests-used-nhs-staff

National Institute for Health and Care Excellence (NICE), 'Dexamethasone', 2021 https://bnf.nice.org.uk/drug/dexamethasone.html#indicationsAndDoses

National Institute for Health and Care Excellence (NICE), 'Hydroxychloroquine sulfate', https://bnf.nice.org.uk/drug/hydroxychloroquine-sulfate.html

National Institute for Health and Care Excellence (NICE), 'Ivermectin', 2021 https://bnf.nice.org.uk/drug/ivermectin.html

National Institute for Health and Care Excellence (NICE), 'Remdesivir', 2021 https://bnf.nice.org.uk/drug/remdesivir.html#indicationsAndDoses

Neumann, G., Noda, T., and Kawaoka, Y., 'Emergence and pandemic potential of swine-origin H1N1 influenza virus', *Nature*, 2009 https://doi.org/10.1038/nature08157

Padhy, B. M., Mohanty, R. R., Das, S., et al., 'Therapeutic potential of ivermectin as add-on treatment in COVID 19: a systematic review and meta-analysis', *Journal of Pharmacy and Pharmaceutical Sciences*, 2020 https://doi.org/10.18433/jpps31457

Pegg, D., 'What was Exercise Cygnus and what did it find?', *Guardian*, 2020 https://www.theguardian.com/world/2020/may/07/what-was-exercise-cygnus-and-what-did-it-find

Perraudin, F., and Duncan, P., 'Britain's coronavirus testing scandal: a timeline of mixed messages', *Guardian*, 2020 https://www.theguardian.com/politics/2020/apr/03/coronavirus-testing-in-uk-timeline-of-ministers-mixed-messages

Pott-Junior, H., Bastos Paoliello, M. M., Miguel, A. Q. C., et al., 'Use of ivermectin in the treatment of COVID-19: a pilot trial', *Toxicology Reports*, 2021 https://doi.org/10.1016/j.toxrep.2021.03.003

Proctor, K., 'Boris Johnson will not hold coronavirus crisis meeting until Monday', *Guardian*, 2020 https://www.theguardian.com/politics/2020/feb/28/boris-johnson-not-to-hold-coronavirus-crisis-meeting-cobra-until-monday

Public Health England, 'Exercise Cygnus report: Tier 1 command post exercise pandemic influenza, 18 to 20 October 2016', 2017 https://assets.publishing.service.gov.uk/government/uploads/system/uploads/attachment_data/file/927770/exercise-cygnus-report.pdf

RECOVERY Collaborative Group, 'Dexamethasone in hospitalized patients with COVID-19', *New England Journal of Medicine*, 2021 https://www.nejm.org/doi/full/10.1056/nejmoa2021436

RECOVERY Collaborative Group, 'Effect of hydroxychloroquine in hospitalized patients with COVID-19', *New England Journal of Medicine*, 2020 https://www.nejm.org/doi/full/10.1056/NEJMoa2022926

RECOVERY Collaborative Group, 'Low-cost dexamethasone reduces death by up to one third in hospitalised patients with severe respiratory complications of COVID-19', 2020 https://www.recoverytrial.net/news/low-cost-dexamethasone-reduces-death-by-up-to-one-third-in-hospitalised-patients-with-severe-respiratory-complications-of-covid-19

RECOVERY Collaborative Group, 'Randomised evaluation of COVID-19 therapy', 2021 https://www.recoverytrial.net/

RECOVERY Collaborative Group, 'Statement from the chief investigators of the randomised evaluation of COVid-19 thERapY (RECOVERY) trial on hydroxychloroquine, 5 June 2020: no clinical benefit from use of hydroxychloroquine in hospitalised patients with COVID-19', 2020 https://www.recoverytrial.net/files/hcq-recovery-statement-050620-final-002.pdf

Science and Technology Committee, 'Third report. Chapter 6: scientific advice and evidence in emergencies (prepared 2 March 2011)', 2011 https://publications.parliament.uk/pa/cm201011/cmselect/cmsctech/498/49809.htm#note157

Smith, B., 'UK was not as prepared for COVID as it could have been, ex-chief medic says', *Civil Service World*, 2020 https://www.civilserviceworld.com/news/article/uk-was-not-as-prepared-for-covid-as-it-could-have-been-ex-chief-medic-says

Spinner C. D., Gottlieb, R. L., Criner, G. J., et al., 'Effect of remdesivir versus standard care on clinical status at 11 days in patients with moderate COVID-19: a randomized clinical trial', *Journal of the American Medical Association*, 2020 https://jamanetwork.com/journals/jama/fullarticle/2769871

*The Week*, 'Jürgen Conings: far-right Belgian anti-vaccination soldier found dead', 2021 https://www.theweek.co.uk/news/world-news/europe/953227/jurgen-conings-far-right-belgian-soldier-found-dead

Todd, B., and Gandossy, T., 'China quarantines US school group over flu concerns', CNN, 2009 https://edition.cnn.com/2009/HEALTH/05/28/us.china.swine.flu/

Tomazini, B. M., Maia, I. S., Cavalcanti, A. B., et al., 'Effect of dexamethasone on days alive and ventilator-free in patients with moderate or severe acute respiratory distress syndrome and COVID-19: the CoDEX randomized clinical trial', *Journal of the American Medical Association*, 2020 https://jamanetwork.com/journals/jama/fullarticle/2770277

Versluis, E., van Asselt, M., and Kim, J., 'The multilevel regulation of complex policy problems: uncertainty and the swine flu pandemic', *European Policy Analysis*, 2019 https://doi.org/10.1002/epa2.1064

Walker, P., '"Part-time PM": Corbyn criticises Johnson's response to floods', *Guardian*, 2020 https://www.theguardian.com/politics/2020/feb/26/part-time-prime-minister-jeremy-corbyn-criticises-boris-johnson-response-to-floods

Wenham, C., 'Modelling can only tell us so much: politics explains the rest', *Lancet*, 2020 https://www.ncbi.nlm.nih.gov/pmc/articles/PMC7180022/

Wise, J., and Coombes, R., 'COVID-19: the inside story of the RECOVERY trial', *BMJ*, 2020 https://doi.org/10.1136/bmj.m2670

World Health Organization Rapid Evidence Appraisal for COVID-19 Therapies Working Group, 'Association between administration of systemic corticosteroids and mortality among critically ill patients with COVID-19: a meta-analysis', *Journal of the American Medical Association*, 2020 https://jamanetwork.com/journals/jama/fullarticle/2770279

World Health Organization Solidarity Trial Consortium, 'Repurposed antiviral drugs for COVID-19: interim WHO Solidarity Trial Results', *New England Journal of Medicine*, 2021 https://www.nejm.org/doi/full/10.1056/nejmoa2023184

World Health Organization, 'What is the pandemic (H1N1) 2009 virus?', 2010 https://www.who.int/csr/disease/swineflu/frequently_asked_questions/about_disease/en/#:~:text=The%20pandemic%20H1N1%20virus%20is,can%20contaminate%20hands%20or%20surfaces

Xue, L., and Zeng, G., 'Global strategies and response measures to the influenza A (H1N1) pandemic', *A Comprehensive Evaluation on Emergency Response in China*, 2018 https://doi.org/10.1007/978-981-13-0644-0_2

## 6. Trump's Divided America and the Children

Abad-Santos, A., 'Trump said he might fire Fauci after the election, but "don't tell anybody"', *Vox*, 2020 https://www.vox.com/21545631/trump-fire-fauci-2020-election

Abutaleb, Y., Dawsey, J., Nakashima, E., et al., 'The US was beset by denial and dysfunction as the coronavirus raged', *Washington Post*, 2020 https://www.washingtonpost.com/national-security/2020/04/04/coronavirus-government-dysfunction/

Adams, R., 'Education inequalities exposed by COVID have no quick fix – survey', *Guardian*, 2021 https://www.theguardian.com/education/2021/mar/25/education-inequalities-exposed-by-covid-have-no-quick-fix-survey

Altman, D., 'Understanding the US failure on coronavirus – an essay', *BMJ*, 2020 https://www.bmj.com/content/370/bmj.m3417

Armstrong, D., and Randall, T., 'US ranks behind EU, Australia, UK in vaccine buys but says it will meet goal', Bloomberg, 2020 https://www.bloomberg.com/news/articles/2020-12-09/which-countries-have-reserved-the-most-covid-19-vaccines-u-s-is-32nd-on-list

Ayanian, J. Z., Cleary, P. D., Weissman, J. S., et al., 'The effect of patients' preferences on racial differences in access to renal transplantation', *New England Journal of Medicine*, 1999 https://www.nejm.org/doi/full/10.1056/NEJM199911253412206

Bach, P. B., Cramer, L. D., Warren, J. L., et al., 'Racial differences in the treatment of early-stage lung cancer', *New England Journal of Medicine*, 1999 https://www.nejm.org/doi/full/10.1056/NEJM199910143411606

Baptiste, N., 'Black people have suffered the most from COVID-19. But they're still suspicious of vaccines', *Mother Jones*, 2020 https://www.motherjones.com/coronavirus-updates/2020/05/black-people-have-suffered-the-most-from-covid-but-theyre-still-suspicious-of-vaccines/

Barclay, E., 'Trump and his staff's refusal to wear a face mask is a catastrophe', *Vox*, 2020https://www.vox.com/2020/10/2/21498414/trump-coronavirus-mask-white-house-kayleigh-mcenany

Bassett, M. T., and Graves, J. D., 'Uprooting institutionalized racism as public health practice', *American Journal of Public Health*, 2018 https://doi.org/10.2105/AJPH.2018.304314

BBC News, 'George Floyd: what happened in the final moments of his life', 2020 https://www.bbc.co.uk/news/world-us-canada-52861726

Belvedere, M., 'Trump says he trusts China's Xi on coronavirus and the US has it "totally under control"', CNBC, 2020 https://www.cnbc.com/2020/01/22/trump-on-coronavirus-from-china-we-have-it-totally-under-control.html

Bhalla, N., '"Education emergency" as third of world's children lack remote learning', Reuters, 2020 https://www.reuters.com/article/us-health-coronavirus-africa-education-t-idUSKBN25N1QT

Centers for Disease Control and Prevention, 'Different COVID-19 vaccines', 2021 https://www.cdc.gov/coronavirus/2019-ncov/vaccines/different-vaccines.html

Cheung, H., 'George Floyd death: why US protests are so powerful this time', BBC News, 2020 https://www.bbc.co.uk/news/world-us-canada-52969905

Courtemanche, C., Le, A., Yelowiz, A., et al., 'School reopenings, mobility and COVID-19 spread: evidence from Texas', National Bureau of Economic Research, 2021 https://www.nber.org/papers/w28753

Cuddy, A., 'COVID: how to protest during a global pandemic', BBC News, 2020 https://www.bbc.co.uk/news/world-54477523

Danner, C., and Stieb, M., 'What we know about the US COVID-19 vaccine distribution plan', *Intelligencer*, 2020 https://nymag.com/intelligencer/2020/12/what-we-know-about-u-s-covid-19-vaccine-distribution-plan.html

Deluca, A., 'Miami-Dade mayor claims without evidence that protests caused COVID-19 spike', *Miami New Times*, 2020 https://www.miaminewtimes.com/news/miami-dade-mayor-carlos-gimenez-black-lives-matter-protests-covid-19-spike-11663729

Devakumar, D., Selvarajah, S., Shannon, G., et al., 'Racism, the public health crisis we can no longer ignore', *Lancet*, 2020 https://www.thelancet.com/pdfs/journals/lancet/PIIS0140-6736(20)31371-4.pdf

European Centre for Disease Prevention and Control, 'COVID-19 in children and the role of school settings in transmission – first update', 2020 https://www.ecdc.europa.eu/en/publications-data/children-and-school-settings-covid-19-transmission

Evelyn, K., 'FBI investigates death of Black man after footage shows officer kneeling on his neck', *Guardian*, 2020 https://www.theguardian.com/us-news/2020/may/26/george-floyd-killing-police-video-fbi-investigation

Farmer, P. E., Nizeye, B., Stulac, S., et al., 'Structural violence and clinical medicine', *PLoS Medicine*, 2006 https://doi.org/10.1371/journal.pmed.0030449

Feagin, J., and Bennefield, Z., 'Systemic racism and US health care', *Social Science and Medicine*, 2014 https://doi.org/10.1016/j.socscimed.2013.09.006

Frieden, T., Koplan, J., Satcher, D., et al., 'We ran the CDC. No president ever politicized its science the way Trump has', *Washington Post*, 2020 https://www.washingtonpost.com/outlook/2020/07/14/cdc-directors-trump-politics/

Garcia, J., and Sharif, M. Z., 'Black Lives Matter: a commentary on racism and public health', *American Journal of Public Health*, 2015 https://doi.org/10.2105/AJPH.2015.302706

Garg, S., Kim, L., and Whitaker, M., 'Hospitalization rates and characteristics of patients hospitalized with laboratory-confirmed coronavirus disease, 2019 – COVID-NET, 14 States, March 1–30, 2020', Centers for Disease Control and Prevention Morbidity and Mortality Weekly Report, 2020 https://www.cdc.gov/mmwr/volumes/69/wr/mm6915e3.htm

Gathright, J., 'Black Washingtonians make up less than half of DC's population, but 80 per cent of coronavirus deaths', DCist, 2020 https://dcist.com/story/20/05/06/black-washingtonians-make-up-less-than-half-of-d-c-s-population-but-80-of-coronavirus-deaths/

Hanushek, E., and Woessmann, L., 'The economic impacts of learning losses', OECD, 2020 https://www.oecd.org/education/The-economic-impacts-of-coronavirus-covid-19-learning-losses.pdf

Healy, B., 'In massive show of solidarity, vigil attendees gather at Franklin Park to protest police brutality', WBUR, 2020 https://www.wbur.org/news/2020/06/02/vigil-attendees-gather-at-franklin-park-to-protest-police-brutality

Jarmanning, A., 'How a day of peaceful protest turned to hours of unrest in Boston', WBUR, 2020 https://www.wbur.org/news/2020/06/04/boston-protests-sunday-peaceful-police

Johnson & Johnson, 'Johnson & Johnson COVID-19 vaccine authorized by US FDA for emergency use – first single-shot vaccine in fight against global pandemic', 2021 https://www.jnj.com/johnson-johnson-covid-19-vaccine-authorized-by-u-s-fda-for-emergency-usefirst-single-shot-vaccine-in-fight-against-global-pandemic

Lewis, D., 'Why schools probably aren't COVID hotspots', *Nature*, 2020 https://www.nature.com/articles/d41586-020-02973-3

Lewis, S., Munro, A., and Smith, G., 'Closing schools is not evidence based and harms children', *BMJ*, 2021 https://doi.org/10.1136/bmj.n521

Lopez, G., 'America still needs more coronavirus testing. Trump's new plan falls short', *Vox*, 2020 https://www.vox.com/2020/4/28/21239729/coronavirus-testing-trump-plan-white-house

Lopez, G., 'Europe's second wave of COVID-19 doesn't excuse Trump's failures', *Vox*, 2020https://www.vox.com/future-perfect/21536607/trump-coronavirus-covid-europe-second-wave

Lopez, G., 'Everyone failed on COVID-19', *Vox*, 2020 https://www.vox.com/future-perfect/22176191/covid-19-coronavirus-pandemic-democrats-republicans-trump

Lopez, G., 'Florida now has more COVID-19 cases than any other state. Here's what went wrong', *Vox*, 2020 https://www.vox.com/future-perfect/2020/7/17/21324398/florida-coronavirus-covid-cases-deaths-outbreak

Lopez, G., 'How Mike Pence enabled Donald Trump's botched COVID-19 response', *Vox*, 2020 https://www.vox.com/future-perfect/2020/10/7/21504186/mike-pence-coronavirus-covid-vice-presidential-debate-trump

Lopez, G., 'How New York governor Andrew Cuomo failed, then succeeded on COVID-19', *Vox*, 2020 https://www.vox.com/future-perfect/21401242/andrew-cuomo-coronavirus-covid-pandemic-new-york

Lopez, G., 'If Trump has learned anything from getting COVID-19, he's not showing it', *Vox*, 2020 https://www.vox.com/future-perfect/2020/10/4/21501511/trump-covid-19-coronavirus-tweet-leaving-walter-reed-hospital

Lopez, G., 'The CDC calls for quarantining even after a negative test. The White House isn't listening', *Vox*, 2020 https://www.vox.com/future-perfect/2020/10/5/21502485/trump-white-house-covid-coronavirus-cluster-kayleigh-mcenany

Lopez, G., 'Trump's attempts to corrupt the CDC, explained', *Vox*, 2020 https://www.vox.com/future-perfect/21436459/cdc-trump-mmrw-covid-19-coronavirus-pandemic

Lopez, G., 'Trump's closing message on COVID-19: my only mistake was mishandling public relations', *Vox*, 2020 https://www.vox.com/2020-presidential-election/2020/11/3/21547973/trump-election-day-covid-coronavirus-pandemic

Malat, J., Clark-Hitt, R., Burgess, D. J., et al., 'White doctors and nurses on racial inequality in health care in the USA: whiteness and colour-blind racial ideology', *Ethnic and Racial Studies*, 2010 https://doi.org/10.1080/01419870903501970

Matza, M., 'COVID vaccine: when will Americans be vaccinated?', BBC News, 2020 https://www.bbc.co.uk/news/world-us-canada-55149138

Maxmen, A., and Tollefson, J., 'Two decades of pandemic war games failed to account for Donald Trump, *Nature*, 2020 https://www.nature.com/articles/d41586-020-02277-6

Mayberry, R. M., Mili, F., and Ofili, E., 'Racial and ethnic differences in access to medical care', *Medical Care Research and Review*, 2000 https://doi.org/10.1177%2F1077558700057001S06

Nabavi, N., and Dobson, J., 'COVID-19 and schools – known unknowns', 2021 https://blogs.bmj.com/bmj/2021/02/03/covid-19-and-schools-known-unknowns/

Newburger, E., 'Dr Fauci says his daughters need security as family continues to get death threats', CNBC, 2020 https://www.cnbc.com/2020/08/05/dr-fauci-says-his-daughters-need-security-as-family-continues-to-get-death-threats.html

Newkirk II, V. R., 'A generation of bad blood', *Atlantic*, 2016 https://www.theatlantic.com/politics/archive/2016/06/tuskegee-study-medical-distrust-research/487439/

Neyman, G., and Dalsey, W., 'Black Lives Matter protests and COVID-19 cases: relationship in two databases', *Journal of Public Health*, 2020 https://doi.org/10.1093/pubmed/fdaa212

North, A., 'The debate over how to handle kids' "lost year" of learning', *Vox*, 2021 https://www.vox.com/22380650/school-remote-distance-learning-pandemic-covid-19

North, A., 'The shift to online learning could worsen educational inequality', *Vox*, 2020 https://www.vox.com/2020/4/9/21200159/coronavirus-school-digital-low-income-students-covid-new-york

North, A., 'Why restaurants are open and schools are closed', *Vox*, 2020 https://www.vox.com/21570207/covid-19-schools-restaurants-new-york-close

Nsikan, A., 'US has only a fraction of the medical supplies it needs to combat coronavirus', *National Geographic*, 2020 https://www.nationalgeographic.com/science/article/us-america-has-fraction-medical-supplies-it-needs-to-combat-coronavirus

Nuriddin, A., Mooney, G., and White, A., 'Reckoning with histories of medical racism and violence in the USA', *Lancet*, 2020 https://doi.org/10.1016/S0140-6736(20)32032-8

Our World in Data, 'United States: coronavirus pandemic country profile', 2021 https://ourworldindata.org/coronavirus/country/united-states

Phelan, J., and Link, B., 'Is racism a fundamental cause of inequalities in health?', *Annual Review of Sociology*, 2015 https://doi.org/10.1146/annurev-soc-073014-112305

Piper, K., 'Biden agreed to waive vaccine patents. But will that help get doses out faster?', *Vox*, 2021 https://www.vox.com/future-perfect/22419842/vaccine-patents-biden-pfizer-moderna-johnson-astrazeneca

Prasad, R., 'Coronavirus: how did Florida get so badly hit by COVID-19?', BBC News, 2020 https://www.bbc.co.uk/news/world-us-canada-53357742

Project TCT, 'The COVID racial data tracker', 2021 https://covidtracking.com/race

Ray, R., 'Why are Blacks dying at a faster rate from COVID?', Brookings, 2020 https://www.brookings.edu/blog/fixgov/2020/04/09/why-are-blacks-dying-at-higher-rates-from-covid-19/

Scott, D., 'Cash bonuses, on-site clinics, paid leave: how US companies are supporting the vaccine drive', *Vox*, 2021 https://www.vox.com/coronavirus-covid19/22444414/covid-19-vaccine-incentives-amazon-target-mcdonalds

Scott, D., 'Why the worst fears about Florida's COVID-19 outbreak haven't been realized (so far)', *Vox*, 2020 https://www.vox.com/2020/4/24/21234641/florida-coronavirus-covid-19-stay-at-home

Skloot, R., *The immortal life of Henrietta Lacks*, Crown, 2010 https://www.google.co.uk/books/edition/The_Immortal_Life_of_Henrietta_Lacks/tzlqDwAAQBAJ?hl=en&gbpv=1&printsec=frontcover

Smith, D., '"Your lives matter": Obama offers words of hope in contrast to Trump's division', *Guardian*, 2020 https://www.theguardian.com/us-news/2020/jun/03/obama-george-floyd-remarks-protests-trump

Smith, W., 'Consequences of school closure on access to education: lessons from the 2013–16 Ebola pandemic', *International Review of Education*, 2021 https://doi.org/10.1007/s11159-021-09900-2

Sprunt, B., 'Despite risks to others, Trump leaves hospital suite to greet supporters', NPR, 2020 https://www.npr.org/sections/latest-updates-trump-covid-19-results/

2020/10/04/920181116/in-brief-drive-by-trump-waves-to-supporters-outside-of-walter-reed?t=1628771726979

Tasamba, J., 'Prolonged closure of schools in Africa sparks fear of setbacks', Anadolu Agency, 2021 https://www.aa.com.tr/en/africa/prolonged-closure-of-schools-in-africa-sparks-fear-of-setbacks/2120982

Temperton, J., 'These children had COVID-19 beat. Then they got seriously ill', *Wired*, 2020 https://www.wired.co.uk/article/coronavirus-children-pims-ts

Temple-West, P., 'US health officials deny Trump claim COVID death toll "exaggerated"', *Financial Times*, 2021 https://www.ft.com/content/d3389c93-3880-4c47-9886-d01a98b57aa6

Thebault, R., Ba Tran, A., and Williams, V., 'The coronavirus is infecting and killing black Americans at an alarmingly high rate', *Washington Post*, 2020 https://www.washingtonpost.com/nation/2020/04/07/coronavirus-is-infecting-killing-black-americans-an-alarmingly-high-rate-post-analysis-shows/

Thompson, D. A., Fry, R., Marchant, E., et al., 'Staff–pupil SARS-CoV-2 infection pathways in schools in Wales: a population-level linked data approach', *BMJ Paediatrics Open*, 2021 https://bmjpaedsopen.bmj.com/content/5/1/e001049

Tollefson, J., 'How Trump damaged science – and why it could take decades to recover', *Nature*, 2020 https://www.nature.com/articles/d41586-020-02800-9

UN News, 'Coronavirus update: UN addresses school disruptions, suspends public access to New York Headquarters', 2020 https://news.un.org/en/story/2020/03/1059121

UN News, '"Emergency" for global education, as fewer than half world's students cannot return to school', 2020 https://news.un.org/en/story/2020/09/1071402

UNESCO, 'One year into COVID-19 education disruption: where do we stand?', 2021 https://en.unesco.org/news/one-year-covid-19-education-disruption-where-do-we-stand

UNESCO, 'UNESCO figures show two thirds of an academic year lost on average worldwide due to COVID-19 school closures', 2021 https://en.unesco.org/news/unesco-figures-show-two-thirds-academic-year-lost-average-worldwide-due-covid-19-school

UNICEF, 'COVID-19: at least a third of the world's schoolchildren unable to access remote learning during school closures, new report says', 2020 https://www.unicef.org/press-releases/covid-19-least-third-worlds-schoolchildren-unable-access-remote-learning-during

UNICEF, 'Latin America and the Caribbean is home to 3 out of 5 children who lost an entire school year worldwide', 2021 https://www.unicef.org/lac/en/press-releases/latin-america-and-caribbean-is-home-of-3-out-5-children-who-lost-an-entire-school-year-in-the-world

United States Census Bureau, 'QuickFacts Florida', 2019 https://www.census.gov/quickfacts/FL

United States Department of Human and Health Services, 'Biden Administration purchases additional doses of COVID-19 vaccines from Pfizer and Moderna', 2021 https://www.hhs.gov/about/news/2021/02/11/biden-administration-purchases-additional-doses-covid-19-vaccines-from-pfizer-and-moderna.html

van Lacker, W., and Parolin, Z., 'COVID-19, school closures and child poverty: a social crisis in the making', *Lancet Public Health*, 2020 https://doi.org/10.1016/S2468-2667(20)30084-0

Vasquez Reyes, M., 'The disproportional impact of COVID-19 on African Americans', *Health and Human Rights*, 2020 https://www.researchgate.net/publication/348736024_The_Disproportional_Impact_of_COVID-19_on_African_Americans

Viglione, G., 'Four ways Trump has meddled in pandemic science – and why it matters', *Nature*, 2020 https://www.nature.com/articles/d41586-020-03035-4

Wagner, J., '"It's real ugly": protesters clash with Minneapolis police after George Floyd's death', CBS Minnesota, 2020 https://minnesota.cbslocal.com/2020/05/26/hundreds-of-protesters-march-in-minneapolis-after-george-floyds-deadly-encounter-with-police/

Walsh, S., Chowdhury, A., Braithwaite, V., et al., 'Do school closures and school reopenings affect community transmission of COVID-19? A systematic review of observational studies', *BMJ Open*, 2021 https://bmjopen.bmj.com/content/11/8/e053371

Walters, Q., '"I'm in a perpetual state of anger": hundreds in Boston protest George Floyd's death', WBUR, 2020 https://www.wbur.org/news/2020/05/29/george-floyd-protest-boston

Wilkinson, A., 'Report: the Trump administration didn't order ventilators or masks until mid-March', *Vox*, 2020 https://www.vox.com/2020/4/5/21208802/coronavirus-trump-ventilators-masks-march

Williams, D. R., 'Miles to go before we sleep: racial inequities in health', *Journal of Health and Social Behaviour*, 2012 https://doi.org/10.1177%2F0022146512455804

Williams, D. R., 'Opinion: stress was already killing black Americans. COVID-19 is making it worse', *Washington Post*, 2020 https://www.washingtonpost.com/opinions/2020/05/13/stress-was-already-killing-black-americans-covid-19-is-making-it-worse/

Williams, D. R., Lawrence, J. A., and Davis, B. A., 'Racism and health: evidence and needed research', *Annual Review of Public Health*, 2019 https://doi.org/10.1146/annurev-publhealth-040218-043750

Williams, D. R., Priest, N., and Anderson, N. B., 'Understanding associations among race, socioeconomic status, and health: patterns and prospects', *Health Psychology*, 2016 https://doi.apa.org/doi/10.1037/hea0000242

Williams, P., and Hemingway, P., 'Donald Trump insists schools reopen, but some parents, teachers and children fear coronavirus', ABC News, 2020 https://www.abc.net.au/news/2020-09-03/trump-insists-schools-open-but-some-parents-fear-coronavirus/12598944

World Health Organization, 'WHO, UNICEF urge safe school reopening in Africa', 2020 https://www.afro.who.int/news/who-unicef-urge-safe-school-reopening-africa

Yglesias, M., 'Trump has consistently mocked adherence to public health guidelines', *Vox*, 2020https://www.vox.com/2020/10/2/21498574/trump-covid-biden-basement

Yglesias, M., 'Trump's flailing incompetence makes coronavirus even scarier', *Vox*, 2020 https://www.vox.com/2020/2/25/21150574/trump-coronavirus-cdc-cuts

## *7. Elimination versus 'Letting Go'*

ABC News, 'Australian medical team touches down in Port Moresby with emergency vaccines', 2021 https://www.abc.net.au/news/2021-03-23/ausmat-team-touches-down-in-port-moresby/100023680

ABC News, 'Italy, EU refuse AstraZeneca request to ship 250,000 doses of vaccine to Australia', 2021 https://www.abc.net.au/news/2021-03-05/italy-eu-block-250000-astrazeneca-doses-to-australia/13218348

Agence France-Presse, 'Dutch court reinstates COVID curfew minutes before its start time', *Guardian*, 2020 https://www.theguardian.com/world/2021/feb/16/dutch-court-orders-government-lift-covid-curfew

Ainge, Roy E., '"Deliberate, malicious": stop spreading COVID misinformation, says New Zealand minister', *Guardian*, 2020 https://www.theguardian.com/world/2020/sep/10/deliberate-malicious-stop-spreading-covid-misinformation-says-new-zealand-minister

Ainge, Roy E., 'Nearly a third of New Zealanders felt badly distressed in COVID lockdown', *Guardian*, 2020 https://www.theguardian.com/world/2020/nov/05/nearly-a-third-of-new-zealanders-felt-badly-distressed-in-covid-lockdown

Ainge, Roy, E., 'New Zealand's three-week streak without local COVID case ends as port worker falls ill', *Guardian*, 2020 https://www.theguardian.com/world/2020/oct/18/new-zealands-three-week-streak-without-local-covid-case-ends-as-port-worker-falls-ill

Ainge, Roy E., 'Trump calls out New Zealand's "terrible" COVID surge, on day it records nine new cases', *Guardian*, 2020 https://www.theguardian.com/world/2020/aug/18/trump-calls-out-new-zealands-big-surge-on-day-it-records-nine-covid-cases

Akindele, B., 'Living with long COVID in Lagos', *Mail and Guardian*, 2021 https://mg.co.za/africa/2021-02-28-living-with-long-covid-in-lagos/

Al Jazeera, 'Dutch court approves government's use of pandemic curfew measures', 2021https://www.aljazeera.com/news/2021/2/26/dutch-court-approves-govt-use-of-pandemic-curfew-measures

Anderson, R. M., Heesterbeek, H., Klinkenberg, D., et al., 'How will country-based mitigation measures influence the course of the COVID-19 epidemic?' *Lancet*, 2020 https://doi.org/10.1016/S0140-6736(20)30567-5

Andrews, P. J., Pendolino, A. L., Ottaviano, G., et al., 'Olfactory and taste dysfunction among mild-to-moderate symptomatic COVID-19 positive health care workers: an international survey', *Laryngoscope Investigative Otolaryngology*, 2020 https://doi.org/10.1002/lio2.507

Assaf, G., Davis, H., McCorkell, L., et al., 'Report: what Does COVID-19 recovery actually look like?', Patient-led Research Collaborative, 2020 https://patientresearchcovid19.com/research/report-1/

Associated Press, 'Ardern reveals the moment she chose COVID elimination strategy', *Guardian*, 2020 https://www.theguardian.com/world/2020/dec/16/ardern-reveals-the-moment-she-c-covid-elimination-strategy

Associated Press, 'Balancing act: Dutch PM eases lockdown amid infection rise', *Independent*, 2021 https://www.independent.co.uk/news/balancing-act-dutch-pm-eases-lockdown-amid-infection-rise-mark-rutte-dutch-netherlands-schools-european-union-b1834683.html

Baker, M., 'New Zealand's elimination strategy for the COVID-19 pandemic: early success but uncertainties and risks remain', Usher Institute COVID-19 Webinar, 2020 https://www.ed.ac.uk/usher/news-events/covid-19-webinars

Baker, M., Kvalsvig, A., Verrall, A. J., et al., 'New Zealand's elimination strategy for the COVID-19 pandemic and what is required to make it work', *New Zealand Medical Journal*, 2020https://journal.nzma.org.nz/journal-articles/new-zealands-elimination-strategy-for-the-covid-19-pandemic-and-what-is-required-to-make-it-work

Baker, M. G., Kvalsvig, A., and Verrall, A. J., 'New Zealand's COVID-19 elimination strategy', *Medical Journal of Australia*, 2020 https://doi.org/10.5694/mja2.50735

Baker, M. G., Wilson, N., and Anglemyer, A., 'Successful elimination of COVID-19 transmission in New Zealand', *New England Journal of Medicine*, 2020 https://www.nejm.org/doi/full/10.1056/NEJMc2025203

Baker, M. G., Wilson, N., and Blakely, T., 'Elimination could be the optimal response strategy for COVID-19 and other emerging pandemic diseases', *BMJ*, 2020 https://doi.org/10.1136/bmj.m4907

Barrett, M., 'What have Norway, Finland and Denmark got right on COVID-19?', *New Statesman*, 2020 https://www.newstatesman.com/world/2020/12/what-have-norway-finland-and-denmark-got-right-on-covid-19

BBC News, 'Coronavirus: Australia's Victoria records huge case jump', 2020 https://www.bbc.co.uk/news/world-australia-53589817

BBC News, 'Coronavirus: how New Zealand relied on science and empathy', 2020 https://www.bbc.co.uk/news/world-asia-52344299

BBC News, 'Coronavirus in Australia: Victoria to ease lockdown as cases fall', 2020 https://www.bbc.co.uk/news/world-australia-54592122

BBC News, 'Coronavirus: Melbourne lockdown to keep a million workers at home', 2020 https://www.bbc.co.uk/news/world-australia-53632980

BBC News, 'Coronavirus: Queensland to close border to New South Wales', 2020 https://www.bbc.co.uk/news/world-australia-53659914

BBC News, 'Coronavirus: Sweden's isolated elderly urged to rejoin society', 2020 https://www.bbc.co.uk/news/world-europe-54643070

BBC News, 'COVID-19: Netherlands suspends use of AstraZeneca vaccine', 2021 https://www.bbc.co.uk/news/world-europe-56397157

BBC News, 'New Zealand election: Jacinda Ardern's Labour Party scores landslide win', 2020 https://www.bbc.co.uk/news/world-asia-54519628

BBC News, '*Ruby Princess*: New South Wales premier apologises over cruise ship outbreak', 2020 https://www.bbc.co.uk/news/world-australia-53802816

Bjorklund, K., and Ewing, A., 'The Swedish COVID-19 response is a disaster. It shouldn't be a model for the rest of the world', *Time*, 2020 https://time.com/5899432/sweden-coronovirus-disaster/

Boffey, D., 'Dutch clubbers hit dancefloor for study into easing lockdown', *Guardian*, 2021 https://www.theguardian.com/world/2021/mar/07/dutch-clubbers-dance-study-easing-lockdown-music-event-coronavirus

Boseley, M., 'TGA admits minors mistakenly given AstraZeneca vaccine and says woman's death an "atypical case"', *Guardian*, 2021 https://www.theguardian.com/australia-news/2021/apr/17/tga-admits-minors-mistakenly-given-astrazeneca-vaccine-and-says-womans-death-an-atypical-case

Brouwers, T., van der Heiden, M., Lakerveld, A., et al., 'COVID-19', Koninklijke Nederlands Vereniging voor Microbiologie, 2021 https://www.knvm.org/vaccinologie/covid-19

Buttler, M., and Rigillo, N., 'Trump's latest attack on Sweden revives coronavirus controversy', Bloomberg, 2020 https://www.bloomberg.com/news/articles/2020-05-01/trump-s-latest-attack-on-sweden-revives-covid-19-controversy

Callard, F., and Perego, E., 'How and why patients made long COVID', *Social Science and Medicine*, 2021 https://doi.org/10.1016/j.socscimed.2020.113426

Carfì, A., Bernabei, R., Landi, F., et al., 'Persistent symptoms in patients after acute COVID-19', *Journal of the American Medical Association*, 2020 https://jamanetwork.com/journals/jama/fullarticle/2768351

Carvalho-Schneider, C., Laurent, E., Lemaignen, A., et al., 'Follow-up of adults with noncritical COVID-19 two months after symptom onset', *Clinical Microbiology and Infection*, 2021 https://doi.org/10.1016/j.cmi.2020.09.052

Chopra, V., Flanders, S. A., O'Malley, M., et al., 'Sixty-day outcomes among patients hospitalized with COVID-19', *Annals of Internal Medicine*, 2021 https://doi.org/10.7326/M20-5661

Claeson, M., and Hanson, S., 'COVID-19 and the Swedish enigma', *Lancet*, 2021 https://doi.org/10.1016/S0140-6736(20)32750-1

Cohen, J., 'Metamorphosis of a Dutch top public health official, Jaap van Dissel, in the face of the COVID-19 pandemic', *Forbes*, 2020 https://www.forbes.com/sites/joshuacohen/2021/03/11/metamorphosis-of-a-dutch-top-public-health-official-jaap-van-dissel-in-the-face-of-the-covid-19-pandemic/?sh=251a12597ad4

Commonwealth of Australia, 'Australian health sector emergency response plan for novel coronavirus (COVID-19)', 2020 https://www.health.gov.au/resources/publications/australian-health-sector-emergency-response-plan-for-novel-coronavirus-covid-19

Commonwealth of Australia, Department of Health, 'Australia's vaccine agreements', https://www.health.gov.au/node/18777/australias-vaccine-agreements

Commonwealth of Australia, Department of Health, 'First confirmed case of novel coronavirus in Australia', 2020 https://www.health.gov.au/ministers/the-hon-greg-hunt-mp/media/first-confirmed-case-of-novel-coronavirus-in-australia

Corder, M., 'Dutch coronavirus infections rise ahead of lockdown easing', ABC News, 2021 https://abcnews.go.com/Health/wireStory/dutch-coronavirus-infections-rise-ahead-lockdown-easing-77247218

Cousins, S., 'Experts criticise Australia's aged care failings over COVID-19', *Lancet*, 2020 https://doi.org/10.1016/S0140-6736(20)32206-6

COVID LIVE, Australia, 2021 https://covidlive.com.au/

Davey, M., 'Melbourne suburbs lockdown announced as Victoria battles coronavirus outbreaks', *Guardian*, 2020 https://www.theguardian.com/australia-news/2020/jun/30/melbourne-hotspot-lockdowns-announced-as-victoria-battles-coronavirus-outbreaks

Davey, M., 'South Australia's COVID lockdown explained: why six days and thirty-six cases isn't low?', *Guardian*, 2020 https://www.theguardian.com/australia-news/2020/nov/18/south-australias-covid-lockdown-explained-why-six-days-and-isnt-36-cases-low

Davies, G., and Roeber, B., 'Sweden has avoided a COVID-19 lockdown so far: Has its strategy worked?', ABC News, 2021 https://abcnews.go.com/International/sweden-avoided-covid-19-lockdown-strategy-worked/story?id=76047258

Davis, H., Assaf, G., McCorkell, L., et al., 'Characterizing long COVID in an international cohort: seven months of symptoms and their impact', *EClinical Medicine*, 2021 https://doi.org/10.1016/j.eclinm.2021.101019

de Graaf, P., '2.6 miljoen nertsen zijn vergast vanwege corona. Wie of wat bracht het virus naar de fokkerijen?', *De Volkskrant*, 2020 https://www.volkskrant.nl/nieuws-achtergrond/2-6-miljoen-nertsen-zijn-vergast-vanwege-corona-wie-of-wat-bracht-het-virus-naar-de-fokkerijen~ba3f1beb/?referrer=https%3A%2F%2Fwww.google.com%2F

de Jong, E., 'New Zealand COVID vaccines to arrive one month early, border staff to be inoculated next week', *Guardian*, 2021 https://www.theguardian.com/world/2021/feb/12/new-zealand-covid-vaccines-to-arrive-one-month-early-border-staff-to-be-inoculated-next-week de Vrieze, J., 'Dutch studies bring back the fun – but are they good science?', *Science*, 2021 https://www.science.org/doi/10.1126/science.372.6541.447

Dennis, A., Wamil, M., Alberts, J., et al., 'Multiorgan impairment in low-risk individuals with post-COVID-19 syndrome: a prospective, community-based study', *BMJ Open*, 2021 https://bmjopen.bmj.com/content/11/3/e048391

Diaz, J., 'New Zealand will give free coronavirus vaccines to residents, neighboring nations', NPR, 2020 https://www.npr.org/sections/coronavirus-live-updates/2020/12/17/947403839/new-zealand-will-give-free-coronavirus-vaccines-to-residents-neighboring-nations

Doherty, B., and Murphy, K., 'Australia declares coronavirus will become pandemic as it extends China travel ban', *Guardian*, 2020 https://www.theguardian.com/world/2020/feb/27/australia-declares-coronavirus-will-become-a-pandemic-as-it-extends-china-travel-ban

DutchNews.nl., 'Dutch health minister, charged with tackling coronavirus, resigns after collapse', 2020 https://www.dutchnews.nl/news/2020/03/dutch-health-minister-charged-with-tackling-coravirus-resigns-after-collapse/

DutchNews.nl, 'Group immunity not main aim of Dutch anti-corona measures, says health chief', 2020 https://www.dutchnews.nl/news/2020/03/group-immunity-not-main-aim-of-dutch-anti-corona-measures-says-health-chief/

Dziedzic, S., 'Australia to supply doses of domestically manufactured COVID-19 vaccines to Melanesian countries, including PNG and Timor-Leste', ABC News, 2021 https://www.abc.net.au/news/2021-04-09/australia-png-covid-vaccine-supply-melanesian-countries/100060206

*Economist*, 'The Swedish exception: why the Swedes are not yet locked down', 2020 https://www.economist.com/europe/2020/04/04/why-swedes-are-not-yet-locked-down

Gallagher, J., 'Long COVID: who is more likely to get it?', BBC News, 2020 https://www.bbc.co.uk/news/health-54622059

Galván-Tejada, C. E., Herrera-Garcia, C. F., Godina-Gonzalez, S., et al., 'Persistence of COVID-19 symptoms after recovery in Mexican population', *International Journal of Environmental Research and Public Health*, 2020 https://doi.org/10.3390/ijerph17249367

Gardner, P., 'For seven weeks I have been through a roller coaster of ill health, extreme emotions and utter exhaustion', *BMJ* Blogs, 2020 https://blogs.bmj.com/bmj/2020/05/05/paul-garner-people-who-have-a-more-protracted-illness-need-help-to-understand-and-cope-with-the-constantly-shifting-bizarre-symptoms/

Garrigues, E., Janvier, P., Kherabi, Y., et al., 'Post-discharge persistent symptoms and health-related quality of life after hospitalization for COVID-19', *Journal of Infection*, 2020 https://doi.org/10.1016/j.jinf.2020.08.029

Giesecke, J., 'The invisible pandemic', *Lancet*, 2020 https://doi.org/10.1016/S0140-6736(20)31035-7

Godin, M., 'Sweden's relaxed approach to the coronavirus could already be backfiring', *Time*, 2020 https://time.com/5817412/sweden-coronavirus/

Government of New South Wales, 'Public health order for Northern Beaches LGA', 2020 https://www.health.nsw.gov.au/news/Pages/20201219_03.aspx#

Government of New Zealand, 'History of the COVID-19 alert system', 2021 https://covid19.govt.nz/alert-system/history-of-the-covid-19-alert-system/

Government of New Zealand, 'New Zealand COVID-19 alert levels summary', 2020 https://covid19.govt.nz/assets/COVID_Alert-levels_v2.pdf

Government of the Netherlands, 'Additional measures to control the spread of coronavirus introduced in more regions', 2020 https://www.government.nl/topics/coronavirus-covid-19/news/2020/09/25/additional-measures-to-control-the-spread-of-coronavirus-introduced-in-more-regions

Government of the Netherlands, 'Avoid busy places and stay 1.5 metres away from others', 2020 https://www.government.nl/topics/coronavirus-covid-19/news/2020/05/06/avoid-busy-places-and-stay-1.5-metres-away-from-others

Government of the Netherlands, 'Coronavirus: additional measures introduced on 23 March 2020', 2020 https://www.government.nl/topics/coronavirus-covid-19/news/2020/03/24/additional-measures-introduced-on-23-march

Government of the Netherlands, 'Coronavirus puts stop to visits to nursing homes', 2020 https://www.government.nl/latest/news/2020/03/19/coronavirus-puts-stop-to-visits-to-nursing-homes

Government of the Netherlands, 'COVID-19: additional measures in schools, the hospitality sector and sport', 2020 https://www.government.nl/topics/coronavirus-covid-19/news/2020/03/15/additional-measures-in-schools-the-hospitality-sector-and-sport

Government of the Netherlands, 'COVID-19: new instructions for inhabitants of North Brabant', 2020 https://www.government.nl/latest/news/2020/03/06/covid-19-new-instructions-for-inhabitants-of-north-brabant

Government of the Netherlands, 'Government adopts advice to cull mink on infected farms', 2020 https://www.government.nl/topics/coronavirus-covid-19/news/2020/06/09/government-adopts-advice-to-cull-mink-on-infected-farms

Government of the Netherlands, 'Lockdown measures tightened in response to concerns about new variants of virus', 2020 https://www.government.nl/topics/coronavirus-covid-19/news/2021/01/20/lockdown-measures-tightened-in-response-to-concerns-about-new-variants-of-virus

Government of the Netherlands, 'Lockdown to minimise contact between people', 2020 https://www.netherlandsworldwide.nl/latest/news/2020/12/16/lockdown-in-the-netherlands-in-order-to-minimise-contact-between-people

Government of the Netherlands, 'Man diagnosed with coronavirus (COVID-19) in the Netherlands', 2020 https://www.government.nl/latest/news/2020/02/27/man-diagnosed-with-coronavirus-covid-19-in-the-netherlands

Government of the Netherlands, 'Negative test declaration now mandatory for all international travel to the Netherlands', 2020 https://www.government.nl/latest/news/2020/12/23/negative-test-declaration-now-mandatory-for-all-international-air-travel-to-the-netherlands#:~:text=From%2000.01%20on%2029%20December,aircraft%20bound%20for%20the%20Netherlands.&text=The%20government%20has%20decided%20to,flights%20bound%20for%20the%20Netherlands

Government of the Netherlands, 'New measures to stop spread of coronavirus in the Netherlands', 2020 https://www.government.nl/latest/news/2020/03/12/new-measures-to-stop-spread-of-coronavirus-in-the-netherlands

Government of the Netherlands, 'New regional measures to control the spread of coronavirus', 2020 https://www.government.nl/topics/coronavirus-covid-19/news/2020/09/18/new-regional-measures-to-control-the-spread-of-coronavirus

Government of the Netherlands, 'No national exams this year', 2020 https://www.government.nl/topics/coronavirus-covid-19/news/2020/03/24/no-national-exams-this-year

Government of the Netherlands, 'Order of vaccination for health and care workers', 2021 https://www.government.nl/topics/coronavirus-covid-19/dutch-vaccination-programme/order-of-vaccination-against-coronavirus/health-and-care-workers

Government of the Netherlands, 'Order of vaccination for people who do not work in healthcare', 2021 https://www.government.nl/topics/coronavirus-covid-19/dutch-vaccination-programme/order-of-vaccination-against-coronavirus/order-of-vaccination-for-people-who-do-not-work-in-healthcare

Government of the Netherlands, 'Partial lockdown needed to bring down infections', 2020 https://www.government.nl/topics/coronavirus-covid-19/news/2020/10/13/partial-lockdown-needed-to-bring-down-infections

Government of the Netherlands, 'Patient with novel coronavirus deceased', 2020 https://www.government.nl/topics/coronavirus-covid-19/news/2020/03/06/patient-with-novel-coronavirus-deceased

Government of the Netherlands, 'Stricter measures to control coronavirus', 2020 https://www.government.nl/topics/coronavirus-covid-19/news/2020/03/23/stricter-measures-to-control-coronavirus

Government of the Netherlands, 'Temporary tightening of partial lockdown', 2020 https://www.government.nl/topics/coronavirus-covid-19/news/2020/11/03/temporary-tightening-of-partial-lockdown

Government of the Netherlands, 'The Netherlands closes its borders to persons from outside Europe', 2020 https://www.government.nl/topics/coronavirus-covid-19/news/2020/03/18/the-netherlands-closes-its-borders-to-persons-from-outside-europe

Government of the Netherlands, 'Travel advice: only travel abroad if essential', 2020 https://www.government.nl/topics/coronavirus-covid-19/news/2020/03/17/travel-advice-only-travel-abroad-if-essential

Government of the Netherlands, 'Urgent advice to wear face masks', 2020 https://www.government.nl/topics/coronavirus-covid-19/news/2020/10/02/urgent-advice-to-wear-face-masks

Graham-McLay, C., 'Ardern urged to review New Zealand COVID measures after election landslide', *Guardian*, 2020 https://www.theguardian.com/world/2020/oct/22/ardern-urged-to-review-covid-measures-after-election-landslide

Graham-McLay, C., 'New Zealand records first new local COVID-19 cases in 102 days', *Guardian*, 2020 https://www.theguardian.com/world/2021/oct/28/new-zealands-south-island-records-first-covid-cases-in-major-city-in-over-a-year

Greenhalgh, T., Knight, M., A'Court, C., et al., 'Management of post-acute COVID-19 in primary care', *BMJ*, 2020 https://doi.org/10.1136/bmj.m3026

*Guardian*, 'Australia's state-by-state COVID restrictions and coronavirus lockdown rules explained', 2021 https://www.theguardian.com/australia-news/2021/feb/15/australia-covid-19-lockdown-rules-coronavirus-restrictions-by-state-nsw-victoria-vic-queensland-qld-western-south-australia-wa-sa-nt-act-travel-border-social-distancing-masks

Hancock, S., 'Netherlands pauses AstraZeneca vaccine rollout for people under sixty', *Independent*, 2021 https://www.independent.co.uk/news/health/astrazeneca-vaccine-netherlands-europe-b1826127.html

Harding, L., '"Weird as hell": the COVID-19 patients who have symptoms for months', *Guardian*, 2020 https://www.theguardian.com/world/2020/may/15/weird-hell-professor-advent-calendar-covid-19-symptoms-paul-garner

Harvey-Jenner, C., 'Living with long COVID: what happens when coronavirus just won't go away', *Cosmopolitan*, 2020 https://www.cosmopolitan.com/uk/body/health/a34412641/long-covid/

Heikkilä, M., 'Swedish king: country has "failed" on coronavirus pandemic', *Politico*, 2020 https://www.politico.eu/author/melissa-heikkila/page/2/

Heneghan, C., Jefferson, T., 'Dying of neglect: the other COVID care home scandal', *Spectator*, 2020 https://www.spectator.co.uk/article/dying-of-neglect-the-other-covid-care-home-scandal

Henley, J., 'Dutch government resigns over child benefits scandal', *Guardian*, 2021 https://www.theguardian.com/world/2021/jan/15/dutch-government-resigns-over-child-benefits-scandal

Henley, J., 'Dutch leaders condemn "criminal" clashes at anti-lockdown protests', *Guardian*, 2021 https://www.theguardian.com/world/2021/jan/25/dutch-leaders-condemn-criminal-clashes-at-anti-lockdown-protests

Henley, J., 'Netherlands election: Mark Rutte claims fourth term with "overwhelming" victory', *Guardian*, 2021 https://www.theguardian.com/world/2021/mar/17/netherlands-election-mark-rutte-on-course-to-win-fourth-term

Henley, J., 'Sweden has highest new COVID cases per person in Europe', *Guardian*, 2021 https://www.theguardian.com/world/2021/apr/13/sweden-has-highest-new-covid-cases-per-person-in-europe

Henley, J., 'Swedish PM warned over "Russian roulette-style" COVID-19 strategy', *Guardian*, 2020 https://www.theguardian.com/world/2020/mar/23/swedish-pm-warned-russian-roulette-covid-19-strategy-herd-immunity

Henriques-Gomes, L., 'Snap five-day COVID lockdown for Victoria announced in bid to contain UK variant', *Guardian*, 2021 https://www.theguardian.com/australia-news/2021/feb/12/snap-five-day-covid-lockdown-for-victoria-announced-in-bid-to-contain-uk-variant

Hinde, N., 'Long COVID isn't just leaving people sick – it's taking everything they've got', *Huffington Post*, 2020 https://www.huffingtonpost.co.uk/entry/long-covid-work-financial-struggle_uk_5f6c7363c5b6e2c91261a97b

Hoekman, L. M., Smits, M. M. V., and Koolman, X., 'The Dutch COVID-19 approach: regional differences in a small country', *Health Policy Technology*, 2020 https://doi.org/10.1016/j.hlpt.2020.08.008

Holligan, A., 'Coronavirus: Dutch shocked to be EU vaccination stragglers', BBC News, 2021 https://www.bbc.co.uk/news/world-europe-55549656

Holligan, A., 'Coronavirus: why Dutch lockdown may be a high-risk strategy', BBC News, 2020 https://www.bbc.co.uk/news/world-europe-52135814

Hollingsworth, J., 'New Zealand and Australia were COVID success stories. Why are they behind on vaccine rollouts?', CNN, 2021 https://edition.cnn.com/2021/04/15/asia/new-zealand-australia-covid-vaccine-intl-dst-hnk/index.html

Hood, L., 'Trust in government soars in Australia and New Zealand during pandemic', *The Conversation*, 2021 https://theconversation.com/trust-in-government-soars-in-australia-and-new-zealand-during-pandemic-154948

Horvath, L., Lim, J. W. J., Taylor, J. W., et al., 'Smell and taste loss in COVID-19 patients: assessment outcomes in a Victorian population', Acta Oto-Laryngologica, 2021 https://doi.org/10.1080/00016489.2020.1855366

Hunt, E., 'Could Covid give New Zealand's struggling tourism sector a chance to go green?', *Guardian*, 2021 https://www.theguardian.com/world/2021/feb/18/could-covid-give-new-zealands-struggling-tourism-sector-a-chance-to-go-green

Hunt, E., 'Words matter: how New Zealand's clear messaging helped beat COVID', *Guardian*, 2021 https://www.theguardian.com/world/2021/feb/26/words-matter-how-new-zealands-clear-messaging-helped-beat-covid

Hurst, D., and Taylor, J., 'Victoria announces Stage 4 coronavirus lockdown restrictions including overnight curfew', *Guardian*, 2020 https://www.theguardian.com/australia-news/2020/aug/02/victoria-premier-daniel-andrews-stage-four-coronavirus-lockdown-restrictions-melbourne-covid-19

Ilanbey, S., and Towell, N., 'Huge new COVID testing blitz to target Melbourne's north-west', *The Age*, 2020 https://www.theage.com.au/politics/victoria/huge-new-covid-testing-blitz-to-target-melbourne-s-north-west-20201109-p56cz4.html

Irwin, R. E., 'Misinformation and de-contextualization: international media reporting on Sweden and COVID-19', *Globalization and Health*, 2020 https://doi.org/10.1186/s12992-020-00588-x

Jones, A., 'How did New Zealand become COVID-19 free?', BBC News, 2020 https://www.bbc.co.uk/news/world-asia-53274085

Kamal, M., Abo Omirah, M., Hussein, A., et al., 'Assessment and characterisation of post-COVID-19 manifestations', *International Journal of Clinical Practice*, 2020 https://doi.org/10.1111/ijcp.13746

Kavaliunas, A., Ocaya, P., Mumper, J., et al., 'Swedish policy analysis for COVID-19', *Health Policy and Technology*, 2020 https://doi.org/10.1016/j.hlpt.2020.08.009

Keay, L., 'Long-term COVID warning: ICU doctor reports having coronavirus symptoms for three months', Sky News, 2020 https://news.sky.com/story/long-term-covid-warning-icu-doctor-reports-having-coronavirus-symptoms-for-three-months-12014361

Knaus, C., 'Victoria records first overseas coronavirus case since hotel quarantine overhaul', *Guardian*, 2021 https://www.theguardian.com/world/2021/apr/10/victoria-records-first-overseas-coronavirus-case-since-hotel-quarantine-overhaul

*Lancet*, 'Facing up to long COVID', 2020 https://doi.org/10.1016/S0140-6736(20)32662-3

Livingstone, H., 'Ardern tells New Zealand border staff: get COVID vaccine now or be redeployed', *Guardian*, 2021 https://www.theguardian.com/world/2021/apr/12/ardern-new-zealand-border-staff-covid-vaccine

Local, The, '"Sweden never had a formal coronavirus strategy": Health Minister tells inquiry', 2021 https://www.thelocal.se/20210410/sweden-never-had-a-formal-coronavirus-strategy-health-minister-to-inquiry/

Lopez-Leon, S., Wegman-Ostrosky, T., Perelman, C., et al., 'More than fifty long-term effects of COVID-19: a systematic review and meta-analysis', *Scientific Reports*, 2021 https://doi.org/10.1038/s41598-021-95565-8

Lowenstein, F., 'We need to talk about what coronavirus recoveries look like', *New York Times*, 2020 https://www.nytimes.com/2020/04/13/opinion/coronavirus-recovery.html

Ludvigsson, J. F., 'The first eight months of Sweden's COVID-19 strategy and the key actions and actors that were involved', *Acta Pædiatr*, 2020 https://doi.org/10.1111/apa.15582

Maguire, D., Maasdopr, J., and Iorio, K., 'Coronavirus Australia live news: Melbourne ordered back into Stage 3 lockdown ahead of NSW–Victoria border closure on Wednesday', ABC News, 2020 https://www.abc.net.au/news/2020-07-07/coronavirus-australia-live-news-melbourne-lockdown-restrictions/12427970

Mandal, S., Barnett, J., Brill, S., et al., 'Long COVID': a cross-sectional study of persisting symptoms, biomarker and imaging abnormalities following hospitalisation for COVID-19', *Thorax*, 2020 http://dx.doi.org/10.1136/thoraxjnl-2020-215818

Mao, F., 'Coronavirus: why has Melbourne's outbreak worsened?', BBC News, 2020 https://www.bbc.co.uk/news/world-australia-53259356

Mao, F., 'Coronavirus: why is Melbourne seeing more cases?', BBC News, 2020 https://www.bbc.co.uk/news/world-australia-53604751

Martin, L., 'Coronavirus: how an aged care crisis seized "ill-prepared" Australia', BBC News, 2020 https://www.bbc.co.uk/news/world-australia-53633356

McClure, T., '"No roadmap": New Zealand mulls reopening options after a year of closed borders', *Guardian*, 2021 https://www.theguardian.com/world/2021/may/14/new-zealand-reopening-options-covid-closed-borders

McGowan, M., 'Victorians describe feeling "intimidated" by police enforcing lockdown laws', *Guardian*, 2020 https://www.theguardian.com/australia-news/2020/apr/21/victorians-describe-feeling-intimidated-by-police-enforcing-lockdown-laws

Milne, R., 'Architect of Sweden's no-lockdown strategy insists it will pay off', *Financial Times*, 2020 https://www.ft.com/content/a2b4c18c-a5e8-4edc-8047-ade4a82a548d

Milne, R., 'Coronavirus: Sweden's king says country's strategy has failed', *Irish Times*, 2020 https://www.irishtimes.com/news/world/europe/coronavirus-sweden-s-king-says-country-s-strategy-has-failed-1.4439515

Milne, R., 'Sweden's distinctive COVID strategy nears an end as lockdown proposed', *Financial Times*, 2021 https://www.ft.com/content/b376ae27-4889-4f54-803d-8507de9dcb7e

Ministry of Health, New Zealand, 'COVID-19 border controls', 2021 https://www.health.govt.nz/our-work/diseases-and-conditions/covid-19-novel-coronavirus/covid-19-response-planning/covid-19-border-controls

Ministry of Health, New Zealand, 'COVID-19: current cases', https://www.health.govt.nz/our-work/diseases-and-conditions/covid-19-novel-coronavirus/covid-19-data-and-statistics/covid-19-current-cases

Ministry of Health, New Zealand, 'COVID-19: use of masks and face coverings in the community', 2021 https://www.health.govt.nz/our-work/diseases-and-conditions/covid-19-novel-coronavirus/covid-19-health-advice-public/covid-19-use-masks-and-face-coverings-community

Ministry of Health, New Zealand, 'New Zealand influenza pandemic plan: a framework for action', 2017 https://www.health.govt.nz/publication/new-zealand-influenza-pandemic-plan-framework-action

Munro, K. J., Uus, K., Almufarrij, I., et al., 'Persistent self-reported changes in hearing and tinnitus in post-hospitalisation COVID-19 cases', *International Journal of Audiology*, 2020 https://doi.org/10.1080/14992027.2020.1798519

Murphy, K., and Visontay, E., 'Federal government had no COVID-19 aged care plan, royal commission hears', *Guardian*, 2020 https://www.theguardian.com/australia-news/2020/aug/10/government-had-no-covid-19-aged-care-plan-inquiry-told-as-catastrophic-failure-alleged-over-st-basils

Murray, J., 'Has Sweden's controversial COVID-19 strategy been successful or not?', *BMJ*, 2020 https://doi.org/10.1136/bmj.m3255

Nath, A., 'Long-haul COVID', *Neurology*, 2020 https://n.neurology.org/content/95/13/559

National Institute for Health Research, 'Living with COVID-19', 2020 https://evidence.nihr.ac.uk/themedreview/living-with-covid19/

National Institute for Public Health and the Environment, The Netherlands, 'Current information about COVID-19', 2021 https://www.rivm.nl/en/novel-coronavirus-covid-19/current-information

NL Times, 'Around 60 per cent of NL residents must get COVID-19 for herd immunity: health institute', 2020 https://nltimes.nl/2020/03/17/around-60-nl-residents-must-get-covid-19-herd-immunity-health-institute

NL Times, 'Controversy over De Jonge call to halt AstraZeneca vaccine for under-sixties', 2021 https://nltimes.nl/2021/04/09/controversy-de-jonge-call-halt-astrazeneca-vaccine-60s

NL Times, 'Coronavirus: full text of Prime Minister Rutte's national address in English', 2020 https://nltimes.nl/2020/03/16/coronavirus-full-text-prime-minister-ruttes-national-address-english

NL Times, 'Curfew had no significant effect on hospital admissions, says acute care leader', 2021 https://nltimes.nl/2021/04/29/curfew-significant-effect-hospital-admissions-says-acute-care-leader

NL Times, 'Dutch PM: "Stop shaking hands" – coronavirus kills fourth, spreads to 100 municipalities', 2020 https://nltimes.nl/2020/03/09/dutch-pm-stop-shaking-hands-coronavirus-kills-4th-spreads-100-municipalities

NL Times, 'Dutch to permanently ban mink farming from April 2021', 2020 https://nltimes.nl/2020/08/27/dutch-permanently-ban-mink-farming-april-2021

NL Times, '"Everyone stay home" if sick, many events banned: Dutch government tightens coronavirus rules', 2020 https://nltimes.nl/2020/03/12/everyone-stay-home-sick-many-events-banned-dutch-government-tightens-coronavirus-rules

NL Times, 'Labor Party member appointed new Medical Care Minister', 2020 https://nltimes.nl/2020/03/20/labor-party-member-appointed-new-medical-care-minister

NL Times, 'Martin van Rijn to step down as Medical Care Minister in July', 2020 https://nltimes.nl/2020/05/28/martin-van-rijn-step-medical-care-minister-july

NL Times, 'Politicians wish Medical Care Minister well after resignation', 2020 https://nltimes.nl/2020/03/20/politicians-wish-medical-care-minister-well-resignation

Orange, R., 'Anger in Sweden as elderly pay price for coronavirus strategy', *Guardian*, 2020 https://www.theguardian.com/world/2020/apr/19/anger-in-sweden-as-elderly-pay-price-for-coronavirus-strategy

Our World in Data, 'Australia: coronavirus pandemic country profile', 2021 https://ourworldindata.org/coronavirus/country/australia

Our World in Data, 'Netherlands: coronavirus pandemic country profile', 2021 https://ourworldindata.org/coronavirus/country/netherlands

Our World in Data, 'New Zealand: coronavirus pandemic country profile', 2021 https://ourworldindata.org/coronavirus/country/new-zealand

Our World in Data, 'Sweden: coronavirus pandemic country profile', 2021 https://ourworldindata.org/coronavirus/country/sweden

Oyinloye, A., 'Living with long COVID: the South African factory manager', africanews, 2020 https://www.africanews.com/2020/12/01/living-with-long-covid-the-south-african-factory-manager//

Pope, R., tweet, 2020 https://twitter.com/preshitorian/status/1242831605330313216

Public Health Agency of Sweden, 'The Public Health Agency of Sweden's regulations and general guidelines relating to everyone's responsibility to prevent COVID-19 infections', https://www.folkhalsomyndigheten.se/the-public-health-agency-of-sweden/communicable-disease-control/covid-19/regulations-and-general-guidelines/

Puntmann, V. O., Carerj, M. L., Wieters, I., et al., 'Outcomes of cardiovascular magnetic resonance imaging in patients recently recovered from coronavirus disease, 2019 (COVID-19)', *Journal of the American Medical Association: Cardiology*, 2020 https://jamanetwork.com/journals/jamacardiology/fullarticle/2768916

Reuters, 'Dutch parliament backs night-time curfew plan to curb COVID-19 spread', Reuters, 2021 https://www.reuters.com/business/healthcare-pharmaceuticals/dutch-parliament-backs-night-time-curfew-plan-curb-covid-spread-2021-01-21/

Reuters, 'New Zealand to donate vaccines for 800,000 to COVAX vaccine facility – Ardern', 2021 https://www.reuters.com/business/healthcare-pharmaceuticals/new-zealand-donate-vaccines-800000-covax-vaccine-facility-ardern-2021-04-15/

Reuters, 'Sweden says Trump criticism of virus strategy "factually wrong"', 2020 https://www.reuters.com/article/uk-health-coronavirus-sweden-trump-idUKKBN21Q1NB

Reynolds, E., 'Sweden says its coronavirus approach has worked. The numbers suggest a different story', CNN, 2020 https://edition.cnn.com/2020/04/28/europe/sweden-coronavirus-lockdown-strategy-intl/index.html

Rijksoverheid, 'COVID-19 vaccinations', 2021 https://coronadashboard.government.nl/landelijk/vaccinaties

Robertson, D., '"They are leading us to catastrophe": Sweden's coronavirus stoicism begins to jar', *Guardian*, 2020 https://www.theguardian.com/world/2020/mar/30/catastrophe-sweden-coronavirus-stoicism-lockdown-europe

Sacks, B., 'COVID is making younger, healthy people debilitatingly sick for months. Now they're fighting for recognition', BuzzFeed, 2020 https://www.buzzfeednews.com/article/briannasacks/covid-long-haulers-who-coronavirus

Savage, M., 'Coronavirus: what's going wrong in Sweden's care homes?' BBC News, 2020 https://www.bbc.co.uk/news/world-europe-52704836

Schaart, E., 'Back-of-the-pack Dutch under fire for slow coronavirus vaccine rollout', *Politico*, 2021 https://www.politico.eu/article/netherlands-coronavirus-vaccination-slow-start-mark-rutte-hugo-de-jonge/

Siganto, T., 'Greater Brisbane is in a three-day lockdown. Here's what you need to know', ABC News, 2021 https://www.abc.net.au/news/2021-01-08/coronavirus-queensland-lockdown-explained/13041766

Smith, B., 'Wealthy countries "are buying up far more COVID-19 vaccines than they need" – and that's bad news for the end of the pandemic', ABC News, 2021 https://www.abc.net.au/news/science/2021-03-28/covid-19-vaccines-covax-pandemic-wealthy-countries/100023022

Smyth, J., 'New Zealand's delayed vaccine rollout threatens early COVID success', *Financial Times*, 2021 https://www.ft.com/content/bb1de4e4-7b42-43a0-b118-bb35719daca1

Sonnweber, T., Boehm, A., Sahanic, S., et al., 'Persisting alterations of iron homeostasis in COVID-19 are associated with non-resolving lung pathologies and poor patients' performance: a prospective observational cohort study', *Respiratory Research*, 2020 https://doi.org/10.1186/s12931-020-01546-2

Sudre, C. H., Murray, B., Varsavsky, T., et al., 'Attributes and predictors of long COVID', *Nature Medicine*, 2021 https://www.nature.com/articles/s41591-021-01292-y

Taquet, M., Luciano, S., Geddes, J. R., et al., 'Bidirectional associations between COVID-19 and psychiatric disorder: retrospective cohort studies of 62,354 COVID-19 cases in the USA', *Lancet Psychiatry*, 2021 https://doi.org/10.1016/S2215-0366(20)30462-4

Taylor, J., 'South Australia to end COVID lockdown early as premier "fuming" over pizza lie', *Guardian*, 2020 https://www.theguardian.com/australia-news/2020/nov/20/south-australia-to-end-covid-lockdown-early-as-premier-fuming-over-pizza-lie

Tenforde, M. W., Kim, S. S., Lindsell, C. J., et al., 'Symptom duration and risk factors for delayed return to usual health among outpatients with COVID-19 in a multistate health care systems network – United States, March–June 2020', *Morbidity and Mortality Weekly Report*, 2020 http://dx.doi.org/10.15585/mmwr.mm6930e1

Thaker, J., 'More than 1 in 3 New Zealanders remain hesitant or sceptical about COVID-19 vaccines. Here's how to reach them', *The Conversation*, 2021 https://theconversation.com/more-than-1-in-3-new-zealanders-remain-hesitant-or-sceptical-about-covid-19-vaccines-heres-how-to-reach-them-156489

Townsend, L., Dyer, A. H., Jones, K., et al., 'Persistent fatigue following SARS-CoV-2 infection is common and independent of severity of initial infection', *PLoS One*, 2020 https://doi.org/10.1371/journal.pone.0240784

*Unherd*, 'Why Sweden is no longer shielding the elderly', 2020 https://unherd.com/thepost/why-sweden-is-no-longer-shielding-the-elderly/

van Mersbergen, C., 'RIVM stelt vaccinatiecijfers bij: 220,000 prikken minder gezet door programmeerfout', DPG Media, 2021 https://www.destentor.nl/dossier-coronavirus/rivm-stelt-vaccinatieenshy-cijenshy-fers-bij-220-000-prikken-minder-gezet-door-programenshy-meerenshy-fout~ac8ae191/?referrer=https%3A%2F%2Fwww.google.com%2F

Venkatesan, P., 'NICE guidelines on long COVID', *Lancet Respiratory Medicine*, 2021 https://doi.org/10.1016/S2213-2600(21)00031-X

Visontay, E., 'Hundreds of Australia's aged care residents will die of COVID because of government failure, expert warns', *Guardian*, 2020 https://www.theguardian.

com/australia-news/2020/aug/12/australias-covid-aged-care-deaths-worst-disaster-that-is-still-unfolding-before-my-eyes

Visontay, E., 'Newmarch House operator tells of COVID-19 "dysfunction" between state and federal officials', *Guardian*, 2020 https://www.theguardian.com/australia-news/2020/aug/11/newmarch-house-operator-tells-of-covid-19-dysfunction-between-state-and-federal-officials

Visontay, E., 'Victoria's lockdown to lift at midnight but some restrictions remain in place', *Guardian*, 2021 https://www.theguardian.com/australia-news/2021/feb/17/victorias-lockdown-to-lift-at-midnight-but-some-restrictions-remain-in-place

Vrangbæk, K., Tynkkynen, L., Janlov, N., et al., 'COVID-19 responses in the Nordic countries: similar systems but different strategies and outcomes', European Observatory on Health Systems and Policies Webinar, 2020 https://www.youtube.com/watch?v=uTqNSRWkv5I

Wageningen University & Research, 'COVID-19 geconstateerd op diverse nertsenbedrijven', 2020 https://www.wur.nl/nl/Onderzoek-Resultaten/Onderzoeks instituten/Bioveterinary-Research/show-bvr/COVID-19-geconstateerd-op-twee-nertsenbedrijven.htm

Wahlquist, C., 'Australia's coronavirus lockdown – the first fifty days', *Guardian*, 2020 https://www.theguardian.com/world/2020/may/02/australias-coronavirus-lockdown-the-first-50-days

Ward, A., 'Sweden's government has tried a risky coronavirus strategy. It could backfire', *Vox*, 2020 https://www.vox.com/2020/4/9/21213472/coronavirus-sweden-herd-immunity-cases-death

Watson, A., and Westcott, B., 'After weeks of drama and setbacks, the Australian Open kicks off', CNN, 2021 https://edition.cnn.com/2021/02/07/tennis/australian-open-coronavirus-sport-intl-hnk/index.html

Willingham, R., 'Victorian coronavirus cases rise by 116 as Premier seeks to extend state of emergency', ABC News, 2020 https://www.abc.net.au/news/2020-08-24/victoria-records-116-new-coronavirus-cases-and-15-deaths/12588284

Wilson, N., Barnard, L. T., Summers, J. A., et al., 'Differential mortality rates by ethnicity in three influenza pandemics over a century, New Zealand', *Emerging Infectious Diseases*, 2012 https://doi.org/10.3201/eid1801.110035

Wise, J., 'Long COVID: WHO calls on countries to offer patients more rehabilitation', *BMJ*, 2021 https://doi.org/10.1136/bmj.n405

Wood, J., and Lydeamore, M., 'Australia's experience and the role of modelling in its responses to COVID-19', Usher Institute COVID-19 Webinar, 2020 https://www.ed.ac.uk/usher/news-events/covid-19-webinars

World Health Organization, 'Statement – update on COVID-19: WHO/Europe calls for action on post-COVID conditions/"long COVID"', 2021 https://www.euro.who.int/en/media-centre/sections/statements/2021/statement-update-on-covid-19-whoeurope-calls-for-action-on-post-covid-conditionslong-covid

Xiong, Q., Xu, M., Li, J., et al., 'Clinical sequelae of COVID-19 survivors in Wuhan, China: a single-centre longitudinal study', *Clinical Microbiology and Infection*, 2021 https://doi.org/10.1016/j.cmi.2020.09.023

Yarmol-Matusiak, E. A., Cipriano, L. E., and Stranges, S., 'A comparison of COVID-19 epidemiological indicators in Sweden, Norway, Denmark, and Finland', *Scandinavian Journal of Public Health*, 2021 https://doi.org/10.1177%2F1403494820980264

Yong, E., 'COVID-19 can last for several months', *Atlantic*, 2020 https://www.theatlantic.com/health/archive/2020/06/covid-19-coronavirus-longterm-symptoms-months/612679/

Yong, E., 'Long-haulers are redefining COVID-19', *Atlantic*, 2021 https://www.theatlantic.com/health/archive/2020/08/long-haulers-covid-19-recognition-support-groups-symptoms/615382/

Zaczeck, Z., 'Quarantine security guard, nineteen, makes bombshell claims she was recruited via WhatsApp and received NO training before working at a COVID-19 hotel – as she lifts the lid on botched scheme that plunged Victoria back into lockdown', *Daily Mail*, 2021 https://www.dailymail.co.uk/news/article-8544851/Melbourne-security-guard-lifts-lid-hotel-quarantine-scheme-sparked-second-wave-COVID-19.html

## *8. Variants and the Global South*

Al Jazeera, 'Return of alcohol ban stirs debate in South Africa', 2020 https://www.aljazeera.com/news/2020/7/14/return-of-alcohol-ban-stirs-debate-in-south-africa web.archive.org/web/

Andrews, M. A., Areekal, B., Rajesh, K. R., et al., 'First confirmed case of COVID-19 infection in India: a case report', *Indian Journal of Medical Research*, 2020 https://journals.lww.com/ijmr/pages/default.aspx

Apuzzo, M., Gebrekidan, S., and Mandavilli, A., 'As coronavirus mutates, the world stumbles again to respond', *New York Times*, 2021 https://www.nytimes.com/2021/01/09/world/europe/coronavirus-mutations.html

Biller D., and Álvares, D., 'Brazil hospitals buckle in absence of national virus plan', Associated Press, 2021 https://apnews.com/article/brazil-rio-de-janeiro-health-coronavirus-pandemic-jair-bolsonaro-04931a387c7fb128266235635a1422a1

Boletim, 'Observatório COVID-19', 2021 https://portal.fiocruz.br/sites/portal.fiocruz.br/files/documentos/boletim_extraordinario_2021-marco-30-red.pdf

Boseley, S., 'Brazilian COVID variant: what do we know about P1?', *Guardian*, 2021 https://www.theguardian.com/world/2021/mar/01/brazil-covid-variant-p1-britain

Brown, J., 'Children eat "plants" as hunger explodes in SA', News24, 2020 https://www.news24.com/news24/southafrica/news/covid-19-children-eat-plants-to-survive-as-hunger-explodes-in-sa-20200717

Burki, T., 'No end in sight for the Brazilian COVID-19 crisis', *Lancet Microbe*, 2021 https://doi.org/10.1016/S2666-5247(21)00095-1

Business Tech, 'Lockdown forced nearly half of small businesses in South Africa to close: study', 2020 https://businesstech.co.za/news/business/455100/lockdown-forced-nearly-half-of-small-businesses-in-south-africa-to-close-study/

Buss, L. F., Prete, C. A. Jr, Abrahim, C. M. M., et al., 'Three quarters attack rate of SARS-CoV-2 in the Brazilian Amazon during a largely unmitigated epidemic', *Science*, 2021 https://www.science.org/doi/10.1126/science.abe9728

Cao, B., Wang, Y., Wen, D., et al., 'A trial of lopinavir–ritonavir in adults hospitalized with severe COVID-19', *New England Journal of Medicine*, 2020 https://www.nejm.org/doi/full/10.1056/NEJMoa2001282

Centers for Disease Control and Prevention, 'Implications of the emerging SARS-CoV-2 variant', 2020 https://stacks.cdc.gov/view/cdc/99303

Chaccour, C., Casellas, A., Blanco-Di Matteo, A., et al., 'The effect of early treatment with ivermectin on viral load, symptoms and humoral response in patients with non-severe COVID-19: a pilot, double-blind, placebo-controlled, randomized clinical trial', *EClinical Medicine*, 2021 https://doi.org/10.1016/j.eclinm.2020.100720

Clarke, S., 'Brazil coronavirus variant: what is it and why is it a concern? An expert explains', *The Conversation*, 2021 theconversation.com/brazil-coronavirus-variant-what-is-it-and-why-is-it-a- concern-an-expert-explains-156234 https://theconversation.com/brazil-coronavirus-variant-what-is-it-and-why-is-it-a-concern-an-expert-explains-156234

CNN Brasil, 'Bolsonaro entra com ação no STF contra restrições de governadores do DF, BA e RS', 2021 https://www.cnnbrasil.com.br/politica/bolsonaro-entra-com-acao-no-stf-contra-restricoes-de-governadores-do-df-ba-e-rs/

Coletta, R. D., '"No que depender de mim nunca teremos lockdown", diz Bolsonaro', *Folha De S. Paulo*, 2021 https://www1.folha.uol.com.br/equilibrioesaude/2021/03/no-que-depender-de-mim-nunca-teremos-lockdown-diz-bolsonaro.shtml

Cooperative Governance & Traditional Affairs, Republic of South Africa, 'Disaster Management Act, 2002', 2020 https://www.cogta.gov.za/index.php/2021/06/28/disaster-management-act-57-2002-amendment-of-regulations-issued-in-terms-of-section-27-2-4/

Cotteril, J., 'Illicit trade thrives as South Africa bans alcohol and tobacco sales', *Financial Times*, 2021 https://www.ft.com/content/b9ba721e-cb82-4cb6-9009-58649b04ee4b

du Toit, P., '"You aren't allowed to sell T-shirts, flip-flops?" Trevor Manuel slams "irrationality" of lockdown regulations', News24, 2020 https://www.news24.com/news24/SouthAfrica/News/you-arent-allowed-to-sell-t-shirts-flip-flops-trevor-manuel-slams-irrationality-of-lockdown-regulations-20200513

*Economist*, 'South Africa bans alcohol sales', 2020 https://www.economist.com/middle-east-and-africa/2020/07/18/south-africa-bans-alcohol-sales

European Centre for Disease Prevention and Control, 'Risk assessment: risk related to spread of new SARS-CoV-2 variants of concern in the EU/EEA', 2020 https://www.ecdc.europa.eu/en/publications-data/covid-19-risk-assessment-spread-new-variants-concern-eueea-first-update

Faria, N. R., Mellan T.A., Whittaker, C., et al., 'Genomics and epidemiology of a novel SARS-CoV-2 lineage in Manaus, Brazil', *Science*, 2021 https://www.science.org/doi/10.1126/science.abh2644

Feni, L., 'Lockdown? Life continues as normal in some parts of the Eastern Cape', SowetanLive, 2020 https://www.sowetanlive.co.za/news/south-africa/2020-04-06-lockdown-life-continues-as-normal-in-some-parts-of-the-eastern-cape/

Ferigato, S., Fernandez, M., Amorim, M., et al., 'The Brazilian government's mistakes in responding to the COVID-19 pandemic', *Lancet*, 2020 https://doi.org/10.1016/S0140-6736(20)32164-4

Fernandes, L. A. C., Silva C. A. F. da, Dameda, C., et al., 'COVID-19 and the Brazilian reality: the role of *favelas* in combating the pandemic', *Frontiers in Sociology*, 2020 https://doi.org/10.3389/fsoc.2020.611990

g1, 'Belém e mais 9 cidades do Pará entram em "lockdown"; estado é o 2º do país a adotar a medida contra o coronavírus', 2020 https://g1.globo.com/pa/para/noticia/2020/05/07/belem-e-mais-9-cidades-do-para-entram-em-lockdown-estado-e-o-20-do-pais-a-adotar-a-medida-contra-o-coronavirus.ghtml

g1, 'Quase 500 pessoas com COVID-19 morreram à espera de um leito de UTI em março no estado de SP', 2021 https://g1.globo.com/sp/sao-paulo/noticia/2021/04/01/quase-500-pessoas-com-covid-19-morreram-a-espera-de-um-leito-de-uti-em-marco-no-estado-de-sp.ghtml

GardaWorld, 'Brazil: quarantine in São Paulo extended through May 10 /update 13', 2020 https://www.garda.com/crisis24/news-alerts/334111/brazil-quarantine-in-sao-paulo-extended-through-may-10-update-13

Guimarães, L., 'Coronavírus: 92% das mães nas favelas dizem que faltará comida após um mês de isolamento, aponta pesquisa', BBC News Brasil, 2020 https://www.bbc.com/portuguese/brasil-52131989

Haffajee, K., 'The day the bottom fell out of South Africa – a triple pandemic has hit us', *Daily Maverick*, 2020 https://www.dailymaverick.co.za/article/2020-07-15-the-day-the-bottom-fell-out-of-south-africa-a-triple-pandemic-has-hit-us/

Hanekom, W., and de Oliviera, T., 'South African scientists who discovered new COVID-19 variant share what they know', *The Conversation*, 2021 https://theconversation.com/south-african-scientists-who-discovered-new-covid-19-variant-share-what-they-know-153313

Hone, T., Mirelman, A. J., Rasella, D., et al., 'Effect of economic recession and impact of health and social protection expenditures on adult mortality: a longitudinal analysis of 5,565 Brazilian municipalities', *Lancet Global Health*, 2019 https://doi.org/10.1016/S2214-109X(19)30409-7

Karrim, A., 'The difficult truth: rise in cases expected after lockdown, says expert', News24, 2020 https://www.news24.com/news24/southafrica/news/the-difficult-truth-rise-in-cases-expected-after-lockdown-says-expert-20200413

Khumalo, S., and Omarjee, L., 'Thanks for reopening the economy, but we need more public investment – Cosatu', Fin24, 2020 https://www.news24.com/fin24/economy/thanks-for-reopening-the-economy-but-we-need-more-public-investment-cosatu-20200816news24.com/fin24/economy/thanks-for-reopening-the-economy-but-we-need- more-public-investment-cosatu-20200816

KPMG, 'Brazil: government and institution measures in response to COVID-19', 2020 https://home.kpmg/xx/en/home/insights/2020/04/brazil-government-and-institution-measures-in-response-to-covid.html

Krenzinger, K., Silva, E. S., Morgado, R., et al., 'Violence against women in Complexo da Maré, Rio de Janeiro', 2018 https://transnationalperspectivesonvawg.files.wordpress.com/2018/03/5_krezinger-et-al-vawg-short-report_online.pdf

Kuhn, J. H., Bao, Y., Bavari, S., et al., 'Virus nomenclature below the species level: a standardized nomenclature for natural variants of viruses assigned to the family *Filoviridae*', *Archives of Virology*, 2013 https://doi.org/10.1007/s00705-012-1454-0

*Lancet*, 'COVID-19 in Brazil: "So what?"', 2020 https://doi.org/10.1016/S0140-6736(20)31095-3

Londoño, E., 'Bolsonaro hails anti-malaria pill even as he fights coronavirus', *New York Times*, 2020 https://www.nytimes.com/2020/07/08/world/americas/brazil-bolsonaro-covid-coronavirus.html

Mahlakoana, T., 'Over 1,500 COJ employees fraudulently received COVID relief, social grant funds', Eyewitness News, 2020 https://ewn.co.za/2020/11/17/probe-finds-over-1-500-coj-employees-fraudulently-received-covid-grant-payments

Manuel, T., 'Let's not forsake our constitution because of the lockdown', News24, 2020 https://www.news24.com/citypress/Voices/trevor-manuel-lets-not-forsake-our-constitution-because-of-the-lockdown-20200510

McIlwaine, C., 'Women in *favelas* are the backbone of responses to the coronavirus crisis in Rio de Janeiro', King's College, London, 2020 https://www.kcl.ac.uk/women-in-favelas-backbone-of-responses-to-coronavirus

Medical Xpress, 'What we know about South Africa's coronavirus variant', 2021 https://medicalxpress.com/news/2021-01-south-africa-coronavirus-variant.html

Meunier, T., 'Full lockdown policies in Western Europe countries have no evident impacts on the COVID-19 epidemic', medRxiv, 2020 https://doi.org/10.1101/2020.04.24.20078717

Moraes, T., and Barberia, L., 'COVID-19: Public policies and society's responses. Quality information for refining public policies and saving lives', Policy Briefing Note 20, Rede de Pesquisa Solidária de Políticas Públicas e Sociedade, São Paulo, 2020https://redepesquisasolidaria.org/en/bulletins/bulletin-20/lacking-a-strategy-the-federal-government-foments-the-countrys-fragmentation-and-fails-to-coordinate-the-fight-against-covid-19-the-demobilization-of-the-ministry-of-health-alongside-disor/

Msomi, N., Mlisana, K., and de Oliveira, T., on behalf of the Network for Genomic Surveillance in South Africa Writing Group, 'A genomics network established to respond rapidly to public health threats in South Africa', *Lancet Microbe*, 2020 https://doi.org/10.1016/S2666-5247(20)30116-6

Mvumvu, Z., 'Here's where the Cuban doctors will be deployed to fight COVID-19', TimesLIVE, 2020 https://www.timeslive.co.za/news/south-africa/2020-04-28-heres-where-the-cuban-doctors-will-be-deployed-to-fight-covid-19/

Nanda, R., 'Lockdown and labour pain: the demand for MNREGA work has never been so strong, says economist Jean Drèze', News18, 2020 https://www.news18.com/news/india/lockdown-labour-pain-the-demand-for-mnrega-work-has-never-been-so-strong-says-economist-jean-dreze-2600383.html

Naveca, F. G., Nascimento, V., Souza, V. C. de, et al., 'COVID-19 in Amazonas, Brazil, was driven by the persistence of endemic lineages and P.1 emergence', *Nature Medicine*, 2021 https://doi.org/10.1038/s41591-021-01378-7

News24, 'From cigarettes to curfews: who is suing government and why', 2020 https://www.news24.com/news24/SouthAfrica/News/from-cigarettes-to-curfews-who-is-suing-government-and-why-20200518

NIDS-CRAM, 'Synthesis Report: Wave 1', 2020 https://cramsurvey.org/wp-content/uploads/2020/07/Spaull-et-al.-NIDS-CRAM-Wave-1-Synthesis-Report-Overview-and-Findings-1.pdf

Omarjee, L., and Magubane, K., 'Ramaphosa announces South Africa's biggest spending plan ever to fight coronavirus', Fin 24, 2020 https://www.news24.com/fin24/economy/south-africa/ramaphosa-announces-r500bn-support-package-adjustment-budget-for-coronavirus-20200421

PANGO Lineages, 'Lineage B.1.1.248', 2021 https://cov-lineages.org/lineage.html?lineage=B.1.1.248

Paraguassu, L., 'Major Brazilian cities set lockdowns as virus spreads', Reuters, 2020 https://www.reuters.com/article/uk-health-coronavirus-brazil-lockdown-idUKKBN22H2V5

Phillips, T., 'Brazilian president Jair Bolsonaro tests positive for coronavirus', *Guardian*, 2020 https://www.theguardian.com/world/2020/jul/07/jair-bolsonaro-coronavirus-positive-test-brazil-president

Ponce, D., 'The impact of coronavirus in Brazil: politics and the pandemic', *Nature Reviews Nephrology*, 2020 https://doi.org/10.1038/s41581-020-0327-0

Pontes, M. R. N., and Lima, and J. P., 'Brazil's COVID-19 response', *Lancet*, 2020 https://doi.org/10.1016/S0140-6736(20)31914-0

Prudenciano, G., 'Brasil tem mais de 6.300 pessoas na fila por leitos de UTI COVID', CNN Brasil, 2021 https://www.cnnbrasil.com.br/saude/brasil-tem-mais-de-6300-pessoas-na-fila-por-leitos-de-uti/

Ranzani, O. T., Bastos, L. S. L., Gelli, J. G. M., et al., 'Characterisation of the first 250,000 hospital admissions for COVID-19 in Brazil: a retrospective analysis of nationwide data', *Lancet Respiratory Medicine*, 2021 https://doi.org/10.1016/S2213-2600(20)30560-9

RECOVERY Collaborative Group, 'Lopinavir–ritonavir in patients admitted to hospital with COVID-19 (RECOVERY): a randomised, controlled, open-label, platform trial', *Lancet*, 2020 https://doi.org/10.1016/S0140-6736(20)32013-4

Republic of South Africa, 'Statement March 23rd. Measures to combat the COVID-19 epidemic', 2020 http://www.dirco.gov.za/docs/speeches/2020/cram0323.pdf

Republic of South Africa, the Presidency, 'Statement by President Cyril Ramaphosa on measures to combat COVID-19 epidemic', 2020 http://www.thepresidency.gov.za/press-statements/statement-president-cyril-ramaphosa-measures-combat-covid-19-epidemic

Ribeiro, G., 'São Paulo extends quarantine until April 22nd', *Brazilian Report*, 2020 https://brazilian.report/liveblog/coronavirus/2020/04/06/state-sao-paulo-extends-quarantine-until-april-22/

Rizzo, L. V., and Wolosker, N., 'Brazil's COVID-19 response', *Lancet*, 2020 https://doi.org/10.1016/S0140-6736(20)31915-2

Schrader, A., 'South Africa bans walking dogs, Spain cracks down on "pet rentals" amid coronavirus crisis', *New York Post*, 2020 https://nypost.com/2020/03/25/south-africa-bans-walking-dogs-spain-cracks-down-on-pet-rentals-amid-coronavirus-crisis/

Shaffer, L., 'Fifteen drugs being tested to treat COVID-19 and how they would work', *Nature*, 2020 https://www.nature.com/articles/d41591-020-00019-9

Siciliano, B., Carvalho, G., Silva, C. M. da, et al., 'The impact of COVID-19 partial lockdown on primary pollutant concentrations in the atmosphere of Rio de Janeiro and São Paulo megacities (Brazil)', *Bulletin of Environmental Contamination and Toxicology*, 2020 https://doi.org/10.1007/s00128-020-02907-9

Silva Bastos, M. H. da, 'Brazil's COVID-19 response', *Lancet*, 2020 https://doi.org/10.1016/S0140-6736(20)31914-0

Simelane, B. C., '"Review 100 per cent taxi capacity" – public health specialists plead with Ramaphosa', *Daily Maverick*, 2020 https://www.dailymaverick.co.za/article/2020-07-15-review-100-taxi-capacity-public-health-specialists-plead-with-ramaphosa/

Singh, O., 'SA's eight stages in the fight against COVID-19: what you need to know', TimesLIVE, 2020 https://www.timeslive.co.za/news/south-africa/2020-04-13-sas-8-stages-in-the-fight-against-covid-19-what-you-need-to-know/

Steinhauser, G., 'South Africa COVID-19 strain: what we know about the new variant', *Wall Street Journal*, 2021 https://www.wsj.com/articles/the-new-covid-19-strain-in-south-africa-what-we-know-11609971229

Takemoto, M. L. S., Menezes, M. O., Andreucci, C. B., et al., 'The tragedy of COVID-19 in Brazil: 124 maternal deaths and counting', *International Journal of Gynecology & Obstetrics*, 2020 https://doi.org/10.1002/ijgo.13300

Tandwa, L., '"Don't give us a reason to arrest you" – Cele, as lockdown arrests rise to 2,289', News24, 2020 https://www.news24.com/news24/SouthAfrica/News/dont-give-us-a-reason-to-arrest-you-cele-as-lockdown-arrests-rise-to-2-289-20200403

Tandwa, L., 'EFF slams relaxation of lockdown regulations on mines, ports and call centres', News24, 2020 https://www.news24.com/news24/southafrica/news/eff-slams-relaxation-of-lockdown-regulations-on-mines-ports-and-call-centres-20200417

Tegally, H., Wilkinson, E., Giovanetti, M., et al., 'Detection of a SARS-CoV-2 variant of concern in South Africa', *Nature*, 2021 https://doi.org/10.1038/s41586-021-03402-9

Tegally, H., Wilkinson, E., Giovanetti, M., et al., 'Major new lineages of SARS-CoV-2 emerge and spread in South Africa during lockdown', medRxiv, 2020 https://www.medrxiv.org/content/10.1101/2020.10.28.20221143v1

Trading Economics, 'Brazil – population living in slums', 2021 https://tradingeconomics.com/brazil/population-living-in-slums-percent-of-urban-population-wb-data.html

Williams, A. C., 'US criticises South Africa's use of Cuban doctors in fight against COVID-19', IOL News, 2020 https://www.iol.co.za/news/politics/us-criticises-south-africas-use-of-cuban-doctors-in-fight-against-covid-19-47384258

World Health Organization, 'Responding to COVID-19 – learnings from Kerala', 2020 https://www.who.int/india/news/feature-stories/detail/responding-to-covid-19---learnings-from-kerala

World Health Organization, 'South Africa alcohol consumption: levels and patterns', 2018 https://www.who.int/substance_abuse/publications/global_alcohol_report/profiles/zaf.pdf?ua=1

World Health Organization, 'WHO press conference on coronavirus disease (COVID-19) – 5 March 2021', https://www.who.int/publications/m/item/covid-19-virtual-press-conference-transcript---5-march-2021

## *9. The Race for a Vaccine*

Antrobus, R. D., Coughlan, L., Berthoud, T. K., et al., 'Clinical assessment of a novel recombinant simian adenovirus ChAdOx1 as a vectored vaccine expressing conserved Influenza A antigens', *Molecular Therapy*, 2014 https://doi.org/10.1038/mt.2013.284

Baden, L. R., Stieh, D. J., Sarnecki, M., et al., 'Safety and immunogenicity of two heterologous HIV vaccine regimens in healthy, HIV-uninfected adults (TRAVERSE): a randomised, parallel-group, placebo-controlled, double-blind, Phase 1/2a study', *Lancet HIV*, 2020 https://doi.org/10.1016/S2352-3018(20)30229-0

BBC News, 'Coronavirus: human trial of new vaccine begins in UK', 2020 https://www.bbc.co.uk/news/health-53061288

BBC News, 'COVID: Australian vaccine abandoned over false HIV response', 2020 https://www.bbc.co.uk/news/world-australia-55269381

BBC News, COVID: South Africa halts AstraZeneca vaccine rollout over new variant', 2021 https://www.bbc.co.uk/news/world-africa-55975052

Beer, J., '*National Geographic*'s new documentary tells the inside story of Pfizer's COVID-19 vaccine', Fast Company, 2021 https://www.fastcompany.com/90607643/national-geographics-new-documentary-tells-the-inside-story-of-pfizers-covid-19-vaccine

Borresen, J., Padilla, R., and Heath, D., 'How mRNA vaccines work', *USA Today*, 2021 https://eu.usatoday.com/in-depth/graphics/2021/03/05/mrna-vaccines-explained/4467256001/

Bucci, E., Andreev, K., Björkman, A., et al., 'Safety and efficacy of the Russian COVID-19 vaccine: more information needed', *Lancet*, 2020 https://doi.org/10.1016/S0140-6736(20)31960-7

Bucci, E. M., Berkhof, J., Gillibert, A., et al., 'Data discrepancies and substandard reporting of interim data of Sputnik V Phase 3 trial', *Lancet*, 2021 https://doi.org/10.1016/S0140-6736(21)00899-0

Bulik, B. S., 'The inside story behind Pfizer and BioNTech's new vaccine brand name, Comirnaty', Fierce Pharma, 2020 https://www.fiercepharma.com/marketing/pfizer-biontech-select-comirnaty-as-brand-name-for-covid-19-vaccine

Bunn, C., '"Getting a clearer picture": Black Americans on the factors that overcame their vaccine hesitancy', NBC News, 2021 https://www.nbcnews.com/news/nbcblk/getting-clearer-picture-black-americans-factors-overcame-their-vaccine-hesitancy-n1263787

Campbell, M., 'The inside story of how the Oxford vaccine was made and the team behind it', *Sunday Times*, 2021 https://www.thetimes.co.uk/article/how-oxford-astrazeneca-vaccine-was-made-vxbzg9929

Cappuccini, F., Stribbling, S., Pollock, E., et al., 'Immunogenicity and efficacy of the novel cancer vaccine based on simian adenovirus and MVA vectors alone and in combination with PD-1 mAb in a mouse model of prostate cancer', *Cancer Immunology, Immunotherapy*, 2016 https://doi.org/10.1007/s00262-016-1831-8

Chappell, K. J., Mordant, F. L., Li, Z., et al., 'Safety and immunogenicity of an MF59-adjuvanted spike glycoprotein-clamp vaccine for SARS-CoV-2: a randomised, double-blind, placebo-controlled, Phase 1 trial', *Lancet Infectious Diseases*, 2021 https://doi.org/10.1016/S1473-3099(21)00200-0

Clinical Trials Arena, 'Imperial College, London, begins dosing in COVID-19 vaccine study', 2020 https://www.clinicaltrialsarena.com/news/imperial-covid19-vaccine-trial-dosing/

Clinical Trials Arena, 'Imperial College, London, to start COVID-19 vaccine trials', 2020 https://www.clinicaltrialsarena.com/news/imperial-covid-vaccine-trials/

Clinical Trials, 'A safety and efficacy study of ChAdOx1 LS2 and MVA LS2', 2018 https://clinicaltrials.gov/ct2/show/NCT03203421

Cohen, J., '"Just beautiful": another COVID-19 vaccine, from newcomer Moderna, succeeds in large-scale trial', *Science*, 2020 https://www.sciencemag.org/news/2020/11/just-beautiful-another-covid-19-vaccine-newcomer-moderna-succeeds-large-scale-trial

Coontz, R., 'Science's top-ten breakthroughs of 2013', *Science*, 2013 https://www.science.org/content/article/sciences-top-10-breakthroughs-2013

Corbett, K. S., Edwards, D. K., Leist, S. R., et al., 'SARS-CoV-2 mRNA vaccine design enabled by prototype pathogen preparedness', *Nature*, 2020 https://www.nature.com/articles/s41586-020-2622-0

Cox, D., 'How mRNA went from a scientific backwater to a pandemic crusher', *Wired*, 2020 https://www.wired.co.uk/article/mrna-coronavirus-vaccine-pfizer-biontech

CSL, 'CSL to manufacture and supply University of Queensland and Oxford University vaccine candidates for Australia', CSL Behring, 2020 https://www.csl.com/news/2020/20200907-csl-to-manufacture-and-supply-uq-and-ou-vaccine-candidates-for-australia

CSL, 'Update on the University of Queensland COVID-19 vaccine', 2020 https://www.csl.com/news/2020/20201211-update-on-the-university-of-queensland-covid-19-vaccine

Danner, C., 'J&J takes over plant where contractor ruined 15 million vaccine doses', *Intelligencer*, 2021 nymag.com/intelligencer/2021/04/j-and-j-takes-over-vaccine-plant-where-15m-doses-were-ruined.html

Dunn, A., 'How the biotech upstart Moderna exploded into one of the most important start-ups of all time, creating a highly effective coronavirus vaccine', *Business Insider*, 2020 https://www.businessinsider.com/moderna-origins-ceo-cofounder-interviews-and-inside-the-lab-2020-5?r=US&IR=T

*Economist*, 'Vaccine diplomacy boosts Russia's and China's global standing', 2021 https://www.economist.com/graphic-detail/2021/04/29/vaccine-diplomacy-boosts-russias-and-chinas-global-standing

Elton, C., 'The untold story of Moderna's race for a COVID-19 vaccine', *Boston Magazine*, 2020 https://www.bostonmagazine.com/health/2020/06/04/moderna-coronavirus-vaccine/

Farge, E., 'Update 1 – GAVI signs COVID-19 vaccine supply deal with J&J', Reuters, 2021 https://www.reuters.com/article/health-coronavirus-covax-idCNL5N2N82A5

Feuerstein, A., 'COVID-19 vaccine trial participant had serious neurological symptoms, but could be discharged today, AstraZeneca CEO says', STAT News, 2020 https://www.statnews.com/2020/09/09/astrazeneca-covid19-vaccine-trial-hold-patient-report/

Fierce Biotech, 'Merck, via its Thémis buy, to move first COVID-19 vaccine into clinical development in Q3', 2020 https://www.fiercebiotech.com/biotech/merck-via-its-themis-buy-to-move-first-covid-19-vaccine-into-clinical-development-q3

Folegatti, P. M., Bittaye, M., Flaxman, A., et al., 'Safety and immunogenicity of a candidate Middle East respiratory syndrome coronavirus viral-vectored vaccine: a dose-escalation, open-label, non-randomised, uncontrolled, Phase 1 trial', *Lancet Infectious Diseases*, 2020 https://doi.org/10.1016/S1473-3099(20)30160-2

Gallagher, J., 'Oxford vaccine: how did they make it so quickly?', BBC News, 2020 https://www.bbc.co.uk/news/health-55041371

Garde, D., 'The story of mRNA: how a once-dismissed idea became a leading technology in the COVID vaccine race', STAT News, 2020 https://www.statnews.com/2020/11/10/the-story-of-mrna-how-a-once-dismissed-idea-became-a-leading-technology-in-the-covid-vaccine-race/

Google Scholar, 'Katalin Karikó', 2021 https://scholar.google.com/citations?hl=en&user=PS_CXoAAAAAJ&view_op=list_works&sortby=pubdate

Gosling, T., 'Russia and China are exploiting Europe's vaccine shortfalls', *Foreign Policy*, 2021 https://foreignpolicy.com/2021/03/31/russia-china-vaccine-diplomacy-slovakia-europe-eu-slow-rollout/

Graham, B. S., and Corbett, K. S., 'Prototype pathogen approach for pandemic preparedness: world on fire', *Journal of Clinical Investigation*, 2020 https://www.jci.org/articles/view/139601

Harris, R., 'Australian COVID vaccine terminated due to HIV "false positives"', *Sydney Morning Herald*, 2020 https://www.smh.com.au/politics/federal/australian-covid-vaccine-terminated-due-to-hiv-false-positives-20201210-p56mju.html

Heath, D., and Garcia-Roberts, G., 'Luck, foresight and science: how an unheralded team developed a COVID-19 vaccine in record time', *USA Today*, 2021 https://eu.usatoday.com/in-depth/news/investigations/2021/01/26/moderna-covid-vaccine-science-fast/6555783002/

Hopkins, J. S., 'How Pfizer delivered a COVID vaccine in record time: crazy deadlines, a pushy CEO', *Wall Street Journal*, 2020 https://www.wsj.com/articles/how-pfizer-delivered-a-covid-vaccine-in-record-time-crazy-deadlines-a-pushy-ceo-11607740483

IAVI, 'IAVI and Merck collaborate to develop vaccine against SARS-CoV-2', 2020 https://www.iavi.org/news-resources/press-releases/2020/iavi-and-merck-collaborate-to-develop-vaccine-against-sars-cov-2

IAVI, 'Merck and IAVI discontinue development of COVID-19 vaccine candidate V590', 2021 https://www.iavi.org/news-resources/press-releases/2021/merck-and-iavi-discontinue-development-of-covid-19-vaccine-candidate-v590

IAVI, 'Participant enrollment begins for Phase I trial of IAVI–Merck COVID-19 vaccine candidate', 2020 https://www.iavi.org/news-resources/features/participant-enrollment-begins-for-phase-i-trial-of-iavi-merck-covid-19-vaccine-candidate

IAVI, 'Watch IAVI experts discuss accelerating COVID-19 vaccine development with Merck', 2020 https://www.iavi.org/news-resources/features/watch-iavi-experts-discuss-accelerating-covid-19-vaccine-development-with-merck

Institut Pasteur, 'Discovery of messenger RNA in 1961', 2021 https://www.pasteur.fr/en/home/research-journal/news/discovery-messenger-rna-1961

Institut Pasteur, 'MV-SARS-CoV-2 vaccine candidate: a new partnership between Institut Pasteur, CEPT, Thémis and MSD', 2020 https://www.pasteur.fr/en/press-area/press-documents/mv-sars-cov-2-vaccine-candidate-new-partnership-between-institut-pasteur-cepi-themis-and-msd

Ives, M., 'Australia scraps COVID-19 vaccine that produced HIV false positives', *New York Times*, 2020 https://www.nytimes.com/2020/12/11/world/australia/uq-coronavirus-vaccine-false-positive.html

Jaffe-Hoffman, M., 'Are COVID-19 vaccines the key to world power?', *Jerusalem Post*, 2021 https://www.jpost.com/health-science/are-covid-19-vaccines-the-key-to-world-power-660994

Jones, I., and Roy, P., 'Sputnik V COVID-19 vaccine candidate appears safe and effective', *Lancet*, 2021 https://doi.org/10.1016/S0140-6736(21)00191-4

Kaiser Permanente Washington Health Research Institute, 'COVID-19 vaccine trial volunteer on the shot heard round the world', 2021 https://www.kpwashington-research.org/news-and-events/blog/2021/covid-19-vaccine-trial-volunteer-shot-heard-round-world

Karikó, K., Buckstein, M., Ni, H., et al., 'Suppression of RNA recognition by Toll-like receptors: the impact of nucleoside modification and the evolutionary origin of RNA', *Immunity*, 2005 https://doi.org/10.1016/j.immuni.2005.06.008

Kolata, G., 'Kati Karikó helped shield the world from the coronavirus', *New York Times*, 2021 https://www.nytimes.com/2021/04/08/health/coronavirus-mrna-kariko.html

Kramer, A. E., 'Russia approves coronavirus vaccine before completing tests', *New York Times*, 2020 https://www.nytimes.com/2020/08/11/world/europe/russia-coronavirus-vaccine-approval.html

Kramer, J., 'They spent twelve years solving a puzzle. It yielded the first COVID-19 vaccines', *National Geographic*, 2020 nationalgeographic.com/science/article/these-scientists-spent-twelve-years-solving-puzzle-yielded-coronavirus-vaccines

LaFraniere, S., and Weiland, N., 'Factory mix-up ruins up to 15 million vaccine doses from Johnson & Johnson', *New York Times*, 2021 https://www.nytimes.com/2021/03/31/us/politics/johnson-johnson-coronavirus-vaccine.html

Leigh, M., 'Vaccine diplomacy: soft power lessons from China and Russia?', Bruegel, 2021 https://www.bruegel.org/2021/04/vaccine-diplomacy-soft-power-lessons-from-china-and-russia/

Levie, H., 'From lab to vaccine vial: the historic manufacturing journey of Johnson & Johnson's Janssen COVID-19 vaccine', Johnson & Johnson, 2021 https://www.jnj.com/innovation/making-johnson-johnson-janssen-covid-19-vaccine

Logan, E., 'Coronavirus vaccine: all the celebrities and public figures who have received it so far', *Glamour*, 2021 https://www.glamour.com/story/coronavirus-vaccine-all-the-celebrities-and-public-figures-who-have-received-it-so-far

Logunov, D. Y., Dolzhikova, I. V., Shcheblyakov, D. V., 'Data discrepancies and substandard reporting of interim data of Sputnik V Phase 3 trial – authors' reply', *Lancet*, 2021 https://doi.org/10.1016/S0140-6736(21)00894-1

Logunov, D. Y., Dolzhikova, I. V., Shcheblyakov, D. V., et al., 'Safety and efficacy of an rAd26 and rAd5 vector-based heterologous prime-boost COVID-19 vaccine: an interim analysis of a randomised controlled Phase 3 trial in Russia', *Lancet*, 2021 https://doi.org/10.1016/S0140-6736(21)00234-8

Logunov, D. Y., Dolzhikova, I. V., Tukhvatullin, A. I., et al., 'Safety and efficacy of the Russian COVID-19 vaccine: more information needed – authors' reply', *Lancet*, 2020 https://doi.org/10.1016/S0140-6736(20)31970-X

Ma, J., 'Coronavirus: 1 million Chinese injected with Sinopharm vaccine under emergency use scheme', *South China Morning Post*, 2020 https://www.scmp.com/news/china/society/article/3110519/china-sinopharms-coronavirus-vaccine-taken-about-1-million

Mannix, L., 'What went wrong with UQ's vaccine – and what we do now?', *Sydney Morning Herald*, 2020 https://www.smh.com.au/national/how-crucial-was-the-uq-csl-vaccine-to-australia-and-what-options-do-we-now-have-20201211-p56mlj.html

Marquez, J. R., 'The COVID-19 data plan: three innovative ways Johnson & Johnson is using data science to fight the pandemic', 2021 https://www.jnj.com/innovation/how-johnson-johnson-uses-data-science-to-fight-covid-19-pandemic

McKie, R., 'Life savers: the amazing story of the Oxford/AstraZeneca COVID vaccine', *Guardian*, 2021 https://www.theguardian.com/world/2021/feb/14/life-savers-story-oxford-astrazeneca-coronavirus-vaccine-scientists

McLellan, J. S., Chen, M., Joyce, M. G., et al., 'Structure-based design of a fusion glycoprotein vaccine for respiratory syncytial virus', *Science*, 2013 https://www.science.org/doi/10.1126/science.1243283

McLellan, J. S., Chen, M., Leung, S., et al., 'Structure of RSV fusion glycoprotein trimer bound to a prefusion-specific neutralizing antibody', *Science*, 2013 https://www.science.org/doi/10.1126/science.1234914

Meredith, S., 'AstraZeneca to work on COVID vaccine combinations with Russia's Sputnik V developers', CNBC, 2020 https://www.cnbc.com/2020/12/11/coronavirus-astrazeneca-to-work-on-vaccine-with-russias-gamaleya.html

Moderna, 'Moderna's COVID-19 vaccine candidate meets its primary efficacy endpoint in the first interim analysis of the Phase 3 COVE study', 2020 https://www.biospace.com/article/releases/moderna-s-covid-19-vaccine-candidate-meets-its-primary-efficacy-endpoint-in-the-first-interim-analysis-of-the-phase-3-cove-study/

Moore, T., 'Story of a vaccine', Sky News, 2020 https://news.sky.com/story/covid-19-the-story-of-oxfords-coronavirus-vaccine-that-could-save-britain-12139898

Morris, S. J., Sebastian, S., Spencer, A. J., and Gilbert, S. C., 'Simian adenoviruses as vaccine vectors', *Future Virology*, 2016 https://doi.org/10.2217/fvl-2016-0070

Munster, V. J., Wells, D., Lambe, T., et al., 'Protective efficacy of a novel simian adenovirus vaccine against lethal MERS-CoV challenge in a transgenic human DPP4 mouse model', *NPJ Vaccines*, 2017 https://doi.org/10.1038/s41541-017-0029-1

O'Hare, R., and Wighton, K., 'Imperial to begin first human trials of new COVID-19 vaccine', Imperial College, London, 2020 https://www.imperial.ac.uk/news/198314/imperial-begin-first-human-trials-covid-19/

Oltermann, P., 'Scientist behind BioNTech/Pfizer vaccine says it can end pandemic', *Guardian*, 2020 https://www.theguardian.com/world/2020/nov/12/scientist-behind-biontech-pfizer-coronavirus-vaccine-says-it-can-end-pandemic

Omarjee, L., 'China and Russia's vaccines flood emerging markets – but what's in it for them?', Fin 24, 2021 https://www.news24.com/fin24/economy/world/china-and-russias-vaccines-flood-emerging-markets-but-whats-in-it-for-them-20210429

Pallesen, J., Wang, N., Corbett, K. S., et al., 'Immunogenicity and structures of a rationally designed prefusion MERS-CoV spike antigen', *Proceedings of the National Academy of Science*, 2017 https://doi.org/10.1073/pnas.1707304114

Pfizer, 'Pfizer-BioNTech announce positive topline results of pivotal COVID-19 vaccine study in adolescents', 2021 https://www.pfizer.com/news/press-release/press-release-detail/pfizer-biontech-announce-positive-topline-results-pivotal

Prakash, T., 'The Quad gives a boost to India's vaccine diplomacy', *Interpreter*, 2021 https://www.lowyinstitute.org/the-interpreter/quad-gives-boost-india-s-vaccine-diplomacy#:~:text=India's%20vaccine%20diplomacy,- Teesta%20Prakash&text=plays%20to%20the%20strengths%20of%20the%20grouping.&text=The %20most%20notable%20takeaway%20from,expanding%20the%20global%20vaccine%20supply

Precision Vaccinations, 'V591 SARS-CoV-2 vaccine', 2020 https://www.precision-vaccinations.com/vaccines/v591-sars-cov-2-vaccine

Rigby, B., 'Vaccine diplomacy: how Russia and China are using their COVID-19 jabs to win friends and influence people', *Telegraph*, 2021 https://www.telegraph.co.uk/global-health/science-and-disease/vaccine-diplomacy-russia-china-using-covid-19-jabs-win-friends/

Salisch, N. C., Stephenson, K. E., Williams, K., et al., 'A double-blind, randomized, placebo-controlled Phase 1 study of Ad26.ZIKV.001, an Ad26-vectored anti-Zika virus vaccine', *Annals of Internal Medicine*, 2021 https://doi.org/10.7326/M20-5306

Scales, D., 'How our brutal science system almost cost us a pioneer of mRNA vaccines', WBUR News, 2020 https://www.wbur.org/news/2021/02/12/brutal-science-system-mrna-pioneer

Scheuber, A., 'Imperial social enterprise to accelerate low-cost COVID-19 vaccine', Imperial College, London, 2020 https://www.imperial.ac.uk/news/198053/imperial-social-enterprise-accelerate-low-cost-covid-19/

Scheuber, A., 'Imperial vaccine tech to target COVID mutations and booster doses', Imperial College, London, 2021 https://www.imperial.ac.uk/news/213313/imperial-vaccine-tech-target-covid-mutations/

Science History Institute, 'Interview with Stéphane Bancel', 2021 https://www.sciencehistory.org/distillations/podcast/interview-with-stephane-bancel

Sheehan, S., Harris, S. A., Satti, I., et al., 'A Phase I, open-label trial, evaluating the safety and immunogenicity of candidate tuberculosis vaccines AERAS-402 and MVA85A, administered by prime-boost regime in BCG-vaccinated healthy adults', *PLoS One*, 2015 https://doi.org/10.1371/journal.pone.0141687

Smith, A., 'Russia and China are beating the US at vaccine diplomacy, experts say', NBC News, 2021 https://www.nbcnews.com/news/world/russia-china-are-beating-u-s-vaccine-diplomacy-experts-say-n1262742

Soriot, P., 'Vaccinating the world', AstraZeneca, 2021 https://www.astrazeneca.com/media-centre/articles/2021/vaccinating-the-world.html

Sputnik V, 'Clinical trials', 2021 https://sputnikvaccine.com/about-vaccine/clinical-trials/

Stieb, M., 'Johnson & Johnson pauses US shipments after error ruining 15 million doses', *Intelligencer*, 2021 https://nymag.com/intelligencer/2021/03/johnson-and-johnson-tosses-15-million-doses-pauses-shipments.html

Taylor, A., 'How did the University of Queensland/CSL vaccine fail due to "false positive" HIV tests? A vaccine expert explains', *The Conversation*, 2020 https://theconversation.com/how-did-the-university-of-queensland-csl-vaccine-fail-due-to-false-positive-hiv-tests-a-vaccine-expert-explains-151911

Terry, M., 'Updated comparing COVID-19 vaccines: timelines, types and prices', BioSpace, 2021 https://www.biospace.com/article/comparing-covid-19-vaccines-pfizer-biontech-moderna-astrazeneca-oxford-j-and-j-russia-s-sputnik-v/

van Doremalen, N., Lambe, T., Sebastian, S., et al., 'A single-dose ChAdOx1-vectored vaccine provides complete protection against Nipah Bangladesh and Malaysia in Syrian golden hamsters', *PLoS Neglected Tropical Diseases*, 2019 https://doi.org/10.1371/journal.pntd.0007462

Verywell Health, 'An overview of the Merck COVID-19 vaccine', 2021 https://www.verywellhealth.com/merck-covid-19-vaccine-5093294

Wallace-Wells, D., 'We had the vaccine the whole time', *Intelligencer*, 2020 https://nymag.com/intelligencer/2020/12/moderna-covid-19-vaccine-design.html

Warimwe, G. M., Gesharisha, J., Carr, B. V., et al., 'Chimpanzee adenovirus vaccine provides multispecies protection against Rift Valley fever', *Scientific Reports*, 2016 https://doi.org/10.1038/srep20617

Wee, S.-L., 'They relied on Chinese vaccines. Now they're battling outbreaks', *New York Times*, 2021 https://www.nytimes.com/2021/06/22/business/economy/china-vaccines-covid-outbreak.html

Wighton, K., 'Imperial COVID-19 vaccine trial expands to additional sites', Imperial College, London, 2020 https://www.imperial.ac.uk/news/200435/imperial-covid-19-vaccine-trial-expands-additional/

Wilkie, M., Satti, I., Minhinnick, A., et al., 'A Phase I trial evaluating the safety and immunogenicity of a candidate tuberculosis vaccination regimen, ChAdOx1 85A prime – MVA85A boost in healthy UK adults', *Vaccine*, 2020 https://doi.org/10.1016/j.vaccine.2019.10.102

Wise, J., 'The story of one dose. Inside the sprawling operational puzzle of bringing the Johnson & Johnson COVID vaccine to the public', *Intelligencer*, 2021 https://nymag.com/intelligencer/2021/04/the-story-of-one-dose.html

World Health Organization, 'Prioritizing diseases for research and development in emergency contexts', 2020 https://www.who.int/activities/prioritizing-diseases-for-research-and-development-in-emergency-contexts

World Intellectual Property Organization (WIPO), 'WO 2018/081318 Al', 2018 patentimages.storage.googleapis.com/68/47/0c/2b5bc4f43c9f74/WO2018081318A1.pdf

Wrapp, D., Wang, N., Corbett, K. S., et al., 'Cryo-EM structure of the 2019-nCoV spike in the prefusion conformation', *Science*, 2020 https://www.science.org/doi/10.1126/science.abb2507

Zimmer, C., 'Russians publish early coronavirus vaccine results', *New York Times*, 2020 https://www.nytimes.com/2020/09/04/health/russia-covid-vaccine.html

## *10. Cooperation Breaks Down*

Adepoju, P., 'Africa prepares for COVID-19 vaccines', *Lancet Microbe*, 2021 https://doi.org/10.1016/S2666-5247(21)00013-6

Akpan, N., 'Coronavirus spikes outside China show travel bans aren't working', *National Geographic*, 2020 https://www.nationalgeographic.com/science/article/why-travel-restrictions-are-not-stopping-coronavirus-covid-19

Al Jazeera, 'Coronavirus: travel restrictions, border shutdowns by country', Al Jazeera.2020https://www.aljazeera.com/news/2020/6/3/coronavirus-travel-restrictions-border-shutdowns-by-country

BBC News, 'Coronavirus: people-tracking wristbands tested to enforce lockdown', 2020 https://www.bbc.co.uk/news/technology-52409893

BBC News, 'Coronavirus: Trump attacks "China-centric" WHO over global pandemic', 2020 https://www.bbc.co.uk/news/world-us-canada-52213439

BBC News, 'Coronavirus: Trump moves to pull US out of World Health Organization', 2020 https://www.bbc.co.uk/news/world-us-canada-53327906

BBC News, 'Coronavirus: what are President Trump's charges against the WHO?', 2020 https://www.bbc.co.uk/news/world-us-canada-52294623

BBC News, 'COVAX: how many COVID vaccines have the US and the other G7 countries pledged?', 2021 https://www.bbc.co.uk/news/world-55795297

BBC News, 'COVAX vaccine-sharing scheme delivers first doses to Ghana', 2021 https://www.bbc.co.uk/news/world-africa-56180161

Beaumont, P., 'Vaccine inequality exposed by dire situation in world's poorest nations', *Guardian*, 2021 https://www.theguardian.com/world/2021/may/30/vaccine-inequality-exposed-by-dire-situation-in-worlds-poorest-nations

Belluz, J., 'Vietnam defied the experts and sealed its border to keep COVID-19 out. It worked', *Vox*, 2021 https://www.vox.com/22346085/covid-19-vietnam-response-travel-restrictions

Belluz, J., Irfan, U., and Resnikc, B., 'A guide to the vaccines and drugs that could fight coronavirus', *Vox*, 2020 https://www.vox.com/science-and-health/2020/3/4/21154590/coronavirus-vaccine-treatment-covid-19-drug-cure

Berkley, S., 'COVAX explained: GAVI', 2020 https://www.gavi.org/vaccineswork/covax-explained

Berkley, S., 'The GAVI COVAX AMC explained', 2020 https://www.gavi.org/vaccineswork/gavi-covax-amc-explained

Borger, J., 'Trump announces US to sever all ties with WHO', *Guardian*, 2020 https://www.theguardian.com/us-news/2020/may/29/trump-who-china-white-house-us

Burns, J., Movsisyan, A., Stratil, J. M., et al., 'International travel-related control measures to contain the COVID-19 pandemic: a rapid review', Cochrane Database of Systematic Reviews, 2021 https://doi.org/10.1002/14651858.CD013717.pub2

CEPI, 'A global coalition for a global problem', 2021 https://cepi.net/about/whoweare/

CEPI, 'COVAX: CEPI's response to COVID-19', 2021 https://cepi.net/covax/

CEPI, 'Creating a world in which epidemics are no longer a threat to humanity', 2021 https://cepi.net/about/whyweexist/

COVAX, 'COVAX facility explainer', 2020 https://cepi.net/wp-content/uploads/2020/10/COVAX_Facility_Explainer.pdf

COVAX, 'COVAX, the ACT-Accelerator vaccines pillar', 2020 https://cepi.net/wp-content/uploads/2020/10/COVAX-Pillar-background.pdf

COVAX, 'COVAX: the vaccines pillar of the Access to COVID-19 Tools (ACT) Accelerator. Structure and principles', 2020 https://cepi.net/wp-content/uploads/2020/11/COVAX_the-Vaccines-Pillar-of-the-Access-to-COVID-19-Tools-ACT-Accelerator.pdf

Dyer, O., 'COVID-19: countries are learning what others paid for vaccines', *BMJ*, 2021 https://www.bmj.com/content/372/bmj.n281

*Economist*, 'Vaccine diplomacy boosts Russia's and China's global standing', 2021 https://www.economist.com/graphic-detail/2021/04/29/vaccine-diplomacy-boosts-russias-and-chinas-global-standing

European Council, '"Impatience with vaccinations is legitimate, but should not blind us," warns President Michel', 2021 https://www.consilium.europa.eu/en/european-council/president/news/2021/03/09/20210309-pec-newsletter-6-vaccines/

Farge, E., Revill, J., Raff, P., Richardson, A., and Heinrich, M., WHO says countries should not order COVID-19 boosters while others still need vaccines', Reuters, 2021 https://www.reuters.com/business/healthcare-pharmaceuticals/

who-says-countries-should-not-order-covid-19-boosters-while-others-still-need-2021-07-12/

Ferhani, A., and Rushton, S., 'The International Health Regulations, COVID-19, and bordering practices: who gets in, what gets out, and who gets rescued?', *Contemporary Security Policy*, 2020 https://doi.org/10.1080/13523260.2020.1771955

Gastrow, C., and Lawrance, B., 'Vaccine nationalism and the future of research in Africa', *African Studies Review*, 2021 https://muse.jhu.edu/article/790099

GAVI, 'Ninety-two low- and middle-income economies eligible to get access to COVID-19 vaccines through GAVI COVAX AMC', 2020 https://www.gavi.org/news/media-room/92-low-middle-income-economies-eligible-access-covid-19-vaccines-gavi-covax-amc

GAVI, 'The GAVI COVAX AMC: an investment opportunity', 2020 https://www.gavi.org/sites/default/files/2020-06/Gavi-COVAX-AMC-IO.pdf

Ghebreyesus, T., tweet, 2020 https://twitter.com/DrTedros/status/1223288481159503873

Gougla, D., Christodoulou, M., Plotkin, S. A., et al., 'CEPI: driving progress toward epidemic preparedness and response', *Epidemiologic Reviews*, 2019 https://doi.org/10.1093/epirev/mxz012

Grépin, K. A., Ho, T.-L., Liu, Z., et al., 'Evidence of the effectiveness of travel-related measures during the early phase of the COVID-19 pandemic: a rapid systematic review', *BMJ Global Health*, 2021 http://dx.doi.org/10.1136/bmjgh-2020-004537

Hoffman, J., and Vazquez, M., 'Trump announces end of US relationship with World Health Organization', CNN, 2020 https://edition.cnn.com/2020/05/29/politics/donald-trump-world-health-organization/index.html

Hurst, D., 'UN urges Australia to act quickly to bring stranded Australians home', *Guardian*, 2021 https://www.theguardian.com/australia-news/2021/apr/16/un-urges-australia-to-act-quickly-to-bring-stranded-australians-home

International Finance Corporation, 'Getting developing countries the COVID-19 supplies they need', 2020 https://www.ifc.org/wps/wcm/connect/news_ext_content/ifc_external_corporate_site/news+and+events/news/covid-19-supplies

International Finance Corporation, 'IFC helps businesses in poorest countries fight pandemic with $4 billion in COVID-19 financing', 2020 https://pressroom.ifc.org/all/pages/PressDetail.aspx?ID=25861

Keeton, C., 'Africa edging towards third wave as 10 million get COVID-19 vaccines', Times Live, Republic of South Africa, 2021 https://www.timeslive.co.za/news/south-africa/2021-04-02-africa-edging-towards-third-wave-as-10-million-get-covid-19-vaccines/

Kennedy, M., 'WHO declares coronavirus outbreak a Global Health Emergency', NPR, 2020 https://www.npr.org/sections/goatsandsoda/2020/01/30/798894428/who-declares-coronavirus-outbreak-a-global-health-emergency

Khan, A., 'What is "vaccine nationalism" and why is it so harmful?', Al Jazeera, 2021 https://www.aljazeera.com/features/2021/2/7/what-is-vaccine-nationalism-and-why-is-it-so-harmful

Kretchmer, H., 'Vaccine nationalism – and how it could affect us all', World Economic Forum, 2021 https://www.weforum.org/agenda/2021/01/what-is-vaccine-nationalism-coronavirus-its-affects-covid-19-pandemic/

Kumar, A., Bernasconi, V., Manak, M., et al., 'The CEPI centralised laboratory network: supporting COVID-19 vaccine development', *Lancet*, 2021 https://doi.org/10.1016/S0140-6736(21)00982-X

Lewis, D., 'Is the coronavirus airborne? Experts can't agree', *Nature*, 2020 https://www.nature.com/articles/d41586-020-00974-w

Lewis, D., 'Mounting evidence suggests coronavirus is airborne – but health advice has not caught up', *Nature*, 2020 https://www.nature.com/articles/d41586-020-02058-1

Maxmen, A., 'What a US exit from the WHO means for COVID-19 and global health', *Nature*, 2020 https://www.nature.com/articles/d41586-020-01586-0

Maxmen, A., 'WHO report into COVID pandemic origins zeroes in on animal markets, not labs', *Nature*, 2021 https://www.nature.com/articles/d41586-021-00865-8

Menon, R., 'The response to the pandemic has been driven by vaccine nationalism', Common Dreams, 2021 https://www.commondreams.org/views/2021/06/21/response-pandemic-has-been-driven-vaccine-nationalism

Mlaba, K., 'Seven leaders from Africa who are leading the charge against COVID-19 vaccine hoarding', Global Citizen, 2021 https://www.globalcitizen.org/en/content/leaders-africa-calling-out-vaccine-nationalism/

Morawska, L., and Milton, D., 'It is time to address airborne transmission of coronavirus disease 2019', *Clinical Infectious Diseases*, 2020 https://doi.org/10.1093/cid/ciaa939

Mullard, A., 'How COVID vaccines are being divvied up around the world', *Nature*, 2020 https://www.nature.com/articles/d41586-020-03370-6

Multilateral Investment Guarantee Agency, 'MIGA's $6.5 billion fast-track facility to help investors and lenders tackle COVID-19', 2020 https://www.miga.org/press-release/migas-65-billion-fast-track-facility-help-investors-lenders-tackle-covid-19

Nebehay, S., 'Exclusive: WHO sweetens terms to join struggling global COVAX vaccine facility – documents', Reuters, 202 https://www.reuters.com/article/uk-health-coronavirus-who-offer-exclusiv-idUKKBN25O1SX

Nyabol, N., 'Vaccine nationalism is patently unjust', *Nation*, 2021 https://www.thenation.com/article/world/coronavirus-vaccine-justice/

Our World in Data, 'Vietnam: coronavirus pandemic country profile', 2021 https://ourworldindata.org/coronavirus/country/vietnam

Plotkin, S. A., Mahmoud, A. A., and Farrar, J., 'Establishing a global vaccine-development fund', *New England Journal of Medicine*, 2015 https://www.nejm.org/doi/full/10.1056/nejmp1506820

Reuters, 'Vietnam says new COVID outbreak threatens stability', Reuters, 2021 https://www.reuters.com/world/asia-pacific/vietnam-says-new-covid-outbreak-threatens-stability-2021-05-09/

Rouw, A., Wexler, A., Kates, J., et al., 'COVID-19 vaccine access: a snapshot of inequality: Kaiser Family Foundation', 2021 https://www.kff.org/policy-watch/global-covid-19-vaccine-access-snapshot-of-inequality/

Russell, T. W., Wu, J. T., Clifford, S., et al., 'Effect of internationally imported cases on internal spread of COVID-19: a mathematical modelling study', *Lancet Public Health*, 2021 https://doi.org/10.1016/S2468-2667(20)30263-2

Safi, M., 'WHO: just twenty-five COVID vaccine doses administered in low-income countries', *Guardian*, 2021 https://www.theguardian.com/society/2021/jan/18/who-just-25-covid-vaccine-doses-administered-in-low-income-countries

Sifferlin, A., 'This new group wants to stop pandemics before they start', *Time*, 2017 https://uk.sports.yahoo.com/news/group-wants-stop-pandemics-start-163111787.html

Spinney, L., 'When will a coronavirus vaccine be ready?', *Guardian*, 2020 https://www.bbc.co.uk/news/av/health-52382236

Sullivan, P., 'WHO denounces vaccine nationalism as global death toll passes 4 million', CBS News, 2021 https://www.cbsnews.com/news/covid-deaths-globally-hit-4-million-who-blasts-abhorrent-vaccine-nationalism/

Torjesen, I., 'COVID-19: pre-purchasing vaccine – sensible or selfish?', *BMJ*, 2020 https://doi.org/10.1136/bmj.m3226

Wintour, P., and Borger, J., 'Member states back WHO after renewed Donald Trump attack', *Guardian*, 2020 https://www.theguardian.com/world/2020/may/19/member-states-back-who-after-renewed-donald-trump-attack

Wolfson, E., 'Trump said he would terminate the US relationship with the WHO. Here's what that means', *Time*, 2020 https://time.com/5847505/trump-withdrawal-who/

Wood, J., and Lydeamore, M., 'Webinar 18: Australia's experience and the role of modelling in its responses to COVID-19', Usher Institute COVID-19 Webinar, 2020 https://www.youtube.com/watch?v=QV_4YrRF44E

World Bank Group, '$1 billion from World Bank to protect India's poorest from COVID-19 (coronavirus)', 2020 https://www.worldbank.org/en/news/press-release/2020/05/15/world-bank-support-protect-poorest-india-coronavirus

World Bank Group, 'How the World Bank Group is helping countries with COVID-19 (coronavirus)', 2021 https://www.worldbank.org/en/news/factsheet/2020/02/11/how-the-world-bank-group-is-helping-countries-with-covid-19-coronavirus

World Bank Group, 'New World Bank support to enable equitable access to COVID-19 vaccines in Senegal', 2021 https://www.worldbank.org/en/news/press-release/2021/06/02/new-world-bank-support-to-enable-equitable-access-to-covid-19-vaccines-in-senegal

World Bank Group, 'One hundred countries get support in response to COVID-19 (coronavirus)', 2020 https://www.worldbank.org/en/news/press-release/2020/

05/19/world-bank-group-100-countries-get-support-in-response-to-covid-19-coronavirus

World Bank Group, 'Saving lives, scaling-up impact and getting back on track.' World Bank Group, 'COVID-19 crisis response approach paper', 2020 https://documents1.worldbank.org/curated/en/136631594937150795/pdf/World-Bank-Group-COVID-19-Crisis-Response-Approach-Paper-Saving-Lives-Scaling-up-Impact-and-Getting-Back-on-Track.pdf

World Bank Group, 'The European Union and the World Bank support Mongolia's efforts to address COVID-19 (coronavirus) impact', https://www.worldbank.org/en/news/press-release/2020/04/10/the-european-union-and-the-world-bank-support-mongolias-efforts-to-address-covid-19-coronavirus-impact

World Bank Group, 'The World Bank approves $26.9 million for Mongolia's COVID-19 emergency response', 2020 https://www.worldbank.org/en/news/press-release/2020/04/02/the-world-bank-approves-269-million-for-mongolias-covid-19-coronavirus-emergency-response

World Bank Group, 'World Bank and African Union team up to support rapid vaccination for up to 400 million people in Africa', 2021 https://www.worldbank.org/en/news/press-release/2021/06/21/world-bank-and-african-union-team-up-to-support-rapid-vaccination-for-up-to-400-million-people-in-africa

World Bank Group, 'World Bank approves $20 million for Senegal to fight COVID-19', 2020 https://www.worldbank.org/en/news/press-release/2020/04/02/world-bank-approves-20-million-for-senegal-to-fight-covid-19

World Bank Group, 'World Bank approves $50.7 million for affordable and equitable COVID-19 vaccine access in Mongolia', 2021 https://www.worldbank.org/en/news/press-release/2021/02/11/world-bank-approves-507-million-for-affordable-and-equitable-covid-19-vaccine-access-in-mongolia

World Bank Group, 'World Bank approves US$20 million grant to support COVID-19 (coronavirus) response in Haiti', 2020 https://www.worldbank.org/en/news/press-release/2020/04/01/world-bank-approves-us20-million-grant-to-support-covid-19-response-in-haiti

World Bank Group, 'World Bank financing for COVID-19 vaccine rollout exceeds $4 billion for fifty countries', 2020 https://www.worldbank.org/en/news/press-release/2021/06/30/world-bank-financing-for-covid-19-vaccine-rollout-exceeds-4-billion-for-50-countries

World Bank Group, 'World Bank Group's operational response to COVID-19 (coronavirus) – projects list', 2021 https://www.worldbank.org/en/about/what-we-do/brief/world-bank-group-operational-response-covid-19-coronavirus-projects-list

World Health Organization, '172 countries and multiple candidate vaccines engaged in COVID-19 vaccine Global Access Facility', 2020 https://www.who.int/news/item/24-08-2020-172-countries-and-multiple-candidate-vaccines-engaged-in-covid-19-vaccine-global-access-facility

World Health Organization, 'COVID-19 – virtual press conference – 1 April 2020', 2020 https://www.who.int/docs/default-source/coronaviruse/transcripts/who-audio-emergencies-coronavirus-press-conference-full-01apr2020-final.pdf?sfvrsn=573dc140_2

World Health Organization, 'COVID-19 virtual press conference – 12 February 2021', 2021 https://www.who.int/publications/m/item/covid-19-virtual-press-conference-transcript---12-february-2021

World Health Organization, 'COVID-19 virtual press conference – 5 June 2020', 2020 https://www.who.int/docs/default-source/coronaviruse/transcripts/who-audio-emergencies-coronavirus-press-conference-full-05jun2020.pdf?sfvrsn=858dc773_2

World Health Organization, 'COVID-19 virtual press conference – 6 April 2020', 2020 https://www.who.int/docs/default-source/coronaviruse/transcripts/who-audio-emergencies-coronavirus-press-conference-full-06apr2020-final.pdf?sfvrsn=7753b813_2

World Health Organization, 'COVID-19 virtual press conference – 7 July 2020', 2020 https://www.who.int/docs/default-source/coronaviruse/transcripts/virtual-press-conference---7-july---covid-19.pdf?sfvrsn=6d4b4eb7_2

World Health Organization, 'International Health Regulations, 2005', 2008 https://apps.who.int/iris/bitstream/handle/10665/43883/9789241580410_eng.pdf;jsessionid=1BDF1434817DDA624A25A9744E6609C4?sequence=1

World Health Organization, 'Modes of transmission of virus causing COVID-19: implications for IPC precaution recommendations', 2020 https://apps.who.int/iris/bitstream/handle/10665/331601/WHO-2019-nCoV-Sci_Brief-Transmission_modes-2020.1-eng.pdf?sequence=1&isAllowed=y

World Health Organization, 'Press briefing by the international team studying the origins of the COVID-19 virus – 30 March 2021', 2021 https://www.who.int/publications/m/item/press-briefing-by-the-international-team-studying-the-origins-of-the-covid-19-virus-30-march-2021

World Health Organization, 'Transmission of SARS-CoV-2: implications for infection prevention precautions', 2020 https://www.who.int/news-room/commentaries/detail/transmission-of-sars-cov-2-implications-for-infection-prevention-precautions

World Health Organization, 'Virtual press conference on COVID-19 – 11 March 2020', 2020 https://www.who.int/docs/default-source/coronaviruse/transcripts/who-audio-emergencies-coronavirus-press-conference-full-and-final-11mar2020.pdf?sfvrsn=cb432bb3_2

World Health Organization, 'WHO Emergencies Coronavirus Emergency Committee second meeting', 2020 https://www.who.int/docs/default-source/coronaviruse/transcripts/ihr-emergency-committee-for-pneumonia-due-to-the-novel-coronavirus-2019-ncov-press-briefing-transcript-30012020.pdf?sfvrsn=c9463ac1_2

World Health Organization, 'WHO issues its first emergency use validation for a COVID-19 vaccine and emphasizes need for equitable global access', 2020 https://www.who.int/news/item/31-12-2020-who-issues-its-first-emergency-use-validation-for-a-covid-19-vaccine-and-emphasizes-need-for-equitable-global-access

World Health Organization, 'WHO lists two additional COVID-19 vaccines for emergency use and COVAX rollout', 2020 https://www.who.int/news/item/15-02-2021-who-lists-two-additional-covid-19-vaccines-for-emergency-use-and-covax-roll-out

York, G., 'Vaccine "hoarding" by wealthy countries triggers growing worries in poorer countries', *Global and Mail*, 2021 https://www.theglobeandmail.com/world/article-vaccine-hoarding-by-wealthy-countries-triggers-growing-worries-in/

## *11. Healing and Recovery*

Barnes, T., 'World Health Organisation fears new "Disease X" could cause a global pandemic', *Independent*, 2018 https://www.independent.co.uk/news/science/disease-x-what-infection-virus-world-health-organisation-warning-ebola-zika-sars-a8250766.html

BBC News, 'Belgian manhunt for heavily armed far-right soldier', 2021 https://www.bbc.co.uk/news/world-europe-57168576

Blaskey, S., 'Former health department employee, Rebekah Jones, granted official whistleblower status', *Tampa Bay Times*, 2021 https://www.tampabay.com/news/health/2021/05/28/former-health-department-employee-rebekah-jones-granted-official-whistleblower-status/

Boffey, D., 'Heavily armed awol Belgian soldier flagged as threat in February', *Guardian*, 2021 https://www.theguardian.com/world/2021/jun/16/heavily-armed-awol-belgian-soldier-jurgen-conings-flagged-threat-february

Burnside, T., and Yan, H., 'Fired Florida data scientist Rebekah Jones turns herself in to jail and tests positive for COVID-19', CNN, 2021 https://edition.cnn.com/2021/01/18/us/rebekah-jones-data-scientist-surrender/index.html

Cassidy, D., and Ceballos, A., 'Rebekah Jones's Twitter account suspended', *Tampa Bay Times*, 2021 https://www.tampabay.com/news/florida-politics/2021/06/07/rebekah-jones-twitter-account-suspended/

Chappell, B., and Treisman, R., 'Data scientist Rebekah Jones, facing arrest, turns herself in to Florida authorities', NPR, 2021 https://www.npr.org/sections/coronavirus-live-updates/2021/01/18/957914495/data-scientist-rebekah-jones-facing-arrest-turns-herself-in-to-florida-authoriti

Chini, M., 'Fugitive soldier spent two hours near home of "target" Marc Van Ranst on Monday', *Brussels Times*, 2021 https://www.brusselstimes.com/news/belgium-all-news/170506/tbtb-fugitive-soldier-spent-two-hours-in-vicinity-of-target-marc-van-ranst-on-monday-jurgen-conings-national-park-hoge-kempen-van-quickenborne-military-base-terrorist-threat-ocam-investigation/

Chini, M., ' "Not smart": Van Ranst joins support group for fugitive soldier Jürgen Conings', *Brussels Times*, 2021 https://www.brusselstimes.com/news/belgium-all-news/171101/not-smart-marc-van-ranst-joins-support-group-for-fugitive-soldier-jurgen-conings-telegram-facebook-henk-van-ess-manhunt/

Cluskey, P., 'When criticism of top COVID-19 advisers turns into a terror campaign', *Irish Times*, 2021 https://www.irishtimes.com/news/world/europe/when-criticism-of-top-covid-19-advisers-turns-into-a-terror-campaign-1.4581385

Franck, T., 'Treasury Secretary Mnuchin says "We can't shut down the economy again" ', CNBC, 2020 https://www.cnbc.com/2020/06/11/treasury-secretary-mnuchin-says-we-cant-shut-down-the-economy-again.html

Luscombe, R., 'Florida scientist says she was fired for refusing to change COVID-19 data "to support reopen plan" ', *Guardian*, 2020 https://www.theguardian.com/us-news/2020/may/20/florida-scientist-dr-rebekah-jones-fired-refusing-change-covid-19-data-reopen-plan

Mackenzie, J., 'Belgium's Van Ranst: COVID scientist targeted by a far-right sniper', BBC News, 2021 https://www.bbc.co.uk/news/world-europe-57358492

Madani, D., 'Florida data scientist in battle with state over COVID dashboard turns herself in', NBC News, 2021, https://www.nbcnews.com/news/us-news/florida-data-scientist-battle-state-over-covid-dashboard-plans-turn-n1254544

Majumder, B., 'Indian Bar Association VS WHO Chief Scientist Soumya Swaminathan: does ivermectin help COVID patients?', *Swarajya*, 2021 https://swarajyamag.com/news-brief/indian-bar-association-vs-who-chief-scientist-soumya-swaminathan-does-ivermectin-help-covid-patients

News.in-24, 'Marc Van Ranst speaks from his hiding place: "My son is very impressed" ', 2021 https://news.in-24.com/news/29158.html

Pilkington, E., 'Armed police raid home of Florida scientist fired over COVID-19 data', *Guardian*, 2020 https://www.theguardian.com/us-news/2020/dec/07/florida-police-raid-data-scientist-coronavirus

'Rebekah Jones for Florida's 1st District', 2021 https://www.rebekahjonescampaign.com/

Sassoon, A. M., 'Florida scientist was fired for "refusing to manipulate" COVID-19 data, she said', *Florida Today*, 2020 https://eu.floridatoday.com/story/news/2020/05/19/florida-scientist-refused-manipulate-covid-19-data-and-fired/5219137002/

Simrin, S., 'Bar association serves legal notice to WHO chief scientist over ivermectin guidelines', *The Print*, 2021 https://theprint.in/india/bar-association-serves-legal-notice-to-who-chief-scientist-over-ivermectin-guidelines/676672/

van der Ploeg, J., 'Viroloog Marc van Ranst laat zich niet intimideren: "Ik ben volledig vrij om te reageren op onzin" ', *De Volkskrant*, 2021 https://webcache.googleusercontent.com/search?q=cache:bvIb7so5Pu8J:https://www.volkskrant.

nl/nieuws-achtergrond/viroloog-marc-van-ranst-laat-zich-niet-intimideren-ik-ben-volledig-vrij-om-te-reageren-op-onzin~b5456dbf/+&cd=1&hl=en&ct=clnk&gl=uk&client=firefox-b-d

Walker, L., 'Jürgen Conings: suicide confirmed as cause of death by public prosecutor', *Brussels Times*, 2021 https://www.brusselstimes.com/belgium/174710/tbtb-jurgen-conings-suicide-confirmed-as-cause-of-death-by-public-prosecutor/

Wamsley, L., 'Fired Florida data scientist launches a coronavirus dashboard of her own', NPR, 2020 https://www.npr.org/2020/06/14/876584284/fired-florida-data-scientist-launches-a-coronavirus-dashboard-of-her-own

# Index